Volume III: Thermal Behaviour, Energy Efficiency in Buildings and Sustainable Construction

Volume III: Thermal Behaviour, Energy Efficiency in Buildings and Sustainable Construction

Editor

Paulo Santos

MDPI

Basel • Beijing • Wuhan • Barcelona • Belgrade • Novi Sad • Cluj • Manchester

Editor
Paulo Santos
Department of Civil Engineering
University of Coimbra
Coimbra
Portugal

Editorial Office
MDPI
St. Alban-Anlage 66
4052 Basel, Switzerland

This is a reprint of articles from the Special Issue published online in the open access journal *Energies* (ISSN 1996-1073) (available at: www.mdpi.com/journal/energies/special_issues/Buildings_and_Sustainable_Construction_2023).

For citation purposes, cite each article independently as indicated on the article page online and as indicated below:

Lastname, A.A.; Lastname, B.B. Article Title. *Journal Name* **Year**, *Volume Number*, Page Range.

ISBN 978-3-7258-0516-7 (Hbk)
ISBN 978-3-7258-0515-0 (PDF)
doi.org/10.3390/books978-3-7258-0515-0

Contents

About the Editor

Paulo Santos

Paulo Santos [ORCID 0000-0002-0134-6762] is an Associate Professor in the Dep. of Civil Engineering of the University of Coimbra, PT. He is a member of the ISISE (Institute for Sustainability and Innovation in Structural Engineering) research centre. His main fields of interest include Thermal Behaviour, Energy Efficiency in Buildings, and Sustainable Construction. He is the author of around 190 scientific publications and has supervised around 55 Doctoral and Master's theses. He has around 2,770 citations and 77,100 reads on Research Gate (Berlin), with an h-index of 26. More information can be found at Web of Science Researcher ID: B-2073-2010, Scopus ID: 55605780628, ORCID ID: 0000-0002-0134-6762, ResearchGate ID: Paulo-Santos-52, and GoogleScolar ID: sYQK7hgAAAAJ. He has participated in around 16 funded European and Portuguese research projects. He was the Principal Investigator of the Tyre4BuildIns: "Recycled tyre rubber resin-bonded for building insulation systems towards energy efficiency" research project. He was a member of the scientific committee and/or organizer of several national and international scientific conferences. He is an Editorial Board Member of 20 international scientific journals, for which he already has 287 verified Web of Science editor records. He has reviewed, by request of the editors, around 125 articles submitted to international journals. He is also a member of the scientific evaluation committee of several national and international research projects proposed for funding. Furthermore, he is a member of iiSBE: International Initiative for a Sustainable Built Environment (2011-present) and Technical Committee TC14: Sustainability and Eco-efficiency of Steel Construction of the ECCS: European Convention for Constructional Steelwork (2008-present).

Preface

As is well known, energy and sustainability are, nowadays, two of the major concerns of mankind. Given the relevant energy consumption share of the buildings sector, it is very important to search for innovative design solutions and for the optimal thermal performance of buildings to reduce energy bills and greenhouse gas emissions while maintaining the comfort levels of the occupants. Additionally, given the environmental burdens of the construction sector, seeking more environmentally responsible processes and a more efficient use of resources is currently attracting more attention.

The third edition of this Special Issue, published in *Energies*, is dedicated to the analysis of recent advances in the following main issues: (1) thermal behaviour improvement in a building's elements (e.g., walls, floors, roofs, windows, doors, etc.), (2) energy efficiency in buildings, and (3) sustainable construction. The main goal is to compile scientific works within these topics, making use of different possible research approaches, such as experimental, theoretical, numerical, analytical, computational, and case studies, as well as mixtures of these. This reprint compiles a set of original research works of an excellent academic standard and of complete scientific soundness.

The Guest Editor would like to express their sincere and deep gratitude for all the scientific contributions from authors, comprising prestigious worldwide scientists, as well as to the reviewers who have significantly contributed to improving the quality of the manuscripts. Moreover, we express acknowledgments to the research project Tyre4BuildIns – "Recycled tyre rubber resin-bonded for building insulation systems towards energy efficiency", supported by FEDER funds through the Competitivity Factors Operational Programme—COMPETE and by national funds through FCT—Foundation for Science and Technology, within the scope of the project POCI-01-0145-FEDER-032061, which contributed three scientific articles to the first volume, one paper to the second volume, and two papers to this Special Issue (third volume). Additionally, the Guest Editor also wishes to thank the following companies for providing their support by being partners of the research project Tyre4BuildIns: Pertecno, Gyptec Ibéria, Volcalis, Sotinco, Kronospan, Hulkseflux, Hilti, and Metabo.

Paulo Santos
Editor

energies

Article

Numerical Simulation and Experimental Validation of Thermal Break Strips' Improvement in Facade LSF Walls

Paulo Santos [1,*], Diogo Mateus [1], Daniel Ferrandez [2] and Amparo Verdu [2]

1 ISISE, Department of Civil Engineering, University of Coimbra, Pólo II, Rua Luís Reis Santos, 3030-788 Coimbra, Portugal
2 Department of Building Technology, Polytechnic University of Madrid, Avenida Juan de Herrera n.6, 28040 Madrid, Spain
* Correspondence: pfsantos@dec.uc.pt

Abstract: Thermal bridges may have a significant prejudicial impact on the thermal behavior and energy efficiency of buildings. Given the high thermal conductivity of steel, in Lightweight Steel Framed (LSF) buildings, this detrimental effect could be even greater. The use of thermal break (TB) strips is one of the most broadly implemented thermal bridge mitigation technics. In a previous study, the performance of TB strips in partition LSF walls was evaluated. However, a search of the literature found no similar experimental campaigns for facade LSF walls, which are even more relevant for a building's overall energy efficiency since they are in direct contact with the external environmental conditions. In this article the thermal performance of ten facade LSF wall configurations were measured, using the heat flow meter (HFM) method. These measurements were compared to numerical simulation predictions, exhibiting excellent similarity and, consequently, high reliability. One reference wall, three TB strip locations in the steel stud flanges and three TB strip materials were assessed. The outer and inner TB strips showed quite similar thermal performances, but with slightly higher thermal resistance for outer TB strips (around +1%). Furthermore, the TB strips were clearly less efficient in facade LSF walls when compared to their thermal performance improvement in load-bearing partition LSF walls.

Keywords: lightweight steel frame; LSF facade walls; thermal resistance; thermal break strips; experimental measurements; numerical simulations

Citation: Santos, P.; Mateus, D.; Ferrandez, D.; Verdu, A. Numerical Simulation and Experimental Validation of Thermal Break Strips' Improvement in Facade LSF Walls. *Energies* **2022**, *15*, 8169. https://doi.org/10.3390/en15218169

Academic Editor: Chi-Ming Lai

Received: 27 September 2022
Accepted: 1 November 2022
Published: 2 November 2022

1. Introduction

Overall, buildings are responsible for about 36% of global CO_2 emissions, making them a key player in the fight against global warming and climate change [1]. The European Union spends approximately 40% of final energy consumption on heating and cooling buildings [2]. For this reason, in recent decades European legislation has been geared towards transforming the existing building stock into a network of nearly zero energy buildings (NZEB) [2], promoting the use of renewable energies and developing renovation strategies to improve the energy efficiency of existing buildings [3].

Currently, approximately 35% of the building stock in the European Union is over 50 years old and needs rehabilitation to meet current energy requirements [4]. Consequently, the European Green Deal presented by the Commission on 11 December 2019 [5], sets out the arrangements for achieving an efficient use of energy and building resources to double the annual renovation rate of the building stock [6]. In line with this initiative, it has been noted that it is important to conduct studies that allow for a reliable thermal characterization of buildings and an exhaustive analysis of the components that make up their envelope [7].

As an alternative for the energy rehabilitation of existing building facades, systems such as External Thermal Insulation Composite Systems (ETICS) have been developed in recent decades [8,9]. These are construction systems that consist of a thermal insulation material applied to the building facade with the help of an adhesive and/or mechanical fastening, and then a reduced thickness rendering system that incorporates a reinforcing mesh [10]. These construction systems make it possible to reduce thermal bridges and moisture condensation on the inside surface of the envelopes, as well as offering the advantage of not reducing the interior space of the rooms in the dwellings and improving the aesthetics of the refurbished facade [11]. Moreover, there are several studies that have found the use of this type of insulation system from the outside saves up to 8% more energy on average than in the case of using an insulation system from the inside [12].

ETICS are also very widely used in Lightweight Steel Framed (LSF) facade walls [13], having the advantage of mitigating the thermal bridges originated by the high thermal conductivity of the steel studs since they are a continuous insulation. There are several types of thermal bridges [13], which could be designated as: (1) repeating thermal bridges (e.g., due to the vertical steel studs inside an LSF wall), where their effect is usually considered by reducing the R-value (increase the U-value); (2) non-repeating (or linear) thermal bridges (e.g., along a wall corner, or along a wall–window joint perimeter [14]), quantified using the linear thermal transmittance (Ψ); and (3) punctual thermal bridges (e.g., due to mechanical steel fasteners crossing the ETICS insulation layer), which are taken into account using the point thermal transmittance (χ). In this study, only the above-mentioned first type of thermal bridges are considered.

The LSF construction system has been used in the last decades due to its multiple advantages such as its low weight compared to its high mechanical resistance [15], a great speed of execution and ease of transport to the building site [16], a high potential for recycling and a high quality as very precise tolerances are achieved in the execution of prefabricated elements [17,18].

However, the existence of thermal bridges through the steel and the lower thermal inertia of the LSF walls can cause problems related to comfort and a higher energy demand [19], with the thermal bridges issue being even more relevant in cold climates [20]. Furthermore, it is sometimes difficult to accurately evaluate the final thermal resistance of the facade as there is a strong contrast between the high thermal conductivity of the steel frame and the insulation materials filling the air cavity [21,22]. Additionally, the size and shape of the stud flanges may have a relevant influence on the thermal performance of LSF walls [23]. Furthermore, the position of the thermal insulation is very important with regards to their effectiveness [24], and the usual fibrous cavity insulation (e.g., mineral wool) is likewise relevant for the noise insulation performance of LSF facade walls [25].

Given the absence of a continuous insulation layer, the steel stud thermal bridges are even more relevant in LSF partition walls [26–28]. To mitigate these thermal bridges, the LSF partition could be split in two steel frames, having a small air gap between them, this way breaking the heat flow across the steel studs. Moreover, since this double-pane LSF partition has an air cavity, an efficient way to increase their thermal resistance even more is to use a reflective foil, e.g., in aluminum [26]. Even though they are not as efficient as in a single pane LSF wall [28], thermal break (TB) strips could also be used along the double-pane steel stud flanges to mitigate the related thermal bridge effects [27].

In fact, in addition to ETICS, TB strips are widely disseminated as a thermal energy loss attenuation strategy in LSF enclosures [29]. In a previous research work, the performance of TB strips in load-bearing and non-load-bearing partition walls was evaluated [30]. The lab measurements were performed using the heat flow meter (HFM) method under nearly steady-state conditions. Several TB materials (recycled rubber, cork/rubber composite and aerogel) were compared, as well as three TB locations (inner, outer and on both stud flanges). It was concluded that outer and inner TB strips display quite analogous thermal performances, but using two TB strips had a comparative noteworthy thermal performance improvement. A systematic experimental campaign to measure the thermal performance improvement of load-bearing LSF facade walls due to the use of thermal break strips was not found in the literature.

This research work is a continuation of the previous one [30], but, instead of partition LSF walls, a new set of ten different facade LSF configurations were evaluated under similar lab conditions and test procedures. The reference facade configuration, in addition to the mineral wool batt insulation, has the usual exterior continuous thermal insulation layer (ETICS) and does not have any TB strips. The remaining nine facade LSF configurations correspond to the three TB strip materials and three TB strip positions. The main goal of this research is to see if the trends and features, related to the use of TB strips, previously measured for the partition LSF walls are similar or not to the ones exhibited now by these facade LSF walls (e.g., will the outer and inner TB strips still have very similar thermal performances? Are the TB strips able to fully mitigate the steel frame thermal bridges effect? Are the TB strips less, equally or more efficient in facade LSF walls?). Notice that the relevance of facade walls in the building's thermal behaviour and energy efficiency is much higher in comparison with partition walls since the temperature gradient (between the indoor conditioned space and the exterior environment) is also considerably bigger than between a conditioned and an unconditioned space.

This paper is organized as explained next. Following this brief introduction, the related materials and methods are presented, including the LSF facade walls' characterization, the experimental tests under laboratory-controlled conditions and the numerical simulations. After, the achieved results for measured and predicted global conductive thermal resistances, the measured *R*-value improvement and the infrared thermography assessment are presented. Later, these results are discussed and compared with the ones provided by the previously assessed partition LSF walls. To conclude, the main outcomes from this study are listed.

2. Methods and Materials

2.1. Characterization of the LSF Facade Walls

Here, the thermal break (TB) strips, as well as the tested facade LSF walls, are characterized including the geometries/dimensions, materials and thermal properties.

2.1.1. Reference Facade LSF Wall

Figure 1 displays the reference facade LSF wall cross-section. Table 1 lists the thermal conductivities of the materials, as well as the thickness of each wall layer. This load-bearing facade wall is supported by C-shaped vertical steel studs (C90 × 43 × 15 × 1.5 mm). The spacing between these vertical steel studs is 400 mm. The mineral wool (MW) batt insulation is 90 mm thick, completely filling the air cavity. On both sides of the steel studs there is an OSB (Oriented Strand Board) structural sheathing panel (12 mm thick). Moreover, in the inner surface there is an extra GPB (Gypsum Plaster Board) sheathing layer (12.5 mm). In the outer surface there is an External Thermal Insulation Composite System (ETICS), having 50 mm of expanded polystyrene (EPS) and a finishing layer (5 mm thick). This reference facade LSF wall has a total thickness equal to 181.5 mm.

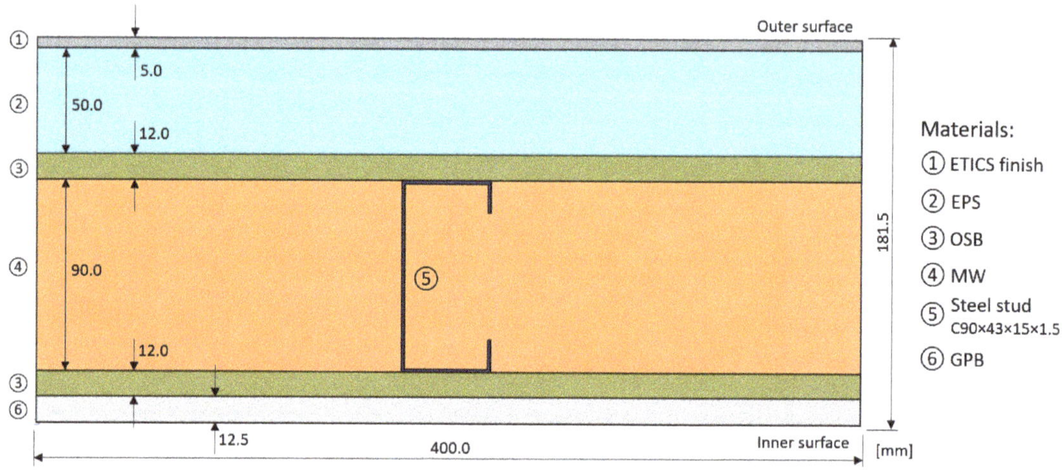

Figure 1. Reference facade LSF wall horizontal cross-section: materials and geometry.

Table 1. Reference facade LSF wall material thickness (*d*) and thermal conductivities (λ).

Material (From Outer to Innermost Layer)	*d* (mm)	λ (W/(m·K))	Ref.
ETICS [1] finishing layer	5.0	0.450	[31]
EPS [2]	50.0	0.036	[32]
OSB [3]	12.0	0.100	[33]
MW [4]	90.0	0.035	[33]
Steel profiles (C90 × 43 × 15 × 1.5 mm)	-	50.000	[34]
OSB [3]	12.0	0.100	[33]
GPB [5]	12.5	0.175	[33]
Global Thickness	**181.5**	-	-

[1] ETICS—External Thermal Insulation Composite System; [2] EPS—expanded polystyrene; [3] OSB—Oriented Strand Board; [4] MW—mineral wool; [5] GPB—Gypsum Plaster Board.

2.1.2. Thermal Break Strips

All the thermal break (TB) strips evaluated have a rectangular cross-section: 50 mm wide by 10 mm thick. The TB strips were placed along the inner, outer and on both sides of the metallic stud flanges, as displayed in Figure 2. They were fixed by compression between the steel studs and the OSB sheathing panels, which were attached to the steel frame using self-drilling screws. Moreover, three materials were tested in the TB strips, namely: recycled rubber and cork composite (R0), recycled rubber (R1) and aerogel (AG). These materials were chosen given their decreasing thermal conductivities, extending from 122 mW/(m·K) down to 15 mW/(m·K), as illustrated in Table 2.

Table 2. Thermal break strips: material and thermal conductivity (λ).

Material	λ (W/(m·K))	Ref.
Recycled rubber (R1)	0.122	[33]
Recycled rubber and cork (R0)	0.088	[35]
CBS [1] aerogel (AG)	0.015	[33]

[1] CBS—cold break strip.

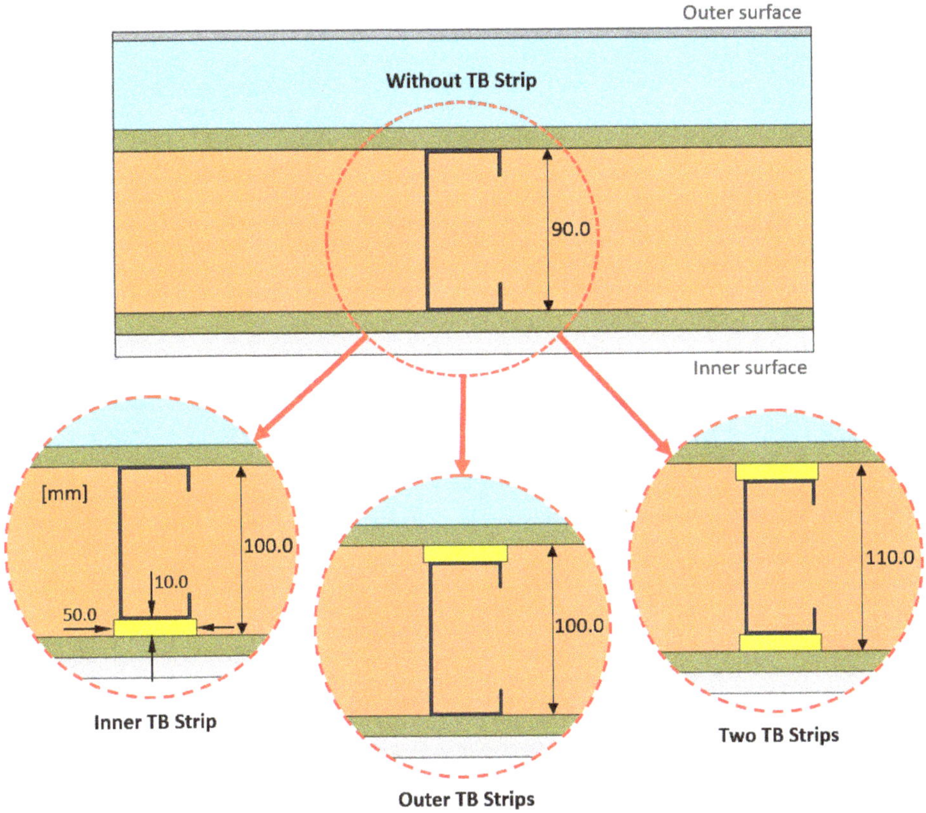

Figure 2. Geometry and location of the thermal break (TB) strips.

Notice that when one or two TBS are attached to the steel stud flanges, the total thickness of the LSF wall is also increased since the steel profiles were always the same. The wall thickness increment is equal to the thickness of the TBS used, i.e., 10 or 20 mm, for one or two TBS, respectively (see Figure 2).

2.2. Experimental Lab Tests

The laboratorial tests were performed using the same experimental setup, as well as the same test procedures and set-points, used in a previous research work to evaluate the thermal performance of partition LSF walls [30]. Therefore, to avoid unnecessary repetitions and for sake of brevity, only the main issues are explained next. More detailed info about this experimental lab test can be found in Reference [30].

2.2.1. Experimental Setup

The thermal performance of the facade LSF elements were assessed making use of the heat flow meter (HFM) method [36]. However, as suggested by Rasooli and Itard [37], this method was modified, having two HF sensors. To ensure a nearly-steady-state difference of temperature, between both surfaces of the tested LSF facade sample, a set of two small climatic boxes was used: one hot (heated using an electric resistance) and another cold (cooled using an attached fridge), as displayed in Figure 3.

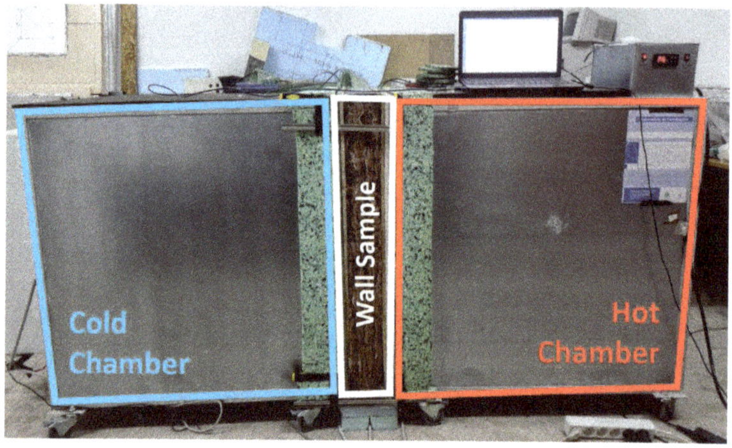

Figure 3. Cold and hot boxes apparatus used in the experiments.

Since the moisture content of the materials, used during the experiments, can strongly conditionate their thermal properties (e.g., thermal conductivity), a stable relative humidity (RH) condition was ensured during and before the lab tests (storage stage).

Two small fans (one for each climatic chamber), reutilized from old computer case fans (12 V, 0.25 A), were used inside hot and cold boxes to promote air circulation and avoid air temperature stratification [37]. Additionally, near the wall sample (10 cm apart) a black radiation shield was placed, one inside the cold box and another one inside the hot box (not illustrated here). This radiation shield was made of PVC and it was placed at vertical position, parallel to the LSF wall test specimen.

The steel frames of the LSF wall test samples (1030 mm by 1060 mm) have three vertical steel studs, where the central one is positioned exactly in the middle, as displayed in Figure 4a. Given the high thermal conductivity of steel, to minimize the lateral heat flow, the LSF wall sample perimeter was covered by two layers (40 mm thick each) of polyurethane foam insulation, ensuring an additional edge thermal resistance of 2.22 $m^2 \cdot K/W$.

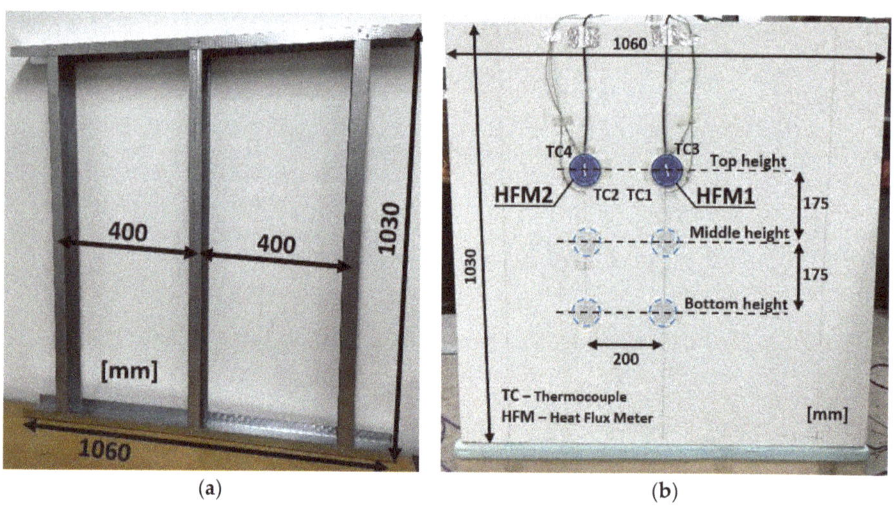

Figure 4. Tested LSF wall sample. (**a**) Steel frame. (**b**) Sensors' position (cold surface).

To increase the precision of the measurement and reduce its time duration (see Reference [38]), instead of the standard use of only one heat flux meter (HFM) placed in the inner side (hot) of the wall sample, another HFM was simultaneously placed in the outer side (cold). Figure 4b illustrates the sensors' position on the cold surface of the tested LSF wall sample. As shown here, to adequately characterize the thermal resistance of the LSF wall samples, the sensors were placed in two vertical alignments: (1) near the central steel stud, where the heat flow is higher (HFM1); and (2) in the middle of the insulation cavity, where the heat flow is lower (HFM2).

In addition to heat flow, the sample surface and air temperatures were also measured, making use of certified (class one precision) type K thermocouples (TCs). To ensure the good temperature measurement accuracy of these TCs, they were previously calibrated for a temperature range [5 °C; 45 °C], with a 5 °C increment, by immersing the TCs in a thermostatic stirring water bath.

Six TCs were used in the hot side and another six in the cold side, with their locations symmetrical with regards to the hot and cold environments. Looking to the cold surface of the wall sample (Figure 4b): two TCs (TC1 and TC2) were used to measure the wall surface temperatures; another two (TC3 and TC4) were used to measure the air temperature between the wall surface and the radiation shield; and the remaining two (TC5 and TC6) measured the environment air temperature inside each chamber (not illustrated).

To record all the data obtained by the HFMs and the TCs during the experiments, one datalogger (PICO TC-08) was used on each surface of the LSF facade sample (cold and hot). Moreover, these two dataloggers were coupled to a computer (see Figure 3, over the "Hot Chamber"), where the PicoLog® version 6.1.10 software (Pico Technology Limited, St Neots, UK) was used to manage the recorded data.

To illustrate some recorded data, Figure 5 displays, for the reference LSF facade wall, the measured surface temperatures and the variation in heat flux densities over time when sensors are placed at top height location (see Figure 4b). As expected, during the first measurement hours, there is some variation in the recorded values, but after this stabilizing period there is clearly a convergence. Moreover, the temperature difference is slightly higher for the thermocouples placed in position 2 (cavity, Figure 5b) and the heat flux density is significantly smaller when compared with the sensors located at the vicinity of the steel stud (position 1, Figure 5a). Thus, the measured local thermal resistance will be much higher at this cavity location, in comparison with the steel studs' location. This relevant thermal resistance reduction was expected, being related to the high thermal conductivity of steel and the consequent thermal bridge phenomenon occurring near this LSF wall location (position 1), even having a 50 mm continuous insulation thickness (ETICS).

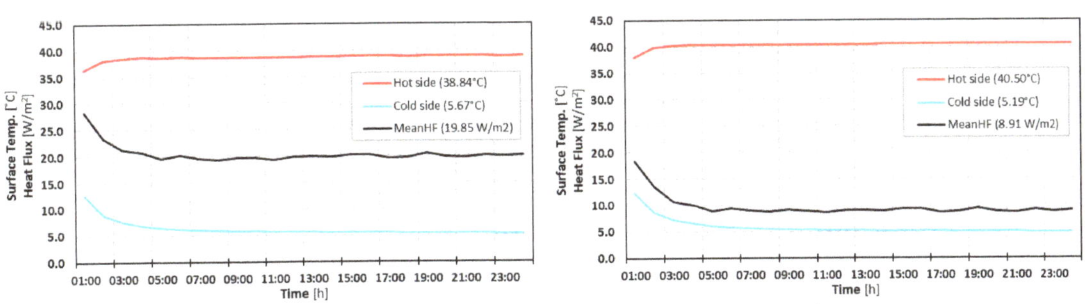

(a) (b)

Figure 5. Measured surface temperatures and variations in heat flux densities over time when sensors are located at top height (see Figure 4b), for the reference LSF facade wall. (**a**) Sensors' position 1—stud. (**b**) Sensors' position 2—cavity.

2.2.2. Set-Points and Test Procedures

To measure the surface-to-surface *R*-value (or thermal resistance) of the assessed LSF facade walls, using the heat flow meter (HFM) method, the test procedures prescribed in some international standards were adopted, including ISO 9869-1 [36], ASTM C 1155-95 [39] and ASTM C 1046-95 [40]. More specifically, the "summation technique" described in the standard ASTM C 1155-95 [39] was adopted, with this methodology having some similarities to the "average method" described in standard ISO 9869-1 [36]. According to the so-called "summation technique" [39], the estimated measured thermal resistance for each time interval, R_e, could be computed making use of the following equation

$$R_e = \frac{\sum_{k=1}^{n}(T_{si,k} - T_{se,k})}{\sum_{k=1}^{n} q_k}, \tag{1}$$

where: T_{si} and T_{se} are the interior (hot) and exterior (cold) surface temperatures (°C), respectively; q is the heat flux density (W/m^2); k is the counter for summation of time-series data; and n is number of time values.

As suggested by this standard [39], the duration of each test was 24 h (minimum). The convergence factor and criterion are also defined in ASTM C 1155-95 [39] and given by the following expression

$$CR_n = \frac{R_e(t) - R_e(t-n)}{R_e(t)} < 0.10, \tag{2}$$

where: CR_n is the convergence factor; $R_e(t)$ is the estimated measured thermal resistance during the evaluated time interval (e.g., 1 h); and $R_e(t-n)$ is the estimated thermal resistance during the previously evaluated time interval. This adopted convergence criterion means that the variation in the estimated measured *R*-value, between a time interval and the previous one, should be smaller than 10%, in relation to the actual time interval.

As already mentioned before, the improvement suggested by Rasooli and Itard [37] was adopted to reduce test duration and increase precision, which consists of the use of two HFMs simultaneously at both cold and hot wall surfaces, as an alternative to measuring only one side, as given by ISO 9869-1 [36].

The adopted set-points for the cold and hot chambers were 5 °C and 40 °C, respectively. The measurements were performed in a quasi-steady-state heat transfer state. This large temperature difference (35 °C) between the cold and hot chambers allowed the reliability and accuracy of the measured *R*-values to be increased [36].

As illustrated in Figure 4b, three height sensor locations were chosen, for each wall configuration, namely: (1) top, (2) middle and (3) bottom, with one test performed for each one. Thus, the overall surface-to-surface thermal resistance of the LSF facade was achieved by averaging the values from the previously mentioned three tests, this way ensuring the repeatability of the experimental measurements.

Making use of the data recorded (heat fluxes and temperatures) for each test and applying the HFM method [36], two distinct conductive local *R*-values were obtained: (1) a lower value for location 1 (Figure 4), i.e., in the vicinity of the steel studs (R_{stud}); and (2) a higher value between the steel studs, i.e., in the middle of the insulation cavity (R_{cav}). The overall surface-to-surface *R*-value of the wall (R_{global}) was obtained by computing an area weighted of both measured conductive *R*-values, as indicated in the following equation

$$\frac{1}{R_{\text{global}}} = \frac{1}{R_{\text{stud}}}\frac{A_{\text{stud}}}{A_{\text{global}}} + \frac{1}{R_{\text{cav}}}\frac{A_{\text{cav}}}{A_{\text{global}}}, \tag{3}$$

where: A_{global} is the total area of the LSF wall (m^2); A_{stud} is the area of influence of the steel stud (m^2); and A_{cav} is the remaining cavity area of the LSF wall (m^2).

The steel stud influence area (A_{stud}) was defined as prescribed by ASHRAE zone method [41], i.e., assuming a zone factor (zf) equal to 2.0 [22]. Therefore, the width of the steel stud influence zone (w) is equal to the flange length (fl) plus two times the thickness of the thicker sheathing layer ($d_{thicker}$).

Notice that these computations to obtain an overall R-value of the tested LSF wall were performed making use of a representative wall zone area defined by the studs' spacing (width) and assuming one meter high (length), i.e., 0.40 m by 1.00 m (0.40 m^2).

2.2.3. Experimental Procedures' Verification

The verification of the reliability and good working conditions of the sensors and dataloggers, in the experimental apparatus, was performed under two different approaches: (1) testing a homogenous extruded polystyrene (XPS) panel (60 mm thick) under the same lab conditions, with known thermal conductivity and resistance; (2) simulating a representative LSF wall cross-section using some finite element models, as described in Section 2.3, and comparing the predicted R-value with the measured thermal resistance.

Regarding the first verification, both measured and manufacturer surface-to-surface thermal resistances were the same (1.784 $m^2 \cdot K/W$). Concerning the second verifications, as displayed later in Section 3.1 (Table 3), there was a very good agreement between the measured and the predicted R-values, the differences ranging between 0% and −3%.

Table 3. Thermal resistances (surface-to-surface) predicted (THERM) and measured.

Wall Code	Wall Description	R-Value ($m^2 \cdot K/W$)	
		THERM	Measured
Ref	Reference LSF Facade Wall	3.200	3.200
	Difference: Absolute ($m^2 \cdot K/W$) and Percentage (%)	0.000	0%
R1in	Inner Recycled Rubber TB Strip	3.404	3.320
	Difference: Absolute ($m^2 \cdot K/W$) and Percentage (%)	−0.084	−2%
R0in	Inner Rubber-Cork TB Strip	3.441	3.396
	Difference: Absolute ($m^2 \cdot K/W$) and Percentage (%)	−0.045	−1%
AGin	Inner Aerogel TB Strip	3.764	3.661
	Difference: Absolute ($m^2 \cdot K/W$) and Percentage (%)	−0.103	−3%
R1out	Outer Recycled Rubber TB Strip	3.384	3.344
	Difference: Absolute ($m^2 \cdot K/W$) and Percentage (%)	−0.040	−1%
R0out	Outer Rubber-Cork TB Strip	3.416	3.427
	Difference: Absolute ($m^2 \cdot K/W$) and Percentage (%)	0.011	0%
AGout	Outer Aerogel TB Strip	3.707	3.695
	Difference: Absolute ($m^2 \cdot K/W$) and Percentage (%)	−0.012	0%
R1 × 2	Inner and Outer Recycled Rubber TB Strip	3.587	3.547
	Difference: Absolute ($m^2 \cdot K/W$) and Percentage (%)	−0.040	−1%
R0 × 2	Inner and Outer Rubber-Cork TB Strip	3.658	3.588
	Difference: Absolute ($m^2 \cdot K/W$) and Percentage (%)	−0.070	−2%
AG × 2	Inner and Outer Aerogel TB Strip	4.224	4.093
	Difference: Absolute ($m^2 \cdot K/W$) and Percentage (%)	−0.131	−3%

TB—thermal break.

Moreover, Figure 6 displays the estimated measured *R*-values' variation over time, at three different sensor height positions, as well as the predicted THERM *R*-value for the reference LSF facade wall. As expected, during the first hours of measurements, there is a convergence phenomenon until the quasi-stationary state is reached (around seven hours). After this period, the convergence criterion is reached (see Equation (2)) and the averaged measured *R*-values become very similar to the predicted THERM value. In fact, using the average of the measured *R*-values during the convergence period, the predicted and measured thermal resistances are the same (3.200 $m^2 \cdot K/W$).

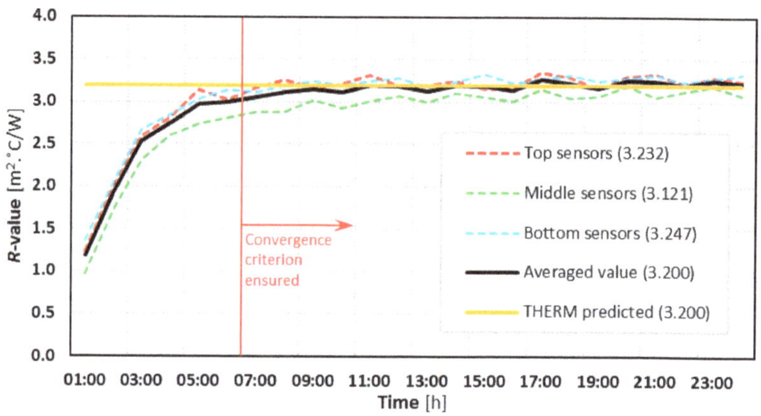

Figure 6. Estimated measured thermal resistances' variation over time, three different sensor heights' locations and the predicted THERM *R*-value for the reference LSF facade wall.

All these verifications allowed the good working conditions of the experimental apparatus to be evidenced, to ensure the reliability of the measurements, as well as the THERM models.

2.3. Numerical Simulations

The bidimensional simulations of the LSF facades' thermal performance were achieved using the Finite Element Method (FEM) commercial software THERM® (version 7.6.1, Lawrence Berkeley National Laboratory, United States Department of Energy: Berkeley, CA, USA). The corresponding model details are briefly explained next.

This algorithm makes use of 2D steady-state conservation of energy equation for isotropic materials

$$\rho C_p \left(v_x \frac{\partial T}{\partial x} + v_y \frac{\partial T}{\partial y} \right) = \left(\frac{\partial^2 T}{\partial x^2} + \frac{\partial^2 T}{\partial y^2} \right) + Q_s, \quad (4)$$

where: ρ is the density (kg/m^3); C_p is the specific heat capacity (J/(kg·K); v is the velocity (m/s); T is the temperature (°C or K); λ is the thermal conductivity (W/(m·K)); and Q_s represents the volumetric heat source (W/m^3).

Given the repetitive features of this LSF facade structure, as previously illustrated in Figure 1, only a typical 2D portion of the facade cross-sections (with 400 mm length) was implemented in the THERM models. The predicted *R*-values were computed in these models along the inner wall surface (Figure 1). In Section 2.1 (Tables 1 and 2) the materials' thermal properties used in the simulations were presented earlier. Additionally, for all models built in this work, a maximum error of 2% on the FEM computations was set.

For each THERM model, two sets of boundary conditions were defined, namely the air temperatures and the surface *R*-values. The colder external (5 °C) and warmer internal (40 °C) air temperatures were defined equal to the set-points defined for the cold and hot climatic chambers during the laboratory measurements, respectively (see Section 2.2.2). Nevertheless, as is well known, since the *R*-values are computed for a unitary temperature difference, they do not rely on the chosen temperature difference between interior and exterior environments.

The surface thermal resistances used in the THERM models were set equal to the average values measured for each LSF wall surface and for each test, considering the air and surface temperature differences, as well as the recorded heat fluxes in the wall surfaces. The measured surface *R*-values, ranging between 0.10 and 0.15 $m^2 \cdot K/W$, were near the value defined in ISO 6946 [42] for the internal surface resistance (R_{si}) when there is a horizontal heat flow, i.e., 0.13 $m^2 \cdot K/W$. These measured values were not even smaller because the fans used were not powerful, having reduced dimensions (only 10 by 10 cm) and oriented to the opposite side of the test-sample wall. Moreover, the black radiation shields provide some protection, uniformizing the air flux near the wall test-sample outer surface.

Several verifications were made to ensure a good accuracy of the THERM models, explicitly: (i) 2D test cases' verification prescribed in ISO 10211 [43]; (ii) computations assuming simplified homogeneous wall layers; and (iii) validation using experimental laboratory measurements.

The two ISO 10211 [43] 2D test cases were modelled with success and the THERM software FEM algorithm for steady-state heat transfer simulations was classified as a high precision algorithm. Furthermore, the authors have a large amount of experience of using this software, as can be seen in these references [21,22,30,33].

Regarding the homogeneous wall layers' verification, the reference LSF facade wall (illustrated in Figure 1), was modelled without the steel studs, and it was assumed the other wall layers were homogenous and continuous. Using this simplified wall model, the analytical solution could be applied since it is well known, being prescribed in ISO 6946 [42]. In this case, the total thermal resistance is calculated as the summation of the layer's thermal resistances. Figure 7b illustrates the obtained results, where, as expected, both analytic and THERM *R*-values are the same when it is assumed there are homogeneous layers within the facade wall (4.283 $m^2 \cdot K/W$).

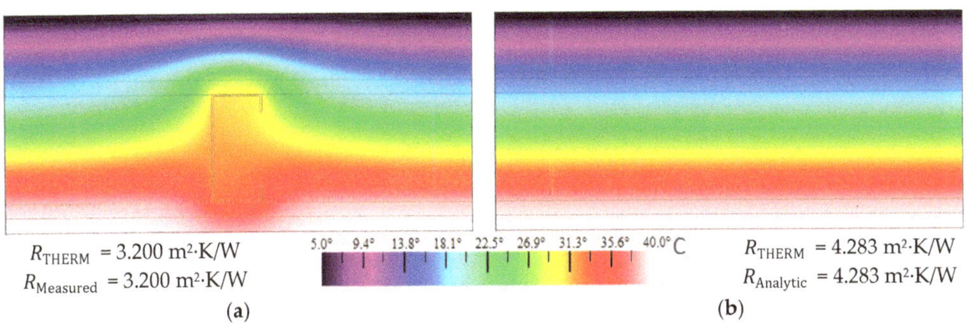

Figure 7. Accuracy verification of the THERM models: temperature distribution and conductive *R*-values. (**a**) Inhomogeneous layers. (**b**) Homogeneous layers.

Concerning the lab measurements' verification, Figure 6a also illustrates the measured overall surface-to-surface *R*-values of the reference facade LSF wall. The measured *R*-value is equal to the predicted THERM value (3.200 $m^2 \cdot K/W$), ensuring an excellent agreement among simulated and predicted surface-to-surface *R*-values, even with all the related uncertainties.

Moreover, even with a continuous exterior thermal insulation, it is very clear that, in the foreseen temperature color distribution (Figure 7), the consequence of the steel stud thermal bridge was originated by the higher heat transfer. This higher heat transfer meaningfully decreases the *R*-value of the facade when related with the simplified model, having homogeneous layers, i.e., without steel studs. This thermal resistance decrease was 1.083 $m^2 \cdot K/W$ (−25%).

3. Results

The achieved results are structured here into three subsections. First, the measured and the predicted thermal resistance values are displayed. Next, the measured thermal resistance improvements due to the use of thermal break (TB) strips are graphically exhibited. Finally, the infrared thermography technic is used to better visualize the steel studs' thermal bridges effect.

3.1. Measured and Predicted R-Values

The obtained measured and predicted results for the evaluated load-bearing LSF facade walls are displayed in Table 3, with these results being organized by the position and material of the TB strips, but starting with the reference LSF facade wall (without any TB strips). As displayed here, the agreement between the predicted and the measured *R*-values is very good. These differences range between −3% and 0%.

In Table 3, the thermal resistance increase that originated from using TB strips along the metallic profile flanges is well visible. This increase is bigger for materials with smaller thermal conductivities (e.g., AG—aerogel), as well as when two TB strips are used.

3.2. Measured Thermal Resistance Improvement

To better visualize these features in a graphical way, Figure 8a shows the thermal performance improvement originated by the TB strips' use on the LSF facades, with the thermal resistance increase ranging from +4% (for the inner recycled rubber, R1in) up to +28% (for the double aerogel TB strips, AG × 2). As expected, the outer and inner TB strips' performances are quite analogous, being a little better when placed in the outer stud flange (+1%).

The major *R*-value increase occurred when two TB strips were used (×2) and for lower thermal conductivity of the TB strip material, i.e., aerogel, AG. Indeed, the aerogel TB strips displayed a considerable increase in the obtained thermal resistances: +15% and +14% for outer and inner strips, respectively, as well as +28% when using two TB strips. However, it was not possible to reach the thermal resistance provided by a homogeneous wall without steel studs (4.283 $m^2 \cdot K/W$), as previously presented in Figure 7b, not even with the greatest thermal performance configuration, i.e., when having two aerogel TB strips (4.093 $m^2 \cdot K/W$).

To compare these facade LSF wall measurements with previous load-bearing partition LSF wall measured *R*-values (see Reference [30]), Figure 8b also exhibits these former values. Starting this comparison by using the reference *R*-values without TB strips, the facade LSF wall has a significantly higher thermal resistance (3.200 $m^2 \cdot K/W$) due to the use of ETICS, when compared with the partition LSF wall (1.558 $m^2 \cdot K/W$), corresponding to a reference *R*-value increase of 1.642 $m^2 \cdot K/W$ (+105%).

Comparing now these measurement results using TB strips for the facade (Figure 8a) with the partition LSF walls (Figure 8b), both plots show a similar tendency, with the thermal resistances being higher for the facade LSF walls, as already justified and previously noticed. However, the thermal performance improvement due to the use of TB strips is now very reduced in the facade LSF walls (Figure 8a), both in absolute values ($m^2 \cdot K/W$) and even more evident in percentage values.

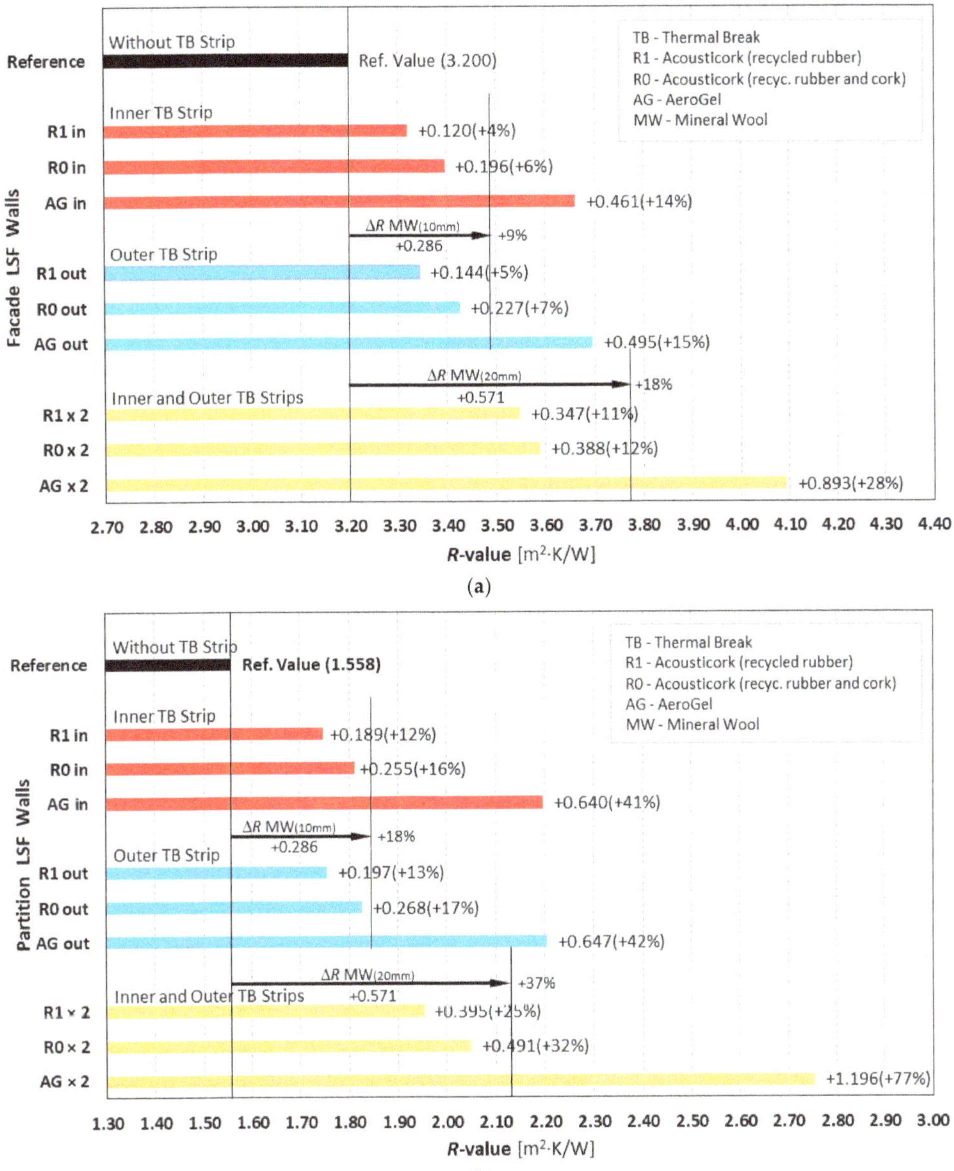

Figure 8. Measured thermal resistances of load-bearing LSF walls. (**a**) Facade LSF walls. (**b**) Partition LSF walls (adapted from [30]).

3.3. Infrared Thermography Assessment

To better visualize the differences regarding the steel studs' thermal bridge effect between the facade and the partition LSF walls, with and without TB strips, some infrared images are displayed in Figure 9. In the partition LSF walls (Figure 9b) the presence of the middle vertical steel stud when there is no TB strip (left image) is well visible, as is the reduction in the thermal bridge effect when two aerogel TB strips are used (right image).

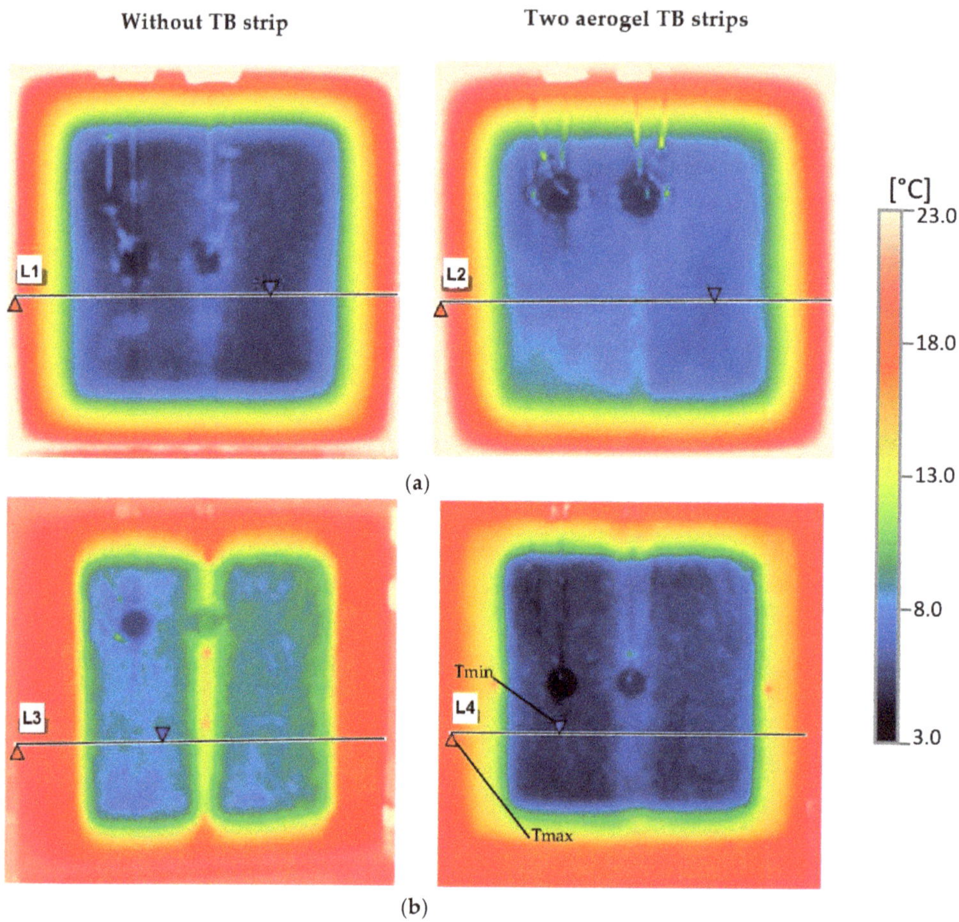

Figure 9. IR images of the LSF walls: cold surface. (**a**) Facade LSF walls. (**b**) Partition LSF walls (adapted from [30]).

However, in the facade LSF walls (Figure 9a), the presence of the vertical central steel stud is not clearly visible, even when there are no TB strips (left image). To see this singularity in more detail, Figure 10 exhibits the surface temperatures along the horizontal lines previously displayed in Figure 9. In this plot the significant peak temperature rises near the central steel stud in the partition LSF wall are even more perceptible (Line 3), as well as the corresponding attenuation due to the use of two aerogel TB strips (Line 4). Again, in the facade wall, with a continuous thermal insulation (ETICS), this steel stud thermal bridge-related temperature increase in the middle of the LSF wall is extremely limited when there are no TB strips (Line 1), and almost negligible when there are two TB strips (Line 2).

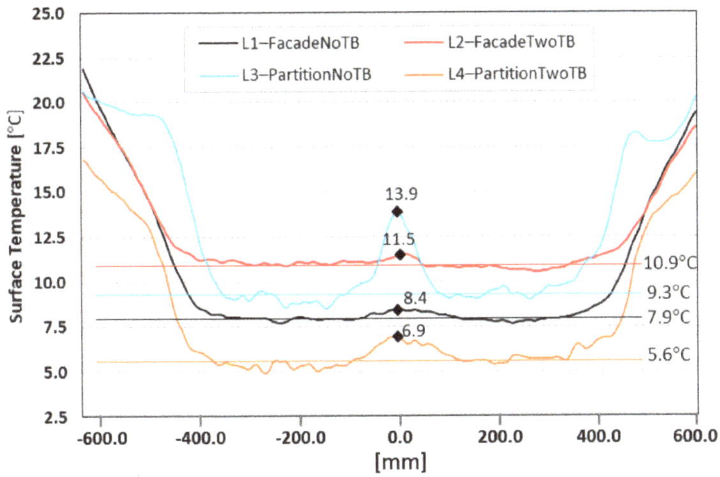

Figure 10. Cold surface temperatures along horizontal lines obtained from thermographic images of the measured facade and partition LSF walls, with and without thermal break (TB) strips made from aerogel.

4. Discussion of Results

To provide an easier comparison of their overall performances, the charts in Figure 11 exhibit the thermal resistance improvements due to the use of TB strips, for both facade (in red color) and partition (in black color) LSF walls.

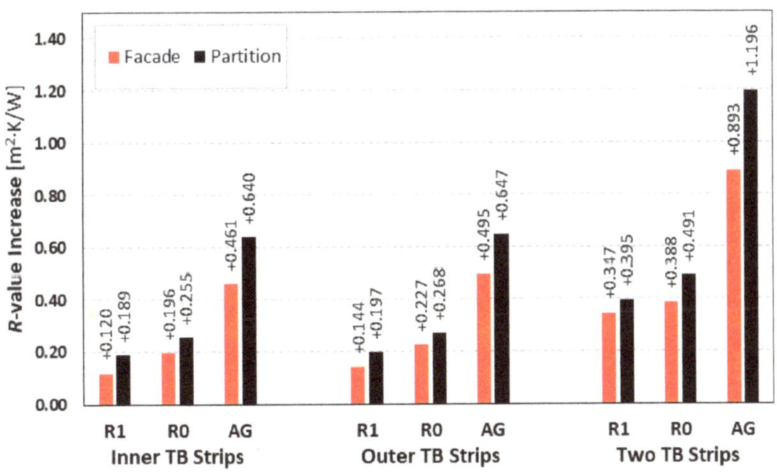

Figure 11. The increase in thermal resistances due to thermal break (TB) strips' use in load-bearing LSF walls: facade vs partition [30].

From the previous plots (Figure 11), it can be concluded that the TB strips are clearly less efficient in facade LSF walls when compared to partition LSF walls. This feature is related with the use of a continuous external thermal insulation layer (ETICS) in facade walls (in this case 50 mm of EPS), decreasing in this way the relevance of the thermal bridges originated by the steel studs. Additionally, the thermal resistance increases due to aerogel (AG) TB strips remain very relevant. In fact, this thermal performance improvement is about three times bigger when compared to recycled rubber (R1). Moreover, the thermal resistance rises due to the TB strips made from rubber/cork composite (R0) are only a little bigger when compared to recycled rubber (R1), i.e., about +0.08 $m^2 \cdot K/W$ for single TB strips and even lower for double TB strips (+0.04 $m^2 \cdot K/W$).

Notice that, in all evaluated LSF walls, the use of TB strips allowed their thermal performance to be improved, by increasing the overall *R*-value of the wall. Obviously, with the higher thermal resistance and the lower thermal transmittance (or *U*-value) and, therefore, less thermal energy being lost or transferred across the wall, the energy efficiency is increased.

5. Conclusions

In this paper, a new set of experimental lab measurements was performed to evaluate the thermal performance improvement in Lightweight Steel Frame (LSF) facade walls due to the use of thermal break (TB) strips. Three different materials were evaluated as TB strips, namely: recycled rubber (R1); recycled rubber/cork composite (R0); and aerogel (AG), for three different TB strip positions along the steel stud flanges: inner, outer and on both sides (double).

Notice that all the measured *R*-values were related with the simulation results achieved using implemented numerical bidimensional finite element models. These comparisons provided excellent thermal resistance agreements, within an error range of ±3%, for the ten evaluated facade LSF walls, ensuring good reliability, as well as a high robustness of the test procedures and implemented measurement apparatus.

This research work has provided new developments of a previous study regarding the experimental assessment of TB strips' performance in non-load-bearing and load-bearing LSF walls [30]. In this case, instead of partition walls, load-bearing facade LSF walls were evaluated, usually with a continuous external thermal insulation composite system (ETICS).

Taking the previous study for load-bearing partition LSF walls as a reference also [30], the main conclusions are summarized as follows:

- The outer and inner TB strips still have very similar thermal performances. However, as happened before for partition walls, the TB strips positioned in the outer steel flange of the facade LSF walls appear to have slightly better performance (around +1% in the measured *R*-value).
- Again, as expected, the best performances were found for double TB strips and for aerogel TB strips.
- However, neither the former load-bearing partition LSF wall, neither this new facade LSF wall, were able to fully mitigate the steel frame thermal bridges' effect, not even for the best thermal performance configuration (two aerogel TB strips).
- Quite similar thermal performances were measured for the other two materials: recycled rubber (R1), and rubber/cork composite (R0).
- Comparing the thermal resistance improvement for these facade LSF walls and the previously assessed partition LSF walls, it was observed that the TB strips were clearly less efficient in facade LSF walls.

This last conclusion can be justified by the existence of a continuous external thermal insulation layer (ETICS) in the facade walls (in this case 50 mm of EPS), decreasing the relevance of the thermal bridges originated by the steel studs. Consequently, the efficiency of the TB strips becomes reduced.

Notice that the improved thermal resistances due to the use of TBS were caused not only to the TB strips themselves, but also provided by the increased total thickness of the LSF wall, particularly the increased thickness of the expansible mineral wool thermal insulation, placed inside the wall cavity.

This research allowed the effectiveness of TB strips for the *R*-values' increase in partition and facade load-bearing LSF walls to be better comprehended, quantified and compared, information that was not available in the literature, particularly for facade LSF walls. Additionally, the measured conductive thermal resistances can be used, for the validation of numerical simulations in LSF walls with identical arrangements, as benchmark values.

Author Contributions: Conceptualization, P.S. and D.M.; Formal analysis, P.S., D.M., D.F. and A.V.; Funding acquisition, P.S.; Investigation, P.S., D.M., D.F. and A.V.; Methodology, P.S. and D.M.; Project administration, P.S.; Resources, P.S.; Supervision, P.S.; Validation, P.S.; Writing—original draft, P.S.; Writing—review & editing, P.S., D.M., D.F. and A.V. All authors have read and agreed to the published version of the manuscript.

Funding: This research was funded by FEDER funds through the Competitivity Factors Operational Programme, COMPETE, and by national funds through FCT, Foundation for Science and Technology, within the scope of the project POCI-01-0145-FEDER-032061.

Cofinanciado por: POCI-01-0145-FEDER-032061

Acknowledgments: The authors also want to thank the support provided by the following companies: Pertecno, Gyptec Ibéria, Volcalis, Sotinco, Kronospan, Hulkseflux, Hilti and Metabo.

Conflicts of Interest: The authors declare no conflict of interest.

References

1. Farouk, N.; El-Rahman, M.A.; Sharifpur, M.; Guo, W. Assessment of CO_2 emissions associated with HVAC system in buildings equipped with phase change materials. *J. Build. Eng.* **2022**, *51*, 104236. [CrossRef]
2. European Union. Directive (EU) 2018/844 of the European Parliament and of the Council of 30 May 2018 amending Directive 2010/31/EU on the energy performance of buildings and Directive 2012/27/EU on energy efficiency. *Off. J. Eur. Union* **2018**, *2018*, 75–91.
3. D'Agostino, D.; Tzeiranaki, S.T.; Zangheri, P.; Bertoldi, P. Assessing Nearly Zero Energy Buildings (NZEBs) development in Europe. *Energy Strateg. Rev.* **2021**, *36*, 100680. [CrossRef]
4. Bustos García, A. Morteros con Propiedades Mejoradas de Ductilidad por Adición de Fibras de Vidrio, Carbono y Basalto. Doctoral Thesis, Universidad Politécnica de Madrid, Madrid, Spain, 2018. (In Spanish).
5. European Commission. The European Green Deal. In *Communication from the European Comission: COM/2019/640 Final, 11.12.2019*; EUR-Lex: Brussels, Belgium, 2019.
6. Cutanda, B.L.; Turrillas, J.C.A. *Administración y Legislación Ambiental*, 11th ed.; Dykinson: Madrid, Spain, 2020; ISBN 978-84-1324-302-3.
7. Buckley, N.; Mills, G.; Reinhart, C.; Berzolla, Z.M. Using urban building energy modelling (UBEM) to support the new European Union's Green Deal: Case study of Dublin Ireland. *Energy Build.* **2021**, *247*, 111115. [CrossRef]
8. Rezai, S.H.; Allard, F.; Abelé, C.; Doya, M. Evaluating External Thermal Insulation Composite Systems (ETICS) regarding the building's global performance. *Energy Procedia* **2015**, *78*, 1562–1567. [CrossRef]
9. Parracha, J.; Borsoi, G.; Flores-Colen, I.; Veiga, R.; Nunes, L.; Dionísio, A.; Gomes, M.G.; Faria, P. Performance parameters of ETICS: Correlating water resistance, bio-susceptibility and surface properties. *Constr. Build. Mater.* **2021**, *272*, 121956. [CrossRef]
10. Fernandes, C.; de Brito, J.; Cruz, C.O. Architectural integration of ETICS in building rehabilitation. *J. Build. Eng.* **2016**, *5*, 178–184. [CrossRef]
11. Luján, S.V.; Arrebola, C.V.; Sánchez, A.R.; Benito, P.A.; Cortina, M.G. Experimental comparative study of the thermal performance of the façade of a building refurbished using ETICS, and quantification of improvements. *Sustain. Cities Soc.* **2019**, *51*, 101713. [CrossRef]
12. Kolaitis, D.I.; Malliotakis, E.; Kontogeorgos, D.A.; Mandilaras, I.; Katsourinis, D.I.; Founti, M.A. Comparative assessment of internal and external thermal insulation systems for energy efficient retrofitting of residential buildings. *Energy Build.* **2013**, *64*, 123–131. [CrossRef]
13. Santos, P.; da Silva, L.S.; Ungureanu, V. *Energy Efficiency of Light-Weight Steel-Framed Buildings*, 1st ed.; Technical Committee 14—Sustainability & Eco-Efficiency of Steel Construction; European Convention for Constructional Steelwork (ECCS): Brussels, Belgium, 2012; ISBN 978-92-9147-105-8.
14. Choi, J.-S.; Kim, C.-M.; Jang, H.-I.; Kim, E.-J. Detailed and fast calculation of wall surface temperatures near thermal bridge area. *Case Stud. Therm. Eng.* **2021**, *25*, 100936. [CrossRef]
15. Liu, C.; Mao, X.; He, L.; Chen, X.; Yang, Y.; Yuan, J. A new demountable light-gauge steel framed wall: Flexural behavior, thermal performance and life cycle assessment. *J. Build. Eng.* **2022**, *47*, 103856. [CrossRef]
16. Perera, D.; Poologanathan, K.; Gillie, M.; Gatheeshgar, P.; Sherlock, P.; Upasiri, I.; Rajanayagam, H. Novel conventional and modular LSF wall panels with improved fire performance. *J. Build. Eng.* **2022**, *46*, 103612. [CrossRef]

17. Alembagheri, M.; Sharafi, P.; Rashidi, M.; Bigdeli, A.; Farajian, M. Natural dynamic characteristics of volumetric steel modules with gypsum sheathed LSF walls: Experimental study. *Structures* **2021**, *33*, 272–282. [CrossRef]
18. Tavares, V.; Soares, N.; Raposo, N.; Marques, P.; Freire, F. Prefabricated versus conventional construction: Comparing life-cycle impacts of alternative structural materials. *J. Build. Eng.* **2021**, *41*, 102705. [CrossRef]
19. Perera, D.; Upasiri, I.; Poologanathan, K.; Gatheeshgar, P.; Sherlock, P.; Hewavitharana, T.; Suntharalingam, T. Energy performance of fire rated LSF walls under UK climate conditions. *J. Build. Eng.* **2021**, *44*, 103293. [CrossRef]
20. Moga, L.; Petran, I.; Santos, P.; Ungureanu, V. Thermo-Energy Performance of Lightweight Steel Framed Constructions: A Case Study. *Buildings* **2022**, *12*, 321. [CrossRef]
21. Santos, P.; Gonçalves, M.; Martins, C.; Soares, N.; Costa, J.J. Thermal Transmittance of Lightweight Steel Framed Walls: Experimental versus Numerical and Analytical Approaches. *J. Build. Eng.* **2019**, *25*, 100776. [CrossRef]
22. Santos, P.; Lemes, G.; Mateus, D. Analytical Methods to Estimate the Thermal Transmittance of LSF Walls: Calculation Procedures Review and Accuracy Comparison. *Energies* **2020**, *13*, 840. [CrossRef]
23. Santos, P.; Poologanathan, K. The Importance of Stud Flanges Size and Shape on the Thermal Performance of Lightweight Steel Framed Walls. *Sustainability* **2021**, *13*, 3970. [CrossRef]
24. Roque, E.; Santos, P. The Effectiveness of Thermal Insulation in Lightweight Steel-Framed Walls with Respect to Its Position. *Buildings* **2017**, *7*, 13. [CrossRef]
25. Roque, E.; Santos, P.; Pereira, A.C. Thermal and sound insulation of lightweight steel-framed façade walls. *Sci. Technol. Built Environ.* **2019**, *25*, 156–176. [CrossRef]
26. Santos, P.; Ribeiro, T. Thermal Performance of Double-Pane Lightweight Steel Framed Walls with and without a Reflective Foil. *Buildings* **2021**, *11*, 301. [CrossRef]
27. Santos, P.; Ribeiro, T. Thermal Performance Improvement of Double-Pane Lightweight Steel Framed Walls Using Thermal Break Strips and Reflective Foils. *Energies* **2021**, *14*, 6927. [CrossRef]
28. Santos, P.; Abrantes, D.; Lopes, P.; Mateus, D. Experimental and Numerical Performance Evaluation of Bio-Based and Recycled Thermal Break Strips in LSF Partition Walls. *Buildings* **2022**, *12*, 1237. [CrossRef]
29. Martins, C.; Santos, P.; da Silva, L.S. Lightweight steel-framed thermal bridges mitigation strategies: A parametric study. *J. Build. Phys.* **2016**, *39*, 342–372. [CrossRef]
30. Santos, P.; Mateus, D. Experimental assessment of thermal break strips performance in load-bearing and non-load-bearing LSF walls. *J. Build. Eng.* **2020**, *32*, 101693. [CrossRef]
31. WEBERTHERM UNO. Technical Sheet: Weber Saint-Gobain ETICS Finish Mortar. 2018. Available online: www.pt.weber/files/pt/2019-04/FichaTecnica_weberthermuno.pdf (accessed on 14 March 2019). (In Portuguese).
32. TincoTerm. Technical Sheet: EPS 100. 2015. Available online: http://www.lnec.pt/fotos/editor2/tincoterm-eps-sistema-co-1.pdf (accessed on 14 March 2021). (In Portuguese).
33. Santos, P.; Lemes, G.; Mateus, D. Thermal Transmittance of Internal Partition and External Facade LSF Walls: A Parametric Study. *Energies* **2019**, *12*, 2671. [CrossRef]
34. Santos, C.; Matias, L. *ITE50—Coeficientes de Transmissão Térmica de Elementos da Envolvente dos Edifícios*; LNEC—Laboratório Nacional de Engenharia Civil: Lisbon, Portugal, 2006. (In Portuguese)
35. *MS-R0 Test Report. Test Report Ref. 5015 PE 1977/08—Determination of Thermal Conductivity (Acousticork MS-R0)*; Departamento de Engenharia Têxtil, Universidade do Minho: Braga, Portugal, 2008.
36. *ISO 9869-1*; Thermal Insulation—Building Elements—In-Situ Measurement of Thermal Resistance and Thermal Transmittance. Part 1: Heat Flow Meter Method. ISO—International Organization for Standardization: Geneva, Switzerland, 2014.
37. Soares, N.; Martins, C.; Gonçalves, M.; Santos, P.; da Silva, L.S.; Costa, J.J. Laboratory and in-situ non-destructive methods to evaluate the thermal transmittance and behaviour of walls, windows, and construction elements with innovative materials: A review. *Energy Build.* **2019**, *182*, 88–110. [CrossRef]
38. Rasooli, A.; Itard, L. In-situ characterization of walls' thermal resistance: An extension to the ISO 9869 standard method. *Energy Build.* **2018**, *179*, 374–383. [CrossRef]
39. *ASTM-C1155–95 (Reapproved-2013)*; Standard Practice for Determining Thermal Resistance of Building Envelope Components from the In-Situ Data. ASTM—American Society for Testing and Materials: Philadelphia, PA, USA, 2013.
40. *ASTM-C1046–95 (Reapproved-2013)*; Standard Practice for In-Situ Measurement of Heat Flux and Temperature on Building Envelope Components. ASTM—American Society for Testing and Materials: Philadelphia, PA, USA, 2013.
41. ASHRAE. *Handbook of Fundamentals (SI Edition)*; ASHRAE—American Society of Heating, Refrigerating and Air-conditioning Engineers: Atlanta, GA, USA, 2017.
42. *ISO 6946*; Building Components and Building Elements—Thermal Resistance and Thermal Transmittance—Calculation Methods. International Organization for Standardization: Geneva, Switzerland, 2017.
43. *ISO 10211*; Thermal Bridges in Building Construction—Heat Flows and Surface Temperatures—Detailed Calculations. ISO—International Organization for Standardization: Geneva, Switzerland, 2017.

energies

Review

Advanced Active and Passive Methods in Residential Energy Efficiency

Hessam Taherian [1] and Robert W. Peters [2,*]

[1] School of Science, Engineering, and Technology, Penn State Harrisburg, Harrisburg, PA 17057, USA
[2] Department of Civil, Construction, and Environmental Engineering, University of Alabama at Birmingham, Birmingham, AL 35294, USA
* Correspondence: rwpeters@uab.edu

Abstract: Energy efficiency in buildings is very important since it contributes significantly to fossil fuel consumption and consequently climate change. Several approaches have been taken by researchers and the industry to address the issue. These approaches are classified as either passive or active approaches. The purpose of this review article is to summarize a number of the technologies that have been investigated and/or developed. In this technical review paper, the more commonly used active and passive building energy conservation techniques are described and discussed. The pros and cons of both the active and passive energy techniques are described with appropriate reference citations provided. This review article provides a description to give an understanding of building conservation approaches. In the active classification, several methods have been reviewed that include earth-to-air heat exchangers, ground-source and hybrid heat pumps, and the use of new refrigerants, among other methods. In the passive classification, methods such as vegetated roofs, solar chimneys, natural ventilation, and more are discussed. Often, in a building, multiple passive and active methods can be employed simultaneously.

Keywords: active building energy systems; passive building energy systems; Trombe wall; white roofs; vegetated roofs; ground-source heat pumps; new refrigerants

Citation: Taherian, H.; Peters, R.W. Advanced Active and Passive Methods in Residential Energy Efficiency. *Energies* **2023**, *16*, 3905. https://doi.org/10.3390/en16093905

Academic Editor: Paulo Santos

Received: 17 February 2023
Revised: 24 April 2023
Accepted: 25 April 2023
Published: 5 May 2023

1. Background

1.1. Decarbonization

Building operations are a critical step towards obtaining global carbon neutrality [1]. Commercial buildings are among the more energy-consuming sectors and they present the fastest growing demand worldwide. Building decarbonization is beyond the scope of this paper. However, in a recent paper addressing the decarbonization of commercial building operations, the results of Xiang et al. [1] showed: (1) the mean carbon intensity of commercial building operations in 16 countries declined approximately 1.94% throughout the period of 2000 to 2019; (2) energy intensity involving different end-uses contributed to decarbonizing commercial buildings with the largest contribution from space heating (-14.33 kg $CO_2/m^3/yr$) and service lighting (-5.29 kg $CO_2/m^3/yr$); and (3) the pace of decarbonization of global commercial buildings is slowing. In another decarbonization study, the results of Ma et al. [2] indicated that: (1) commercial buildings operational carbon emissions continue to increase by 17.7% with economic growth effects and energy use being the key drivers contributing to the increase; (2) since 2009, operational carbon emissions have decoupled from economic growth effects in most megalopolises; and (3) the operational decarbonization of megalopolises commercial buildings has been gradually accelerating. The National Governors Association [3] supports state decarbonization efforts to enhance energy reliability and resiliency, reduce air pollution, improve public health, and foster economic development. Energy efficiency plays an extremely important role in this effort. Benefits of increasing energy efficiency include: cost savings, increased reliability and resiliency, reduced emissions, increased health benefits, economic and workforce development, and energy affordability. Barriers to increased energy efficiency include: high up-front costs, utility regulatory disincentives, workforce gaps, valuing energy savings, and lack of consumer awareness.

1.2. Introduction

Residential and commercial buildings account for nearly 40% of energy use [4]. Space conditioning accounts for 40% to 60% of building end-use. In the U.S., heating, ventilation, and air conditioning (HVAC) make up about 50% of energy use in buildings, which is about 20% of their total energy consumption [5]. Sun et al. [6] and Ben Romdhane et al. [7] have shown that phase change materials can be used in a variety of ways to store and release energy by using their endothermic/exothermic properties, resulting in improved energy consumption, heat storage, and improved thermal comfort. Additionally, phase change materials provide assistance to energy shortages, and carbon emissions. Li et al. [5] provide an excellent discussion of active and passive energy methods. Active systems require the use of a fan and heat pumps/boilers to move the energy throughout the building. Passive energy techniques are more energy-efficient for application in buildings. Passive techniques reduce building energy consumption, abate the carbon footprint, and alleviate building energy bills [8], although not all strategies are cost-effective. Ponmurugan et al. [9] emphasize that reduction in heat gains into buildings through walls, ceilings, and roofing plays a vital role in passive energy techniques. Additionally, passive heating and cooling systems can reduce space-heating and space-cooling loads in buildings [10].

This paper addresses both active and passive energy conservation methods in buildings. This paper does not address every energy conservation method, but we address the major energy conservation methods. Development of effective and efficient energy techniques is critical to controlling room air temperatures and improving building energy consumption [11].

2. Active Building Energy Conservation Methods

Any heat exchange method for providing indoor comfort to buildings that does not employ a prime mover (pump, fan, etc.) to force the flow of heat from the source to the point of use is considered a passive approach. By that definition, for example, solar collectors can be considered active systems if they are a component in a forced flow system and otherwise passive systems. The presence of heat pumps or similar air conditioning devices brings the system to the active category. Active systems typically offer better overall efficiency, while passive systems are attributed to lower costs. Another major difference between an active and a passive system is that the latter typically requires less maintenance. In some buildings, a combination of active and passive systems can be applied.

2.1. Modifications to the Conventional Vapor-Compression Refrigeration Systems

Analyses of various modifications to the conventional vapor-compression refrigeration cycle have been reported in the literature that is reviewed below. A conventional vapor-compression refrigeration cycle uses a vapor compressor to increase the enthalpy (increased pressure and temperature) of an appropriate working fluid (refrigerant) before it is run through a heat exchanger (condenser) to reject heat to the surroundings. The refrigerant which is slightly subcooled (compressed liquid phase) is passed through a throttling device that causes a pressure drop and consequently a significant drop in the refrigerant's temperature. The cold, low-pressure vapor–liquid refrigerant mixture enters the evaporator where it receives heat from the indoor environment (cooling effect) while it turns into vapor prior to entering the suction side of the compressor. Conventional vapor-compression refrigeration systems offer a typical coefficient of performance (COP) of 2 to 3. It has been shown that by using the soil around the building both as a source of thermal energy and as a storage system by hooking up the heat pump system to a ground loop, the system's COP and consequently its energy efficiency improves significantly due to less electricity use.

The ground loop is usually either vertical (boreholes) or horizontal (shallow). Additional modifications have also proven to be effective. These additional modifications include the use of assistive equipment such as liquid dry coolers [12–14], solar collectors [15], and a combination of solar collectors and biomass generators shown in Figure 1 [16]. The mentioned modifications in most cases increase the system COP by 1 to 2.5. Adding a liquid dry cooler to the ground loop of a ground-source heat pump increases the system's thermal efficiency (reduces seasonal electricity use) when the outdoor temperature is favorable to bypass the ground loop. Bypassing the ground loop allows the soil to recover from continuous heat removal/rejection which otherwise would diminish its potential to supply the needed energy. The benefit of adding a solar thermal collector array to the ground loop is apparent because the thermal energy collected can assist with the performance of the heat pump. A solar-assisted heat pump system is evidently only effective during the heating season, and COP values of as high as 6 have been reported for these systems [17]. However, when photovoltaic/thermal (PV/T) collectors are employed, the benefits can be collected in both the heating and cooling seasons [18]. In PV/T systems, while the top layer of the collector is made of solar cells that generate electricity, the lower layer acts as a thermal collector that can heat a liquid or air for heating the building. The electricity generated by the PV cells is used to supplement the electricity use by the compressor and fans of the air conditioning equipment. Researchers in Canada reported using recovered waste heat from a data center as the heat source for activating an absorption chiller system for cooling [19].

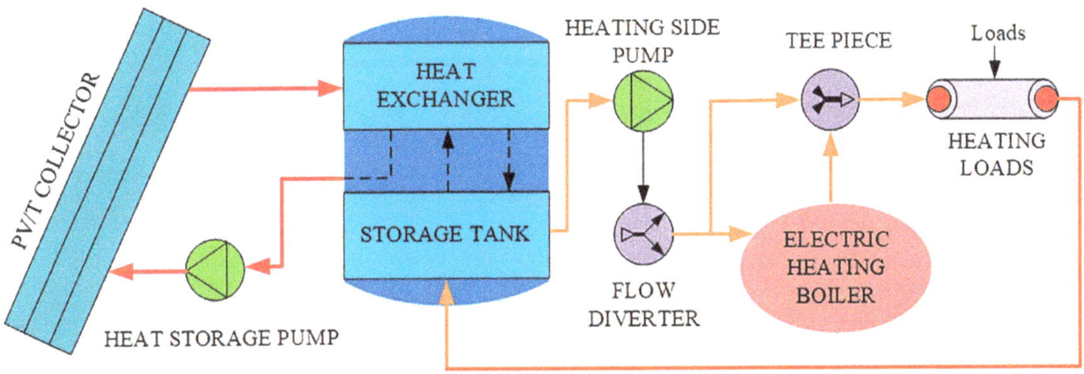

Figure 1. Schematic diagram of a combined solar collector and biomass boiler system [16].

2.2. *Advances in Heat-Activated Cooling Systems*

Absorption water chillers have been in use for many years. The principle of operation is similar to the vapor-compression refrigeration system with the difference that the process of increasing the pressure of the refrigerant occurs in a combination of an absorber-generator system in which one fluid has a considerable capacity for absorbing a second fluid. The pair of fluids that are used for this application is commonly, ammonia-water or lithium bromide-water. This category of refrigeration systems is called "thermally-activated cooling" [20]. A schematic diagram of a solar-assisted absorption cooling system is presented in Figure 2.

These systems use less electricity because they do not have to compress a gas in a compressor. Instead, they employ a pump to pump a liquid from the absorber to the generator. The pump uses significantly less electricity than a compressor. The required energy is supplied by a heat source. Various heat sources can be used to activate the process as long as the temperature of the source is at least 80 °C. The heat from the direct combustion of natural gas [21], exhaust heat from a microturbine system which is also used to generate electricity [22], the heat from an internal combustion engine's jacket coolant loop [23] and the heat generated by operating a phosphoric acid fuel cell [24] have been reported to have been used for this purpose. In comparison to vapor-compression refrigeration systems, absorption refrigeration systems do not achieve high COP values and are limited to COP values of less than 1.5 in practice.

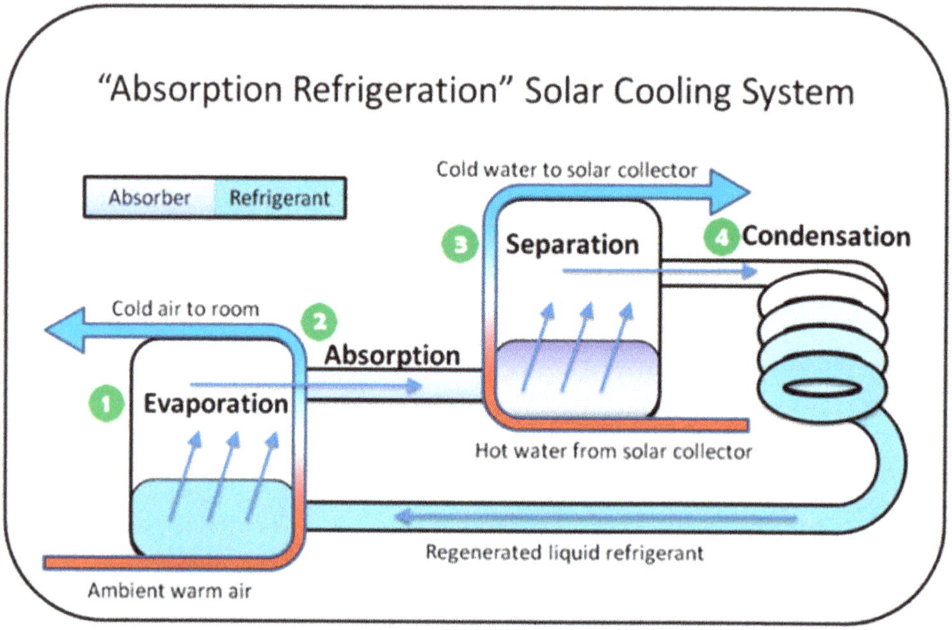

Figure 2. Schematic diagram of an absorption cooling system with heat supply by solar collectors.

2.3. Non-Conventional Systems

Besides the evaporative cooling method which is only feasible for dry climates, the vapor-compression and absorption refrigeration systems are considered conventional cooling systems. Several cooling techniques are outside the realm of conventional systems for which examples are given below. Adsorption cooling systems use a solid desiccant material to adsorb a liquid into the solid desiccant and then desorb the liquid as vapor in a generator in a somewhat similar manner to the process in an absorption refrigeration system. Shanghai Jiao Tong University researchers reported a COP of greater than 0.5 on a 9-kW adsorption refrigeration system [25]. Similar to absorption chillers, various heat sources have been used for the desorption process of the adsorption systems. Desiccant materials have also been used in modified evaporative cooling units for buildings. In desiccant evaporative cooling systems, the outdoor air is passed through a desiccant wheel or a liquid desiccant tower to reduce its humidity to near 0% relative humidity. The ultra-dry air then enters an evaporative cooler where it absorbs water vapor to the saturation point before it enters the interior conditioned area [26,27]. In liquid desiccant air conditioning systems, the desiccant material ($CaCl_2$) used is regenerated in a heat exchanger that uses a range of heat sources including solar energy [28] or a heat pump [29].

Proximity to large lakes offers the potential to use deep lake water's low temperatures (12–13 °C) for cooling of buildings as reported by Fung et al. [30] and Kuyuk et al. [31]. The proposed system requires a significant initial investment, however the return on investment is quick due to the utilization of a natural reservoir (sink) for the heat removed from the buildings.

Other methods exist that are considered assisting technologies to the conventional cooling systems. Earth-to-air heat exchangers (EAHE) [32], have gained popularity to reduce the energy use by air conditioning systems. Outdoor air flows underground through a pipe structure that optimizes heat gain/loss to the soil. The conditioned air is then used as the outdoor fresh air for the building's air handling unit. The COP of the system is expected to increase due to the pre-conditioning of the outdoor air [33].

Radiative sky cooling [34,35] is another assistive technology that is gaining popularity. The clear night sky has favorable conditions to cool down the liquid that flows in an unglazed radiator due to the very low apparent sky temperature. The apparent sky temperature can be as low as 4 °C when the ambient temperature is 30 °C according to some simplified models [36].

2.4. New Refrigerants

In conventional vapor-compression refrigeration systems, the search for less harmful refrigerants is a continuous battle. Many thermodynamically near-perfect refrigerants have been ruled out as potentially harmful to the environment. In searching for or developing new refrigerants, researchers must balance the global warming potential and the thermodynamic properties that improve the efficiency (COP) of the refrigeration system. Carbon dioxide and propane are some of the natural refrigerants that have been the subject of some studies. Using CO_2 as a refrigerant has some inherent issues to be considered. The system needs a cascade transcritical compressor because the typical condensing pressure of CO_2 is 71 bar [37,38]. Nonetheless, in comparison to traditional refrigerants, carbon dioxide is considered less harmful to the ozone layer.

The other front on which scientists are focusing their attention is developing new refrigerants that have new formulations or creating new ones that are made of blending two or more existing refrigerants. An extremely crucial point that is to be considered is that any new refrigerant must be compatible with the refrigeration systems that are in operation around the world. Any attempt at introducing a new refrigerant that requires major modifications to the existing refrigeration systems to the market will fail due to the overwhelming cost associated with it. Among the new refrigerants that researchers are studying are R-454b and R-1234yf. R-454b is an environmentally sustainable replacement for the popular R-410A refrigerant [39]. Similarly, R-1234yf is a replacement for the popular refrigerant R-134a [40]. The importance of introducing new drop-in refrigerants is in their ability to replace older refrigerants that are usually more harmful to the environment when released into the atmosphere during the repair.

2.5. Summary of Active Heating and Cooling Systems for Buildings

The vapor compression refrigeration cycle is still the most widely used method for cooling despite its heavy dependency on grid electricity. Moreover, since it is prone to refrigerant leaks, new refrigerants that are generally less harmful to the environment must be developed. On another front, the development of technologies that can reduce electricity use is imperative. Thermally activated cooling technologies if supplied by waste heat or by renewable sources are a solution. Non-conventional cooling and heating systems are also being developed that require less electricity to operate, use less harmful refrigerants (or none), or are better compatible with renewable energy sources.

2.6. Pros and Cons of Active Energy Conservation Methods

The advantages and limitations of active energy conservation methods are listed in Table 1.

Table 1. Pros and Cons of Various Active Energy Conservation Techniques.

Technology	Estimated Payback Period	Pros	Cons
Ground-source heat pumps–vertical ground loop [12–14]	5–10 years	• High energy storage in the soil • Higher deep soil temperatures	• High initial cost • Longer construction time • Requires special equipment to dig the boreholes
Ground-source heat pumps–horizontal ground loop [12–14]	4–9 years	• Less additional initial cost • Simpler equipment to maintain • Less overall energy use compared to an air-source heat pump	• Moderate additional initial cost • Requires large open area to bury the pipes • Lower heat storage capacity • Lower soil temperature due to the shallow depth
Ground-source heat pumps–hybrid [14]	4–8 years	• Better energy efficiency compared to all other heat pump types • Increase the energy storage capacity of soil	• Additional equipment to maintain • Need a complex control strategy to optimize
Solar-assisted heat pumps [15–18]	2–5 years	• Renewable source of assisted heat • High overall energy efficiency • Less adverse environmental effects	• Require large area for collector installation • Additional equipment to maintain • Higher initial cost • Not applicable to cooling
Absorption refrigeration systems with renewable heat source [19–22]	2–3 years	• Can use various heat sources including renewables • Requires less electricity	• Contain harmful chemicals • Not available in small capacities • Low efficiency compared to vapor-compression refrigeration cycle
Desiccant cooling–solid desiccant [25–27]	3 years	• Less maintenance compared to liquid desiccant • High efficiency	• More equipment to maintain compared to conventional methods • Reactivation requires a heat source
Desiccant cooling–liquid desiccant [28,29]	3–4 years	• High efficiency • Reactivation can be done by renewable sources of heat • Can be used to precisely control the humidity	• High maintenance and more equipment • Higher initial cost and maintenance cost • Usually contain corrosive material • Reactivation requires a heat source

Table 1. *Cont.*

Technology	Estimated Payback Period	Pros	Cons
Deep lake water cooling [30,31]	5–9 years	• High energy efficiency • Free source of cold • No harm to environment	• Only applicable where the building is near a deep lake • Additional piping and equipment
Earth-to-air heat exchangers [32,33]	1.5–2 years	• Low maintenance • Long life • Applicable for heating and cooling • Can be easily bypassed	• Additional initial cost of burial • Use of land • Small temperature differences
Radiative sky cooling [34,35]	1.5–4.5 years	• Low maintenance • Long life • Can be bypassed	• Require unobstructed view of sky • Require a relatively large surface area • Not applicable to heating

3. Passive Building Energy Conservation Methods

3.1. Shading

Global warming and heat island effects have caused increased cooling energy consumption in buildings [41,42]. Probably the most effective way to cool a building in summer is to keep the heat energy from building up in the first place [43]. Kamal [43] and Abdel-Aziz et al. [44] state that the most important passive cooling strategy is shading. Shading involves blocking the sun before its energy can get into the building. The primary source of heat buildup (heat gain) is sunlight absorbed by the building through the roof, windows, and walls, while secondary sources include heat-generating appliances in the building and air leakage. Shading can reduce the peak-cooling load in buildings, thereby reducing the size of the heating, ventilation, and air conditioning (HVAC) equipment. Shading minimizes solar radiation incidence [43–45]. Trees and vegetation are effective ways to shade and reduce heat gain. Energy savings are reported to range from 10% to 40% [43]. Tree-shading has resulted in up to 2.7 °C reduction and up to 3.0 °C temperature difference compared to unshaded areas; and trees affect ambient air temperature from 0.5 °C to 1.0 °C average overnight and day and up to 2.0 °C lower than the outside temperature [42]. Abdel-Aziz et al. [44] reported that shading can help to reduce peak summer temperatures by 1 °C to 5 °C. Homeowners in older neighborhoods with established trees were observed to use less energy for air conditioning than homeowners in a recently developed site [46].

Shading the effects of direct solar radiation can be achieved by Kamal [43]: (a) shade provided by the effect of recesses in the building envelope; (b) shading provided by static or movable blinds or louvers; (c) transient shading provided by orientation of the building; (d) permanent or transient shading provided by surrounding buildings, screens, or vegetation; and (e) shading of roofs by rolling reflective canvas, earthen pots, vegetation, etc. Several shading overhang configurations are available for shading devices, including movable opaques (curtains, awnings, etc.), louvers, and fixed overhangs [43,45].

Trees have many benefits for urban environments, including carbon sequestration, removal of air pollutants, noise reduction, and reduced energy consumption [41]. Akamphon and Akamphon [41], as well as other researchers [47,48] noted that trees can reduce a building's energy cost by 25% to 50%. Abdel-Aziz et al. [44] noted that a medium-sized deciduous tree (with leaves on it) can reduce solar irradiance by 80% and 40% if leafless. Trees reduce the amount of radiant energy absorbed and stored by a building, use evapotranspiration converting liquid water in plants to vapor thereby cooling the air, and reduce wind speed which reduces infiltration of outside air and causes effective ventilation and convective cooling of building surfaces [42,49]. Shading performance is affected by building surface, tree location, tree size, canopy density, season, solar angle (time of day), and microclimate conditions [44,45,49]. Trees are capable of lowering air temperature, improving air quality, and reducing building energy consumption [42]. Hwang et al. [50] noted that trees can reduce residential cooling and heating energy consumption by: (1) casting shade onto building surfaces and manmade ground covers, (2) modifying airflow around buildings, and (3) allowing ambient air temperature through evapotranspiration.

Balogun et al. [42] investigated the effect of tree-shading on energy demand for two similar buildings in Akure, Nigeria, one that was shaded and the other unshaded. The two buildings were of the same architectural design, built at the same time with the same materials, had the same orientation, and were located 60 m apart from each other. The unshaded building warmed earlier and faster than the tree-shaded building. Their results indicated that tree-shading can save up to USD 218 on energy costs on a monthly basis.

Trees facing west produce the largest energy savings. Trees with high shading coefficients and moderate crown size (such as jackfruit, mango, mahogany, and Indian cork trees) maximize shading [41].

3.2. Natural Ventilation

Natural ventilation is a common passive technology, but it is dependent on the ambient environment, such as air temperature and wind speed. (Figure 3). If the air contains too much moisture, ventilation can bring additional moisture into the building causing discomfort to the building inhabitants. Mushtaha et al. [51] addressed three passive energy techniques: shading devices, natural ventilation, and thermal insulation. Implementation of such techniques can reduce 59% of the building's total energy consumption.

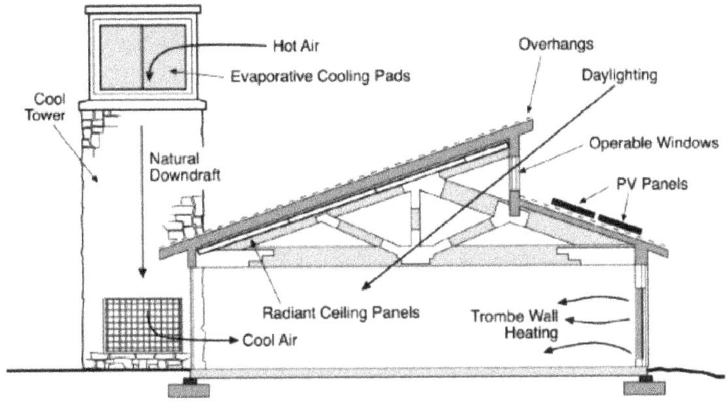

Source: NREL and NPS drawings.

Figure 3. A proposed natural ventilation system for homes.

3.3. *Windcatcher*

A windcatcher is a building's architectural feature used for natural ventilation. Windcatchers are passive systems that require no energy for their operation. They generally take the form of small towers installed on the top of buildings. The tower draws air from the outside into the building, providing natural ventilation in hot arid, and humid climates. They are generally built facing away from the wind direction to minimize ingress of blown dust and sand. They can significantly influence to reduce cooling loads and supply the necessary ventilation rate of buildings [52].

3.4. *Solar Chimneys*

Solar chimneys are inexpensive and easy passive solar heating and cooling systems in a building. Solar chimneys are hollow containers that connect the inside part of a building with the outside environment. They are tall structures constructed facing the sun, typically with a dark-colored surface to absorb solar radiation. As the chimney heats up and rises, it draws additional air in at the bottom of the chimney. They are very effective in climates that are hot and humid. There may be multiple solar chimneys to increase surface area. Low emissivity coatings and glazings help reduce heat losses to the outside environment. The solar chimneys should be insulated from the building so as to not transmit heat gains into occupied spaces. In cooler conditions, solar chimneys can be used to direct absorbed heat back into the building.

Among renewable technologies, a solar chimney is an efficient renewable energy system that has been frequently adopted in buildings to reduce the energy consumption by HVAC systems through enhanced natural ventilation [53]. Solar chimneys can be adopted for both energy savings and fire safety.

3.5. *Window Glazing*

Insulated glazing on windows reduces heat conduction through the windows. Double-paned windows offer significant energy savings compared to single-paned windows. Glass panes have poor insulation properties. Single-paned windows have an *R*-value (insulation) of about 0.9. Double-paned windows (two panes with a gap between the panes) have an *R*-value of about 2.1; a triple-paned window has an *R*-value of about 3.2. Higher *R*-vales can be obtained when a less conductive gas (such as argon) replaces air as the gas in the gap; *R*-values as high as 10 have been reported. While these *R*-values are less than those of walls (*R*~12 to 15), significant reductions in heat transfer can be realized. In a separate study, Choi et al. [54] investigated *U*-values of walls, roof, and floor elements.

3.6. Trombe Walls

A Trombe wall is an equator-facing wall painted a dark color to absorb thermal energy from incident sunlight and is covered with glass on the outside with an insulating air gap between the wall and the glaze. Trombe walls are representative of passive solar building design. During cold winter months when leaves are off trees and the sun's path is low, heat will travel to colder spaces; the brick wall will absorb heat and will release the heat into the space through radiation. As heat enters the room, it will rise as the colder air within the room drops. As the cold air drops, the Trombe wall heats it, creating a convection cycle thereby creating an even heat distribution. Trombe walls also function during the summer months. Trombe walls generally have roof overhangs built at their eaves. Because the sun is higher in the summer, the overhang blocks most of the sun's rays. National Renewable Energy Laboratory [55] provides a good discussion regarding Trombe walls. Liu et al. [56] noted that higher Trombe wall-to-wall ratio limits and more significant potential for energy savings than office buildings.

Advantages of a Trombe wall include:

- Trombe walls can significantly reduce heating costs (due to Trombe walls ability to capture energy and radiate it for longer periods of time, releasing heat during evening hours).
- Trombe walls provide comfortable heat, radiating heat into the building space and creating a convection cycle).
- The system is a passive energy technique, containing no moving parts and requires no maintenance.
- Trombe walls are based on simple and inexpensive construction.
- Convection heats the room from top to bottom, allowing the entire building source to heat evenly.
- Trombe walls significantly reduce heating bills.
- Even in spaces too large to heat entirely by Trombe walls, Trombe walls cans help supplement oil, gas, or electric heating systems, thereby reducing energy costs and resource consumption.

Limitations of Trombe walls include:

- Spaces that are not well insulated may not realize the benefits of solar radiation.
- Trombe walls are not very attractive.
- Trombe walls can be a source of heat loss during extended overcast days.
- Trombe walls do not work everywhere, particularly if the sun's path is blocked by trees, mountains, or other buildings.

3.7. High-Albedo Roofs

White roofs (high-albedo roof coatings) can provide cooling energy savings of 10% to 79%, as reported in residential buildings in Florida and California [57]. White roofs involve reflective materials to reduce the transmittance of energy through the roof. Roofs can reach temperatures exceeding 150 °F. White roofs can commonly reduce the energy demand by 10% to 30%. Figure 4 is a graphical presentation of the difference in heat absorption between a "Dark roof" and a "Cool roof". Insulation thickness was noted to play a significant role in preventing heat loss from buildings [58]. Insulation thickness also delays the transfer of heat due to thermal inertia of the insulation layer. Applying a high-albedo coating to building rooftops results in significant energy reductions [59]. Passive design and landscape variables (rooftop albedo and shading vegetation) are considered important green building techniques [60].

Figure 4. A cool roof compared to a dark roof. Source: EPA, 2022.

Dirt can accumulate on the high-albedo roofing surfaces to alter the performance of the roofing system. Bretz and Akbari [57] note that most of the albedo degradation occurs in the first year of application, with a 70% decrease in performance with the first two months of exposure; after the first year, the degradation slowed, with data showing small losses in albedo after the second year. Cleaning the roofs with soap is effective for restoring the original albedo, but it is not cost-effective. Reflective coatings have also been used on the University of California, Davis exterior walls, resulting in significant energy reductions [61].

3.8. Vegetated Roofs

Getter and Rowe [62] provide a good review of the role of extensive green roofs in sustainable development. Green roofs are classified as being either intensive or extensive systems. Intensive roofs employ a variety of plant species (that may include trees and shrubs) that require deep substrate layers and they require intensive maintenance. Intensive roofs generally are limited to flat roofs. Extensive roofs generally require minimal maintenance. They are generally not accessible to the general public. Due to their shallower substrates, plant species are generally limited to herbs, grasses, mosses, and drought-tolerant species such as sedum. Extensive green roofs can be installed on sloped roof surfaces.

Green roofs have a roof barrier installed on top of a roofing membrane. Above this barrier is a drainage layer that allows excess water to flow away from the roof. On top of the drainage layer is a fabric filter to prevent silt and particulates from clogging the drainage layer. An optional water retention fabric can be placed on top of the fabric filter to allow extra water to be retained for the benefit of the plant species. On top of that is a growing substrate to support plant growth. The depth, density, and humidity of the growing medium are three components of the growing medium contributing to its efficiency and performance.

There are four main components of vegetated roofs: (a) waterproofing membrane and filter membranes; (b) drainage films; (c) growing medium; and (d) vegetation. Saadatian et al. [63] note that some vegetated roofs need water for irrigation, increasing water consumption.

Green roofs (vegetated roofs) help mitigate the urban heat island effect, provide thermal comfort for the building occupants, add aesthetic environmental values, and most importantly reduce the energy consumption of buildings [63]. In a pilot study involving a vegetated roof of area 1388.0 m^2 and containing approximately 20,000 sedum plants conducted on a campus building at the University of Alabama at TBirmingham, Peters et al. [64] gathered utility bill information both prior to and after implementation of the green roof, involving electricity, natural gas, and chilled water. After the implementation of the pilot green roof, utility bill (energy) savings of ~20% to 25% were achieved (compared to the case prior to the implementation of the vegetated roof system). Saadatian et al. [63] also note that green roofs can reduce the roof temperature to 27 °C (compared to a temperature of 80 °C on a conventional black roof). Further, in terms of air quality, plants produce more oxygen and sequester carbon dioxide. Further, Tsoka et al. [65] noted that a continuous shading canopy of trees offers significant energy savings of about 54%.

In a study conducted in Iran, the reduction in energy consumption associated with green roofs ranged from 6.6% to 9.2%, resulting in a payback period of 25 to 57 years [61]. Significant results obtained from their study include: proper selection of green roof type depends on the climate; using green roofs causes a significant decrease in energy consumption and its associated energy costs; increasing the soil layer thickness increases the required energy for cooling and decreases the required energy for heating; the desired and positive effects of green roofs on energy consumption are better for buildings with fewer floors; and decreasing energy consumption should not be the only factor used to justify the use of green roofs.

Benefits associated with vegetated roofs include [62,63,66,67]:

- reduced volume of stormwater runoff;
- delayed stormwater runoff;
- increased stormwater runoff water quality;
- increased life span of roofing membranes;
- energy conservation and reduced urban heat island effect;
- increased biodiversity and providing habitats for wildlife;
- improved aesthetic value;
- mitigation of air pollution;
- noise reduction;
- insulation benefits; and
- application of LEED (Leadership in Energy and Environmental Design) credits.

Roofs represent 21% to 26% of urban areas [62,68]; therefore, vegetated roof systems provide an excellent opportunity to counteract the destruction of natural habitats in cities.

3.9. *Summary of Passive Energy Methods*

By incorporating passive energy design options in air-conditioned residential buildings, Abdul Mujeebu and Bano [69] showed a 34% reduction in the energy performance index could be achieved, compared to a 32% reduction for a green roof. Latent heat storage units can effectively reduce building energy consumption, reduce indoor temperature fluctuations, and improve indoor thermal comfort [6]. Furthermore, careful attention to building envelope design focusing on passive energy options can improve the indoor thermal environment by 5.82 °C, reduce thermal discomfort by 80.75%, and save up to 77% on energy requirements [70]. Passive cooling techniques can maintain the indoor temperature within a comfortable range while reducing the building energy load [71]. Passive building strategies that favor building energy conservation also have a positive impact on the building's resilience [72]. Passive energy systems can be incorporated into retrofitting buildings [73].

3.10. *Pros and Cons of Passive Energy Conservation Methods*

The advantages and limitations of passive energy conservation methods are listed in Table 2.

Table 2. Pros and cons of various passive energy conservation techniques.

Technology	Estimated Payback Period	Pros	Cons
Shading [42–45,74,75]	2.0 years	• Can provide a significant re-duction in summertime energy consumption • Provides evaporative cooling • In the summertime, trees block unwanted solar radiation entering the building, reducing the cooling load • Can improve building aesthetics • Trees can act as windbreaks to lower ambient wind speed	• High maintenance cost • May be difficult to clean windows • May attract birds and associated droppings
Natural Ventilation [5,51]	2.5 years	• Access to continuous airflow providing good humidity control • Delivers fresh air to the building interior • Reduces risk of condensation, residue, and mold • Lower operating and maintenance costs • Better health and wellness • Expels air impurities and odors • Easier access to better day-light • Eco-friendly	• Variable conditions • Dependent on natural forces • Addition to existing buildings can be expensive • Possibility of intrusions • Potentially adverse noise levels, air pollution, insect vectors, and security
Windcatcher [52,76]	6 months–1 year	• Windcatchers are an efficient cooling passive system that can help reduce the cooling loads in buildings • Adding evaporative cooling to windcatchers improves their function and indoor conditions • Windcatchers can decrease temperature down to 14 °C • Is a sustainable method of ventilation in buildings • Reduces electrical power consumption • Particularly useful in hot, arid, and humid climates • Effective for small rooms in buildings	• Windcatchers do not operate well for low air flow rates • Tall windcatchers operate more efficiently than smaller wind-catchers • Insects can go through wind-catchers resulting in uncomfortable living conditions
Solar Chimneys [77–80]	4.29 years	• Do not rely on direct sunlight • Can store heat and generate solar power around the clock • Versatile potential for commercial use • Do not require complex water cooling system • Cost-effective and energy efficient • Reliable • Low maintenance • Environmentally friendly • No electricity or gas is needed to power the chimney	• High construction costs • Efficiency concerns limits application to large systems • Technology is not widely known or used • While photovoltaic systems con-vert ~8% to 15% of energy collected into electricity, solar chimneys only convert ~1% to 2% into electricity

Table 2. *Cont.*

Technology	Estimated Payback Period	Pros	Cons
Window Glazing [81–83]	6.0 years	• Glazed windows reduce the amount of energy escaping due to two or more panes of glass • Improved noise insulation/soundproofing • Improved security/safety • Reduces interior fading • Increases property values	• Glazed windows trap heat • Difficult to repair if condensation occurs (necessitating replacement) • High initial costs • May not be a good match for buildings with older architectural styles
Trombe Walls [84–86]	2.56–2.85 years	• Are a very flexible solar option • Can be used to both cool and heat a building • Provides improved energy savings • Eliminates sun-drenching • Reduces glare on furnishings • Effective for nighttime heating • Provides reduced condensation • Provides thermal mass in a concentrated area without taking up much space • Provides comfortable, quiet rooms with stable temperatures • Are an attractive option	• Adds to construction costs primarily due to the increased foundations weight • Can become serios heat sinks (losing heat at night unless well-insulated • Operating louvres or dampers to close the vents is an operational hassle • Reduce daylighting and access to panoramic views • Covering them at night can be a hassle
Cool Roofs [58,87–89]	5.7 years	• Light colored (white) reflect sunlight reducing the need for air conditioning (reductions of 10% to 30% in electricity demand) • Lowers air conditioning bills • Results in better indoor comfort • Eco-friendly • Reduced roofing wastes in landfills • Decreased heat island effect	• White roofs may contribute to global warming (limiting the trans-port of moisture to the atmosphere which limits cloud formation resulting in less rainfall and increased drought conditions) • High installation costs • Dirt accumulation decreases system performance • Causes higher heating costs • May not be suitable for colder climates
Green Roofs [62–64,66,67,90–93]	6.2–18 years	• Helps insulate buildings, absorb rainfall, and reduce urban air temperatures • Provides biodiversity and habitat for animals • Reduces building energy consumption • Delayed stormwater runoff • Increased stormwater runoff water quality • Improves roof longevity • Adds beauty and value to building (aesthetics) • Mitigation of air pollution • Noise reduction	• High installation costs • Require more maintenance than traditional roofs

4. Summary and Conclusions

This paper does not address all the active and passive energy techniques being used today; rather, this paper describes and summarizes the more commonly used techniques applied in the field today. The paper presents a review of new and advanced technologies for heating and cooling of buildings that can reduce energy use while providing the required comfort for the occupants. As stated in the disclaimer, this paper does not center its attention on building decarbonization approaches. The paper has two sections for active and passive methods, respectively. A table at the end of each major section provides the advantages and disadvantages of each of the described technologies.

On the topic of active methods, several variations of and modifications to a conventional heat pump system have been discussed. There are various levels of initial investment associated with each of the variations, and the associated energy savings have been identified. Ground-source heat pumps (also called geothermal heat pumps) have shown promise and are being widely considered by HVAC designers and contractors for installation on small- to medium-sized buildings. The other methods that were discussed in the paper are still in the development stage and are not as common as ground-source heat pumps yet. Some of the techniques are limited in their application depending on the size of the HVAC load. A section was dedicated to the development of new refrigerants because they are under constant development to make them more environmentally friendly.

When it comes to passive techniques, there are multiple approaches that can be adopted by the designers. Among the methods discussed, shading and natural ventilation require the least amount of additional initial investment. Nonetheless, they can offer a significant reduction in the HVAC load and consequently contribute to lower energy bills. Some of the methods are still in the development phase to make them economically and technologically feasible when it comes to their relation to the building structure. Architectural and esthetic considerations are important when a technology is to be adopted. Trombe walls and solar chimneys fall into this category. Trombe walls can be used with smaller buildings, while a solar chimney installation only makes sense in larger buildings. The windcatcher concept has been around for more than a thousand years and perhaps it is time to adapt it to modern-day buildings to save energy. Its effective implementation requires a careful balance between architecture and zone layout, and it might be limited to use in low-rise buildings. Low-emissivity and multiple-glazing windows have documented proof that they work. The extra cost is a factor to be considered. Modifications to the roof of a building, such as green roofs and high-albedo roofs, can also provide energy savings. While high-albedo roofs are easier to implement, green roofs are more effective and at the same time require quite high maintenance.

As pointed out by Ma et al. [2] and the National Governors Association [3], attention on integrated decarbonation and building energy conservation approaches is accelerating.

Energy conservation of existing buildings plays a major role in reducing global carbon emissions and building energy consumption [94]. This is extremely important because building energy consumption accounts for nearly 40% of the global energy consumption [93,94], and the U.S. accounts for 20% of the world's CO_2 emissions [93]; it is key to the world's energy conservation and low-carbon development [94]. Energy use is currently driven by a number of factors, such as: population growth, economic changes, building size, service demands, and efficiency of energy use [93]. The main energy-consuming components in buildings are lighting and HVAC systems [93]. The retrofitting of existing buildings to reduce energy consumption provides an excellent opportunity for reducing global energy consumption. Building maintenance structure retrofits include transforming roofing systems and windows. Active measures relate to heating, ventilation, and air conditioning (HVAC) systems and lighting systems. Renewable energy measures refer to solar photovoltaic systems and ground source heat pump systems. Huang et al. [94] indicated that building energy conservation focuses on three aspects: energy conservation influencing factors and energy conservation barriers, energy conservation measures, and energy conservation optimization methods. As noted by DOE [93], 34 states and more than 253 municipalities have now instituted green building policies. The international Energy Agency [95] noted that without concerted efforts to bolster efficiency, energy intensity in industry may increase.

From the discussion provided, it is apparent that no single building active or passive energy conservation technique will maximize energy conservation in a building, rather an integrated approach involving a combination of building active or passive energy conservation techniques must be employed to maximize energy conservation efficiency for buildings.

Author Contributions: Conceptualization, H.T. and R.W.P.; methodology, H.T.; software, H.T.; formal analysis, H.T. and R.W.P.; investigation, H.T. and R.W.P.; data curation, H.T.; writing—original draft preparation, H.T. and R.W.P.; writing—review and editing, H.T. and R.W.P.; visualization, H.T. and R.W.P.; supervision, R.W.P. All authors have read and agreed to the published version of the manuscript.

Funding: This research received no external funding.

Acknowledgments: The authors gratefully acknowledge the support of Pennsylvania State University Harrisburg and the University of Alabama at Birmingham.

Conflicts of Interest: The authors declare no conflict of interest.

References

1. Xiang, X.; Mam, M.; Ma, X.; Chen, L.; Cai, W.; Feng, W.; Ma, Z. Historical Decarbonization of Global Commercial Building Operations in the 21st Century. *Appl. Energy* **2022**, *322*, 119401. [CrossRef]
2. Ma, M.; Feng, W.; Huo, J.; Xiang, X. Operational Carbon Transition in the Megalopolises' Commercial Buildings. *Build. Environ.* **2022**, *226*, 109705. [CrossRef]
3. National Governors Association. *State Energy Efficiency Policy in a New Era: A Toolkit for Governors*; National Governors Association (nga.org): Washington, DC, USA, 2021.
4. Ibem, E.O.; Udezi, B.E.; Oti, O.M.; Fakorede, O.A. Assessment of Architects' Knowledge of Passive Design Strategies in Terminal Buildings among Architectural Firms in Lagos State. In Proceedings of the 1st International Conference on Sustainable Infrastructural Development, IOP Conference Series: Materials Science and Engineering, Ota, Nigeria, 24–28 June 2019; Volume 640. [CrossRef]
5. Li, X.; Shen, C.; Yu, C.W.F. Building Energy Efficiency: Passive Technology or Active Technology? *Indoor Built Environ.* **2017**, *26*, 729–732. [CrossRef]
6. Sun, W.; Feng, J.; Zhang, Z.; Fang, X. Research Progress of Phase Change Heat Storage Technology for Passive Energy Conservation in Buildings. *Huagong Jinzhan/Chem. Ind. Energy Prog.* **2020**, *38*, 1824–1834. [CrossRef]
7. Ben Romdhane, S.; Amamou, A.; Ben Khalifa, R.; Said, N.M.; Younsi, Z.; Jemni, A. A Review on Thermal Energy Storage Using Phase Change Materials in Passive Building Applications. *J. Build. Eng.* **2020**, *32*, 101563. [CrossRef]
8. Elaouzy, Y.; El Fadar, A. Energy, Economic and Environmental Benefits of Integrating Passive Design Strategies into Buildings: A Review. *Renew. Sustain. Energy Rev.* **2022**, *167*, 112828. [CrossRef]
9. Ponmurugan, M.; Ravikumar, M.; Selvendran, R.; Merlin Medona, C.; Arunraja, K.M. A Review of Energy Conserving Materials for Passive Cooling in Buildings. *Mater. Today Proc.* **2022**, *64*, 1689–1693. [CrossRef]
10. Rempel, A.R.; Lim, S. Numerical Optimization of Integrated Passive Heating and Cooling Systems Yields Simple Protocols for Building Energy Decarbonization. *Sci. Technol. Built Environ.* **2019**, *25*, 1226–1236. [CrossRef]
11. Kumar, V.V. Energy consumption of Residential Buildings in Extreme Climates with Phase Change Material-Aluminum Radiation Reflector Cool Roof. *Energy Sources Part A Recover. Util. Environ. Eff.* **2022**, *44*, 9703–9715. [CrossRef]
12. Hou, G.; Taherian, H.; Li, L. A Predictive TRNSYS Model for Long-Term Operation of a Hybrid Ground Source Heat Pump System with Innovative Horizontal Buried Pipe Type. *Renew. Energy* **2020**, *151*, 1046–1054. [CrossRef]
13. Hou, G.; Taherian, H.; Li, L.; Fuse, J.; Moradi, L. System Performance Analysis of a Hybrid Ground Source Heat Pump with Optimal Control Strategies Based on Numerical Simulations. *Geothermics* **2020**, *86*, 101849. [CrossRef]
14. Hou, G.; Taherian, H. Performance Analysis of a Hybrid Ground Source Heat Pump System Integrated with Liquid Dry Cooler. *Appl. Therm. Eng.* **2019**, *159*, 113830. [CrossRef]
15. Al-Yasiri, Q.; Szabó, M.; Arıcı, M. A Review on Solar-Powered Cooling and Air Conditioning Systems for Building Applications. *Energy Rep.* **2022**, *8*, 2888–2907. [CrossRef]
16. Xu, L.; Hou, G.; Taherian, H.; Song, Y.; Wang, Y.; Moradi, L. Modeling and Assessment of a Novel Solar-Biomass Based Distributed Multi-Generation System. *J. Renew. Sustain. Energy* **2022**, *14*, 036301. [CrossRef]
17. Tehranian, M.; Taherian, H. Solar-Assisted Heat Pump Combined with Seasonal Storage Tank to Meet Multi-Family Buildings Heat Demand. In Proceedings of the Solar Conference, Denver, CO, USA, 9–12 October 2017; American Solar Energy Society: Denver, CO, USA, 2017.

18. Herrando, M.; Ramos, A. Photovoltaic-Thermal (PV-T) Systems for Combined Cooling, Heating and Power in Buildings: A Review. *Energies* **2022**, *15*, 3021. [CrossRef]
19. Amiri, L.; Madadian, E.; Bahrani, N.; Ghoreishi-Madiseh, A. Techno-Economic Analysis of Waste Heat Utilization in Data Centers: Application of Absorption Chiller Systems. *Energies* **2021**, *14*, 2433. [CrossRef]
20. Deng, J.; Wang, R.; Han, G. A Review of Thermally Activated Cooling Technologies for Combined Cooling, Heating and Power Systems. *Prog. Energy Combust. Sci.* **2011**, *37*, 172–203. [CrossRef]
21. Torrella, E.; Sánchez, D.; Cabello, R.; Larumbe, J.; Llopis, R. On-Site Real-Time Evaluation of an Air-Conditioning Direct-Fired Double-Effect Absorption Chiller. *Appl. Energy* **2009**, *86*, 968–975. [CrossRef]
22. Ochoa, G.; Acevedo-Peñaloza, C.; Forero, J. Thermo-Economic Assessment of a Gas Microturbine-Absorption Chiller Trigeneration System under Different Compressor Inlet Air Temperatures. *Energies* **2019**, *12*, 4643. [CrossRef]
23. Han, X.; Li, J.; Xiangqiang Kong, X.; Sun, T.; Zhang, C.; Yin, L. Thermodynamic Performance Study on a Novel Absorption Compression Cascade Refrigeration Activated by an Internal Combustion Engine. *Int. J. Energy Res.* **2021**, *45*, 9595–9612. [CrossRef]
24. Darwish, M. Building Air Conditioning System Using Fuel Cell: Case Study for Kuwait. *Appl. Therm. Eng.* **2007**, *27*, 2869–2876. [CrossRef]
25. Liu, Y.; Wang, R.; Xia, Z. Experimental Study on a Continuous Adsorption Water Chiller with Novel Design. *Int. J. Refrig.* **2005**, *28*, 218–230. [CrossRef]
26. Taherian, H.; Salarian, H.; Ghadamian, H.; Khalaji Assadi, M. A Study of Liquid Desiccant System Performance. In Proceedings of the 6th International Symposium on Heating, Ventilating and Air Conditioning, Nanjing, China, 6–9 November 2009. Available online: https://www.researchgate.net/publication/255961670_A_Study_of_Liquid_Desiccant_System_Performance?enrichId=rgreq-65aada6b-bb94-4d5d-89c2-2687553c78d8&enrichSource=Y292ZXJQYWdlOzI1NTk2MTY3MDtBUzoxOTM1NDExMjI2NTAxMTlAMTQyMzE1NTIwNzMxMg%3D%3D&el=1_x_2 (accessed on 6 April 2023).
27. Lee, Y.; Park, S.; Kang, S. Operational Optimization of a Hybrid Desiccant Cooling System for Building Air Conditioning System. *Heat Mass Transf.* **2022**, *59*, 39–54. [CrossRef]
28. Sharma, A.; Kaushal, R. Performance Evaluation of a Solar-Powered Flat-Plate Multichannel Liquid Desiccant Air Conditioning System. *Heat Transf.* **2021**, *50*, 4103–4121. [CrossRef]
29. Lee, J.; Jeong, J. Hybrid Heat-Pump-Driven Liquid-Desiccant System: Experimental Performance Analysis for Residential Air-Conditioning Applications. *Appl. Therm. Eng.* **2021**, *195*, 117236. [CrossRef]
30. Fung, A.; Taherian, H.; Rahman, M.Z.; Selim, M. Feasibility Study of Deep-Lake Water-Cooling System at Ryerson University. *ASHRAE Trans.* **2015**, *121*, 393–401.
31. Kuyuk, A.; Ghoreishi-Madiseh, S.; Hassani, F.; Saspito, A.; Amiri, L.; Templeton, J. Performance and Economic Assessment of Large-Scale Deep-Lake Cooling Systems: A Canadian Example. *Energy Procedia* **2019**, *158*, 43–48. [CrossRef]
32. Fatahian, H.; Taherian, H.; Salarian, H.; Fatahian, E. Thermal Performance of a Ground Heat Echanger in a Cold Climate: A CFD Study. *Eur. Phys. J. Plus* **2021**, *136*, 265. Available online: https://ui.adsabs.harvard.edu/link_gateway/2021EPJP..136..265F/doi:10.1140/epjp/s13360-021-01265-7 (accessed on 6 April 2023). [CrossRef]
33. Peretti, C.; Zarrella, A.; De Carli, M.; Zecchin, R. The Design and Environmental Evaluation of Earth-to-Air Heat Exchangers (EAHE). A Literature Review. *Renew. Sustain. Energy Rev.* **2013**, *28*, 107–116. [CrossRef]
34. Hosseinzadeh, E.; Taherian, H. An Experimental and Analytical Study of a Radiative Cooling System with Unglazed Flat Plate Collectors. *Int. J. Green Energy* **2012**, *9*, 766–779. [CrossRef]
35. Sun, X.; Sun, Y.; Zhou, Z.; Alam, M.; Bermel, P. Radiative Sky Cooling: Fundamental Physics, Materials, Structures, and Applications. *Nanophotonics* **2017**, *6*, 997–1015. [CrossRef]
36. Evangelisti, L.; Guattari, C.; Asdrubali, F. On the Sky Temperature Models and their Influence on Buildings Energy Performance: A Critical Review. *Energy Build.* **2019**, *183*, 607–625. [CrossRef]
37. Dilshad, S.; Kalair, A.; Khan, N. Review of Carbon Dioxide (CO_2) Based Heating and Cooling Technologies: Past, Present, and Future Outlook. *Int. J. Energy Res.* **2020**, *44*, 1408–1463. [CrossRef]
38. Hussman Corporation. *CO2 Transcritical Systems Training Manual*; Hussman Corporation: Bridgeton, MO, USA, 2018.
39. The CHEMOURS Company. Opteon™ XL41 (R-454b) Refrigerant. 2022. Available online: https://www.opteon.com/en/products/refrigerants/xl41 (accessed on 6 April 2023).
40. Honeywell International Inc. Solstice® yf (R-1234yf). 2022. Available online: https://www.honeywell-refrigerants.com/europe/product/solstice-yf-refrigerant/ (accessed on 6 April 2023).
41. Akamphon, S.; Akamphon, K. Cost and Benefit Tradeoffs in Using a Shade Tree for Residential Building Energy Saving. *Environ. Asia* **2014**, *7*, 19–24. [CrossRef]
42. Balogun, A.A.; Morakinyo, T.E.; Adegun, O.B. Effect of Tree-Shading on Energy Demand of Two Similar Buildings. *Energy Build.* **2014**, *81*, 305–315. [CrossRef]
43. Kamal, M.A. Shading: A Simple Technique for Passive Cooling and Energy Conservation in Buildings. *Archit.–Time, Space and People, Monthly Magazine*, January 2011; pp. 18–23.
44. Abdel-Aziz, D.M.; Shboul, A.A.; Al-Kurdi, N.Y. Effects of Tree Shading on Building's Energy Consumption—The Case of Residential Buildings in a Mediterranean Climate. *Am. J. Environ. Eng.* **2015**, *5*, 131–140.

45. Maleki, B.A. Shading: Passive Cooling and Energy Conservation in Buildings. *Int. J. Tech. Phys. Probl. Eng.* **2011**, *3*, 72–79. Available online: https://www.coa.gov.in/show_img.php?fid=100 (accessed on 6 April 2023).
46. Carver, A.D.; Unger, D.R.; Parks, C.L. Modeling Energy Savings from Urban Shade Trees: An Assessment of the CITYgreen® Energy Conservation Module. *Energy Manag.* **2004**, *34*, 650–655. [CrossRef]
47. Huang, Y.; Akbari, H.; Hara, T.; Rosenfeld, A. The Potential of Vegetation in Reducing Summer Cooling Loads in Residential Buildings. *J. Appl. Meteorol.* **1987**, *26*, 1103–1116. [CrossRef]
48. Meier, A.K. Strategic Landscaping and Air-Conditioning Savings: A Literature Review. *Energy Build.* **1991**, *15*, 479–486. [CrossRef]
49. Simpson, J.R. Improved Estimates of Tree-Shade Effects on Residential Energy Use. *Energy Build.* **2002**, *34*, 1067–1076. [CrossRef]
50. Hwang, W.H.; Wiseman, P.E.; Thomas, V.A. Enhancing the Energy Conservation Benefits of Shade Trees in Dense Residential Developments Using an Alternative Tree Placement Strategy. *Landsc. Urban Plan.* **2017**, *158*, 62–74. [CrossRef]
51. Mushtaha, E.; Salameh, T.; Kharrufa, S.; Mori, T.; Aldawoud, A.; Hamad, R.; Nemer, T. The Impact of Passive Design Strategies on Cooling Loads of Buildings in Temperate Climate. *Case Stud. Therm. Eng.* **2021**, *28*, 101588. [CrossRef]
52. Sangdeh, P.K.; Nasrollahi, N. Windcatchers and their Applications in Contemporary Architecture. *Energy Built Environ.* **2022**, *3*, 56–72. [CrossRef]
53. Shi, L.; Zhang, G. An Empirical Model to Predict the Performance of Typical Solar Chimneys Considering Both Room and Cavity Configurations. *Build. Environ.* **2016**, *103*, 250–261. [CrossRef]
54. Choi, B.-E.; Shin, J.-H.; Lee, J.-H.; Kim, S.-S.; Cho, Y.-J. Establishment of Passive Energy Conservation Measure of Fenestration System in Nonresidential Building of Korea. *Int. J. Polym. Sci.* **2017**, *2017*, 8681737. [CrossRef]
55. National Renewable Energy Laboratory. Building Research: Building a Better Trombe Wall. In *NREL Facts*; Undated; National Renewable Energy Laboratory: Golden, CO, USA. Available online: https://www.nrel.gov/docs/legosti/fy98/22834.pdf (accessed on 6 April 2023).
56. Liu, Y.; Yu, Z.; Song, C.; Wang, D. Heating Load Reduction Characteristics of Passive Solar Buildings in Tibet, China. *Build. Simul.* **2022**, *15*, 975–994. [CrossRef]
57. Bretz, S.E.; Akbari, H. Long-Term Performance of High-Albedo Roof Coatings. *Energy Build.* **1997**, *25*, 159–167. [CrossRef]
58. Ramamurthy, P.; Sun, T.; Rule, K.; Bou-Zeid, E. The Joint Influence of Albedo and Insulation on Roof Performance: An Observational Study. *Energy Build.* **2015**, *93*, 249–258. [CrossRef]
59. Akbari, H.; Bretz, S.; Kurn, D.M.; Hanford, J. Peak Power and Cooling Energy Savings of High-Albedo Roofs. *Energy Build.* **1997**, *25*, 117–126. [CrossRef]
60. Douthat, T.; Morgan, J.D.; Alibašić, H.; Grant, A. Passive Building Characteristics, and Summertime Residential Energy Use: A Spatial Analysis of Energy Efficiency in Gainesville, FL. *Build. Environ.* **2020**, *169*, 106542. [CrossRef]
61. Celniker, C.; Chen, S.; Meier, A.; Levinson, R. Targeting Buildings for Energy-Saving Cool-Wall Retrofits: A Case Study at the University of California, Davis. *Energy Build.* **2021**, *249*, 111014. [CrossRef]
62. Getter, K.L.; Rowe, D.B. The Role of Extensive Green Roofs in Sustainable Development. *HortScience* **2006**, *41*, 1276–1285. [CrossRef]
63. Saadatian, O.; Spian, K.; Salleh, E.; Lim, C.H.; Riffat, S.; Saadatian, E.; Toudeshki, A.; Sulaiman, M.Y. A Review of Energy Aspects of Green Roofs. *Renew. Sustain. Energy Rev.* **2013**, *23*, 155–168. [CrossRef]
64. Peters, R.W.; Sherrod, R.D.; Winslett, M. Chapter 2: Energy Savings Resulting from Installation of an Extensive Vegetated Roof System on a Campus Building in the Southeastern United States. In *New Developments in Renewable Energy*; Arman, H., Yuksel, I., Eds.; Intech Publishers: London, UK, 2013; pp. 21–49. [CrossRef]
65. Tsoka, S.; Leduc, T.; Rodler, A. Assessing the Effect of Urban Street Trees on Building Cooling Energy Needs: The Role of Foliage Density and Planting Pattern. *Sustain. Cities Soc.* **2021**, *65*, 102633. [CrossRef]
66. Refahi, A.H.; Talkhabi, H. Investigating the Effective Factors on the Reduction of Energy Consumption in Residential Buildings with Green Roofs. *Renew. Energy* **2015**, *80*, 595–603. [CrossRef]
67. Kirby, J.T.; Peters, R.W.; Watts, S.; Winslett, M. *Completion Report: Experimental Investigation of Runoff Mitigation from Vegetated Roofs*; Alabama Water Resources Research Institute: Auburn, AL, USA, 2011.
68. Wong, E. Green Roofs and the Environmental Protection Agency's Heat Island Reduction Initiative. In Proceedings of the 3rd North American Greed Roof Conference: Greening Rooftops Sustainable Communities, Washington, DC, USA, 4–6 May 2005; pp. 32–44.
69. Abdul Mujeebu, M.; Bano, F. Integration of Passive Energy Conservation Measures in a Detached Residential Building Design in Warm Humid Climate. *Energy* **2022**, *255*, 124587. [CrossRef]
70. Kini, P.G.; Garg, N.R.; Kamath, K. An Assessment of the Impact of Passive Design Variations of the Building Envelope Using Thermal Discomfort Index and Energy Savings in Warm and Humid Climate. *Energy Rep.* **2022**, *8*, 616–624. [CrossRef]
71. Bhamare, D.K.; Rathod, M.K.; Banerjee, J. Passive Cooling Techniques for Building and their Applicability in Different Climate Zones—The State of Art. *Energy Build.* **2019**, *198*, 467–490. [CrossRef]
72. White, L.; Wright, G. Assessing Resiliency and Passive Survivability in Multifamily Buildings. In Proceedings of the Thermal Performance of the Exterior Envelopes of Whole Buildings, Clearwater Beach, FL, USA, 9–12 December 2019. Available online: https://www.ashrae.org/file%20library/conferences/specialty%20conferences/2020%20building%20performance/papers/d-bsc20-c018.pdf (accessed on 6 April 2023).

73. Coelho, G.B.A.; Henriques, F.M.A. Performance of Passive Retrofit Measures for Historic Buildings that House Artefacts Viable for Future Conditions. *Sustain. Cities Soc.* **2021**, *71*, 102982. [CrossRef]
74. Kamal, M.A. A Study on Shading of Buildings as a Preventive Measure for Passive Cooling and Energy Conservation in Buildings. *Int. J. Civ. Environ. Eng.* **2010**, *10*, 19–22.
75. Abdel-Aziz, D.M. Effects of Tree Shading on Building's Energy Consumption. *Archit. Eng. Technol.* **2014**, *3*, 135. [CrossRef]
76. Jomehzadeh, F.; Nejat, P.; Calautit, J.K.; Yusof, M.B.M.; Zaki, S.A.; Hughes, B.R.; Yazid, M.N.A.W.M. A Review on Windcatcher for Passive Cooling and Natural Ventilation in Buildings, Part 1: Indoor Air Quality and Thermal Comfort Assessment. *Renew. Sustain. Energy Rev.* **2017**, *70*, 736–756. [CrossRef]
77. Klimes, L.; Charvát, P.; Hejčík, J. Comparison of the Energy Conversion Efficiency of a Solar Chimney and a Solar PV-Powered Fan for Ventilation Applications. *Energies* **2018**, *11*, 912. [CrossRef]
78. Clean Power Professionals Group. How Solar Chimneys Can Help Renewables Grow. 2022. Available online: https://energycentral.com (accessed on 6 April 2023).
79. Tiondi, E. 5 Advantages to Using a Solar Chimney. 2010. Available online: https://www.doityourself.com/ (accessed on 6 April 2023).
80. Shi, L.; Ziem, A.; Zhang, G.; Li, J.; Setunge, S. Solar Chimney for a Real Building Considering Both Energy-Saving and Fire Safety—A Case Study. *Energy Build.* **2020**, *221*, 110016. [CrossRef]
81. U.S. Department of Energy, Energy Efficiency & Renewable Energy. *Guide to Energy-Efficient Windows 2010 (October)*; U.S. Department of Energy: Washington, DC, USA, 2010.
82. Gorantla, K.; Shaik, S.; Kontoleon, K.J.; Mazzeo, D.; Maduru, V.R.; Shaik, S.V. Sustainable Reflective Triple Glazing Design Strategies: Spectral Characteristics, Air Conditioning Cost Savings, Daylight Factors, and Payback Periods. *J. Build. Eng.* **2021**, *42*, 103089. [CrossRef]
83. ConnectUS. 9 Advantages and Disadvantages of Double Glazed Windows. 26 September 2016. Available online: https://connectusfund.org/ (accessed on 6 April 2023).
84. Kostikov, S.; Grinkrug, M.; Yiqiang, J. Comparative Technical and Economic Analysis of the Trombe Wall Use in the Heat Supply System at Different Climatic Conditions. *J. Phys. Conf. Ser.* **2020**, *1614*, 012064. [CrossRef]
85. Gromicko, N.; Trombe Walls. International Association of Certified Home Inspectors, Boulder, CO, USA. Undated. Available online: https://www.nachi.org/trombe-walls.htm (accessed on 6 April 2023).
86. National Renewable Energy Laboratory. *Building a Better Trombe Wall*; U.S. Department of Energy: Golden, CO, USA, 1998.
87. Al Fayad, F.; Maref, W.; Awad, M.M. Review of White Roofing Materials and Emerging Economies with Focus on Energy Performance Cost-Benefit, Maintenance, and Consumer Indifference. *Sustainability* **2021**, *13*, 9967. [CrossRef]
88. Energy Roofing. Cool Roof Advantages and Disadvantages. Undated. Available online: https://energy-roofing.com (accessed on 6 April 2023).
89. Homes.com. The Pros and Cons of Cool Roofs. 2016. Available online: https://www.homes.com (accessed on 6 April 2023).
90. Florida Independent. The Pros and Cons of Having a Green Roof. 2022. Available online: https://floridaindependent.com (accessed on 6 April 2023).
91. General Services Administration. Green Roofs. Washington, DC, USA. Undated. Available online: https://www.gsa.gov (accessed on 6 April 2023).
92. American Society of Landscape Architects. Professional Practice—Increasing Energy Efficiency: Residential Green Roofs. Undated. Available online: https://www.asla.org (accessed on 6 April 2023).
93. U.S. Department of Energy, Energy Efficiency & Renewable Energy. *Energy Efficiency Trends in Residential and Commercial Buildings 2010 (August)*; U.S. Department of Energy: Washington, DC, USA, 2010.
94. Huang, H.; Wang, H.; Hu, Y.-J.; Li, C.; Wang, X. The Development Trends of Existing Building Energy Conservation and Emission Reduction—A Comprehensive Review. *Energy Rep.* **2022**, *8*, 13170–13188. [CrossRef]
95. International Energy Agency. *Buildings—Energy Efficiency 2020—Analysis—IEA*; International Energy Agency: Paris, France, 2020.

energies

Article

Thermal Performance of Lightweight Steel Framed Facade Walls Using Thermal Break Strips and ETICS: A Parametric Study

Paulo Santos *, **Paulo Lopes** and **David Abrantes**

ISISE, Department of Civil Engineering, University of Coimbra, 3030-788 Coimbra, Portugal
* Correspondence: pfsantos@dec.uc.pt

Abstract: The thermal performance of lightweight steel framed (LSF) facade walls depends on many factors, such as the steel studs, the batt insulation, the external thermal insulation composite systems (ETICS), and the sheathing layers. Moreover, the high thermal conductivity of steel could negatively affect their thermal performance due to the consequent thermal bridge effect. Furthermore, in LSF walls, the batt insulation is usually bridged by the steel studs. Thus, some analytical calculation procedures defined in standards (e.g., ISO 6946) are not valid, further complicating their thermal performance quantification. In this research, a parametric study to evaluate the thermal performance of facade LSF walls is presented. Seven relevant parameters are assessed, most of them related to the use of thermal break strips (TBS) and ETICS. The 2D numerical models used to predict the conductive *R*-values were experimentally validated, and their precision was successfully verified. As earlier found in a previous research work for partition LSF walls, it is also more effective for facades to increase the TBS thickness rather than their width, with the *R*-value increments being slightly smaller for facade LSF walls. These features were more pronounced for double TBS and for the smaller stud spacing (400 mm). The major thermal performance improvements were found when increasing the ETICS insulation thickness and decreasing their thermal conductivity.

Keywords: thermal performance; lightweight steel framed; facade walls; thermal break strips; ETICS; parametric study

Citation: Santos, P.; Lopes, P.; Abrantes, D. Thermal Performance of Lightweight Steel Framed Facade Walls Using Thermal Break Strips and ETICS: A Parametric Study. *Energies* **2023**, *16*, 1699. https://doi.org/10.3390/en16041699

Academic Editor: Elena Lucchi

Received: 31 December 2022
Revised: 31 January 2023
Accepted: 5 February 2023
Published: 8 February 2023

1. Introduction

A thermal bridge, or cold bridge, can be defined as a localized zone of a building envelope which has a significantly higher heat transfer when compared to the adjacent areas [1]. Thermal bridges in buildings, regarding the way they are computed, can be categorized into three types: repeating thermal bridges (e.g., vertical steel studs equally spaced in a LSF wall), linear (nonrepeating) thermal bridges (e.g., along a projecting balcony, a wall-to-wall connection or a wall to floor edges), and point thermal bridges (e.g., the mechanical metallic fasteners to fix the insulation panels in ETICS).

Regarding the origin, there are usually two types of thermal bridges: geometric and construction thermal bridges. An example of a geometric thermal bridge is a wall-to-wall corner, given the higher outer surface area when compared to the inner one. Construction thermal bridges originate from a higher thermal conductivity of a specific localized material, such as a concrete column within a masonry ceramic brick wall.

Thermal bridges may lead to several negative consequences in buildings, such as increased heat transfer through the building envelope, localized colder zones in the inner sheathing surfaces, thermal discomfort of the occupants, moisture and condensation problems, mold growth, and decrease of materials' durability [1]. The ASIEPI research project [2] concluded that "the total impact of thermal bridges on the heating energy need is in general considerable and can be as high as 30%". Moreover, Ge and Baba [3] simulated the dynamic effect of thermal bridges on the energy performance of reinforced concrete residential building located in a hot climate and concluded that the presence of thermal bridges is able to increase the annual cooling load by 20%.

Given the high thermal conductivity of steel, the relevance of thermal bridges may be even higher if this issue is not adequately addressed at design stage and their mitigation strategies implemented during construction stage [4]. This is a very hot research topic, which has been addressed using several approaches, such as in situ or lab measurements [5–7], parametric studies [8–10], analytical evaluation [11], and numerical simulations [4,7].

Several thermal bridge mitigation strategies in LSF building elements have been developed and studied, including the use of steel profiles having indented flanges [10], the use of ETICS (external thermal insulation composite systems) [8,9,12,13], the placement of thermal break strips (TBS) along the steel studs flanges [5–7,9,10], and the application of steel studs with slotted webs [9,14].

There is no perfect thermal mitigation strategy since each one has its inherent drawbacks and advantages. The most frequent thermal bridge mitigation strategy is the use of ETICS. Being a continuous thermal insulation, the thermal performance improvement is more effective. Moreover, the building's net floor area is not reduced due to the additional ETICS thickness, since it is usually located outside the floor area [13]. On the other hand, since the very reduced insulation material's quantity used is placed only where it is needed more, i.e., along the steel stud flanges, the use of TBS is a highly cost-effective strategy to mitigate thermal bridges in LSF walls [5]. Nevertheless, the use of TBS may have a drawback due to the higher distance between the steel frame and the sheathing panels, equal to the TBS thickness, which will lead to a smaller mechanical resistance to lateral loading shear actions [15].

The steel stud flanges' indentation allows a reduction in the contact surface area between the sheathing panels and the steel, thereby reducing the heat transfer across the wall in the vicinity of the steel profile, without increasing the LSF wall thickness [10]. Moreover, an extra advantage is that the initial flange air gap can be filled with insulation material, similarly to a TBS, but without increasing the LSF wall thickness. Regarding the slotted steel studs strategy, the key disadvantage is related to the reduction in their load-bearing capacity [10].

In a previous study [16], the authors performed a parametric study about the use of thermal break strips (TBS) in load-bearing partition LSF walls, to mitigate the steel frame thermal bridges and improve their thermal performance. In this parametric study, five different parameters were assessed: (1) the steel stud distance, (2) the TBS position and number on the steel stud's flanges, (3) the TBS material's thermal conductivities, (4) the thicknesses, and (5) the width of TBS cross-section geometry. It was concluded that increasing the TBS width did not always lead to a thermal resistance improvement, and that increasing the TBS thickness is more effective than increasing their width, with this latter conclusion related to the volumetric expansibility of the mineral wool batt insulation.

A similar parametric study for load-bearing facade LSF walls was not found in the literature. Therefore, in the present research, a parametric study is performed for load-bearing facade LSF walls. In addition to the five abovementioned parameters, in the present parametric study, two more parameters are evaluated: (6) the thickness of the ETICS thermal insulation layer, and (7) their thermal conductivity. In this parametric study, bidimensional numerical models were used, which were experimentally validated, and their accuracy was successfully verified using three different additional strategies.

In this paper, after this brief introduction, Section 2 presents the materials and methods, including a description of the facade LSF wall used as reference, the evaluated parameters and their values, the characterization of the materials, and the numerical simulations. Next, the achieved results are presented and discussed in Section 3, starting by the reference facade LSF wall, followed by the results for a single TBS and for two TBSs, and finally the results about the relevance of ETICS insulation thickness and their thermal conductivity values. To finalize, the key conclusions of this study are presented in Section 4.

2. Materials and Methods

Here, the evaluated facade LSF walls are described, with emphasis on the reference one. Furthermore, the assessed parameters in this study are specified, including the geometry and dimensions of the TBS and ETICS, the steel stud spacing, and the materials used with their respective thermal properties. Next, the numerical simulations are described, including the domain discretization, the boundary conditions, the model validation, and accuracy verifications.

2.1. Reference Facade LSF Wall

Figure 1 displays the horizontal cross-section for the reference load-bearing facade LSF wall, used in this parametric study. The structure of the evaluated walls was made using vertical steel studs with a web of 90 mm, a flange of 43 mm, a lip return of 15 mm, and a sheet thickness of 1.5 mm. The wall cavity was fully filled with a mineral wool (MW) batt insulation, 90 mm thick. On both sides of the vertical studs, there was an oriented strand board (OSB) structural sheathing panel, 12 mm thick. Furthermore, in the inner surface there was an extra gypsum plasterboard (GPB), 12.5 mm thick, while, on the outer surface, there was an external thermal insulation composite system (ETICS), with expanded polystyrene (EPS) insulation, 50 mm thick, and an ETICS finishing layer, 5 mm thick.

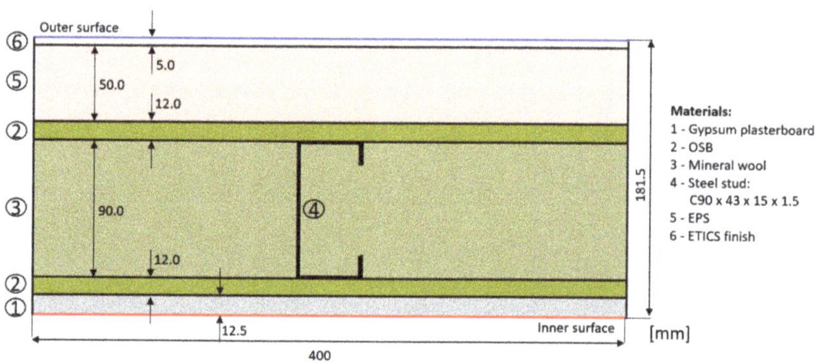

Figure 1. Horizontal cross-section of reference load-bearing facade LSF wall.

2.2. Evaluated Parameters

The parameters assessed in this parametric study, as well as the values to be used for each one, are listed in Table 1 and schematically illustrated in Figure 2. These parameters were (1) the spacing of the steel studs, (2) the dimensions (thickness and width), thermal conductivity, and number of the TBSs, and (3) the thickness and thermal conductivity of the ETICS insulation material. Note that, as illustrated in Figure 2, the addition of a TBS (single or double) increased the total LSF wall thickness and, given the high MW expansion capacity, it was assumed always a fully filled air cavity with this batt insulation.

Table 1. Parameters of the facade LSF walls to be evaluated and assigned values.

Parameter	Values
Steel stud spacing [mm]	**400** *, 600
Thermal Break Strips:	
-Thickness [mm]	5, 7.5, **10** *, 12.5, 15
-Width [mm]	30, 40, **50** *, 60, 70
-Conductivity [mW/(m·K)]	7.5, 15, **30** *, 60, 120
-Number	**Zero** *, one [2], two [3]

Table 1. *Cont.*

Parameter	Values
ETICS [1]:	
-Insulation thickness [mm]	30, 40, **50** *, 60, 70
-Insulation conductivity [mW/(m·K)]	7.5, 15, 30, **36** *[4], 60, 120

* Reference values are in bold. [1] ETICS—External thermal insulation composite system; [2] one—outer flange; [3] two—both inner and outer flanges; [4] thermal conductivity value of the reference ETICS insulation material (EPS—expanded polystyrene).

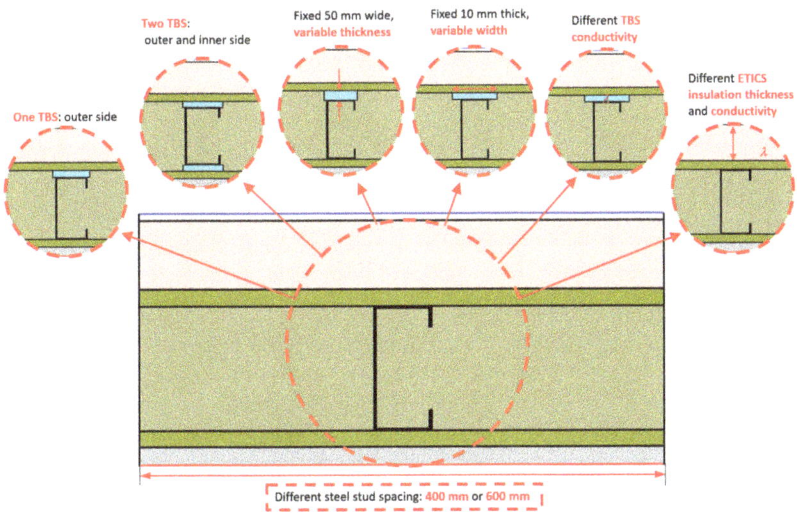

Figure 2. Evaluated parameters of the facade LSF walls.

2.3. Material Characterization

Table 2 displays the materials used in the LSF walls, including the thickness, t, and the thermal conductivity, λ. Note that the TBSs were previously characterized in Table 1, regarding their evaluated dimensions (thickness and width) and assessed thermal conductivities.

Table 2. Thickness (t) and thermal conductivity (λ) of the materials used in the reference facade LSF wall.

Material	t [mm]	λ [W/(m·K)]	Ref.
Gypsum plaster board (GPB)	12.5	0.175	[17]
Oriented strand board (OSB)	12.0	0.100	[18]
Mineral wool (MW)	90.0	0.035	[19]
Steel studs (C90 × 43 × 15 × 1.5)	90.0	50.000	[20]
ETICS [2] insulation (EPS [1])	50.0	0.036	[21]
ETICS [2] finish	5.0	0.450	[22]

[1] EPS—expanded polystyrene; [2] ETICS—external thermal insulation composite system.

2.4. Numerical Simulations

To perform the numerical simulations of the evaluated facade LSF walls, the software THERM (version 7.6.1) [23] was used. In this subsection, some relevant details related to the implemented THERM finite element method (FEM) models are explained, including the discretization of the domain, the employed boundary conditions, the model validation, and accuracy verifications.

2.4.1. Domain Discretization

The materials' thermal properties and dimensions were previously presented in Tables 1 and 2. It was only needed to model a representative part of the LSF walls' cross-section, as displayed previously in Figure 1. This strategy allowed decreasing the calculation effort and time. Moreover, the numerical computation error of the implemented FEM models was capped at 2%.

2.4.2. Boundary Conditions

To perform the numerical simulations, two sets of boundary conditions were set: (1) the environment air temperatures, and (2) the surface thermal resistances. For this parametric study, the external environment temperature was set as 0 °C, as for the usual outdoor winter season temperature in mild climates. Moreover, the interior environment temperature was set to 20 °C, as for the usual indoor comfort temperature in the winter season. Note that the computed thermal resistance (or *R*-value) does not depend on the adopted temperatures indoor or outdoor environment temperatures, as prescribed by standard ISO 6946 [24]. This same standard recommends some default values for the surface thermal resistances. In this study, these standardized values were adopted, specifically 0.04 $m^2 \cdot K/W$ for external surface thermal resistance (R_{se}) and 0.13 $m^2 \cdot K/W$ for internal surface thermal resistance (R_{si}).

2.4.3. Model Accuracy Verifications and Validation

Three precision verifications and one validation procedure were realized to ensure the reliability of the bidimensional THERM software [23] models.

Regarding the models' accuracy verifications, the references were as follows: (1) the Annex C, ISO 10211 standard [25] 2D test cases; (2) the analytical calculation procedures defined in standard ISO 6946 [24]; (3) the 3D FEM simulation results provided by ANSYS software [26].

Regarding the model validation, the results were compared with some experimental measurements, under controlled laboratory conditions, as depicted in the next subsection.

(1) ISO 10211 Test Cases Verification

For this verification, the two bidimensional test cases depicted in Annex C of standard ISO 10211 [25] were modeled, and the results obtained were within the bounds permitted, ensuring the precision of the THERM algorithm and the models. For sake of brevity and to avoid repetitions, the results are not displayed here. However, they can be accessed in previous studies from the authors [13,27,28].

(2) ISO 6946 Analytical Calculus Verification

Since standard ISO 6946 [24] provides some analytical calculation procedures for walls having homogeneous layers, a simplified LSF wall THERM model was built, assuming no steel frame. For this accuracy verification, the THERM model was very similar to the one previously illustrated in Figure 1, for the reference facade LSF wall, but having no steel stud. Moreover, the material's properties were also the same (see Table 2), as well as the boundary conditions (see Section 2.4.2).

The computed thermal transmittances for the numerical and analytical approaches are displayed in Table 3. Both obtained *U*-values were the same, highlighting the excellent accuracy and reliability of the THERM models.

Table 3. Thermal transmittance calculated for the simplified reference facade LSF wall assuming homogeneous layers (without steel stud).

LSF Wall Type	*U*-Value [W/(m^2·K)]	
	Numerical (THERM)	**Analytical (ISO 6946)**
Facade	0.225	0.225

(3) 3D FEM Verification

In this precision verification, a comparison between the 2D model results, obtained using the THERM software [23], and the 3D model results, computed using the ANSYS software [26], was performed. Two different facade LSF walls were assessed: (1) the reference LSF wall without any TBS (previously illustrated in Figure 1); (2) a similar LSF wall containing an outer TBS, having a thermal conductivity equal to 30 mW/(m·K), a width equal to 50 mm, and 10 mm thickness. All other thicknesses and thermal conductivities were early displayed in Table 2, and the implemented boundary conditions were the same as described in Section 2.4.2.

The computed conductive *R*-values and the isothermal colour distribution are displayed in Figure 3a for the reference facade LSF wall (without TBS) and in Figure 3b for a similar wall having a TBS along the outer steel stud flange. The greatest difference between 2D and 3D models occurred for the reference LSF facade (+0.5%), while, for the other LSF wall configuration having a TBS, the difference was almost negligible (−0.001 m^2·K/W, 0.0%), again demonstrating the exceptional accuracy of the THERM models used in this parametric study.

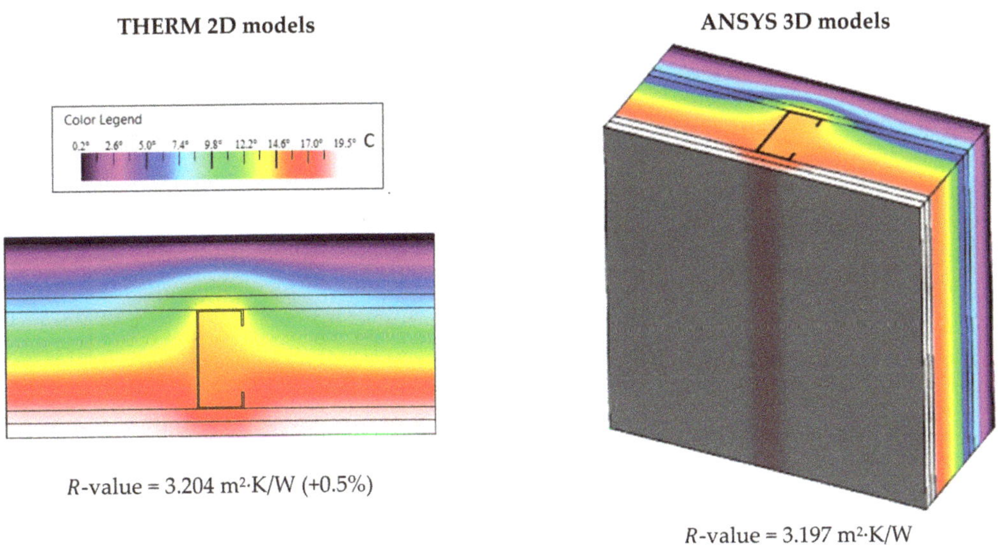

(a)

Figure 3. *Cont.*

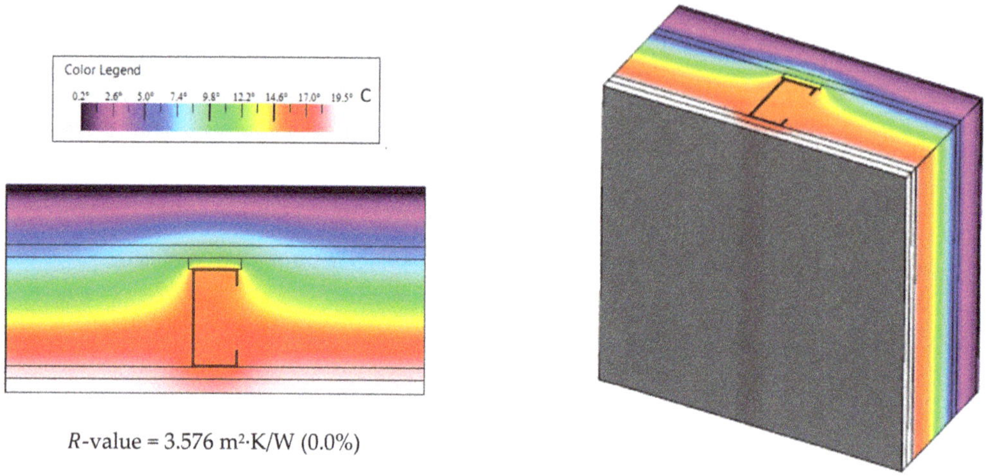

R-value = 3.576 m²·K/W (0.0%)

R-value = 3.577 m²·K/W

(**b**)

Figure 3. Accuracy verification of the LSF facade wall model: temperature distribution and *R*-values: (**a**) reference LSF facade; (**b**) LSF facade with an outer thermal break strip.

(4) Lab Measurement Validation

The THERM model for the reference facade LSF wall, in addition to the three previous explained accuracy verifications, was also validated against lab measurements, under controlled conditions. These laboratory measurements were performed using the same test procedures and setup as those described in previous research from the same authors [16], for the thermal performance evaluation of load-bearing partition LSF walls. Therefore, for the sake of conciseness and to avoid unnecessary repetitions, these test procedures and lab setup are not explained here again.

Nevertheless, the achieved results for the reference facade LSF wall are displayed in Table 4, which lists the measured conductive thermal resistances for the three height sensor locations and the equivalent average conductive *R*-value (3.200 $m^2 \cdot K/W$). Note that the conductive thermal resistance computed by the THERM model is also shown in this table (3.204 $m^2 \cdot K/W$). This very reduced difference between the measured and the predicted thermal resistances (only +0.1%) allowed ensuring the precision of these THERM models, but also permitted their validation.

Table 4. Reference facade LSF wall conductive *R*-values, for both experimental and numerical methods.

Test	Sensors Location	*R*-Value [$m^2 \cdot K/W$]
1	Top	3.247
2	Middle	3.121
3	Bottom	3.232
Measurement Average		**3.200**
Computed in THERM		**3.204**
Percentage Deviation		+0.1%

3. Results and Discussion

In this section, the computed results are displayed and analyzed. First, the *R*-values for the reference facade LSF wall are presented. Then, the thermal performance of this LSF facade wall is evaluated when using a single TBS, placed in the outer steel stud flange. Next, the use of two TBS is assessed. Lastly, the influence of the ETICS insulation thermal conductivity and thickness, when there is no TBS, is also analyzed.

3.1. Reference Facade LSF Wall

The reference surface-to-surface thermal resistance for the facade LSF wall with commercial C90 × 43 × 15 × 1.5 steel studs, placed 400 mm apart (Figure 1), is 3.204 $m^2 \cdot K/W$ (Figure 3a). For the 600 mm steel stud spacing, the *R*-value is improved to 3.499 $m^2 \cdot K/W$ (+9%). This thermal performance improvement of the LSF facade wall was expected, due to the decreased amount of steel, originating from the increased steel studs' spacing.

3.2. One Thermal Break Strip

3.2.1. The Influence of TBS Thickness and Thermal Conductivity

The surface-to-surface thermal resistances obtained for the facade LSF walls, having one TBS with variable thickness and 50 mm width, for two different steel stud spacings ((a) 400 mm; (b) 600 mm), are displayed in Figure 4. The current charts show the same trend as those described in the previous study [16], but it can be noted that, on these facade LSF walls (having ETICS), the TBS performance improvement was smaller than in the previous partition LSF walls. This could be explained by the reduced relevance of the steel studs' thermal bridges due to the ETICS insulation continuous layer.

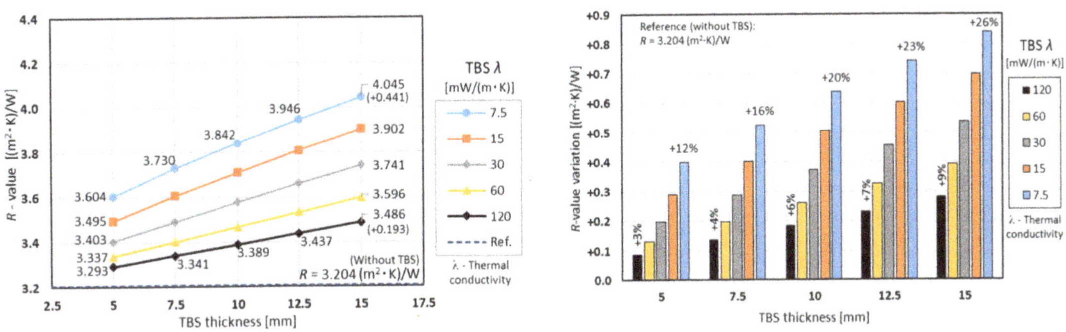

(a)

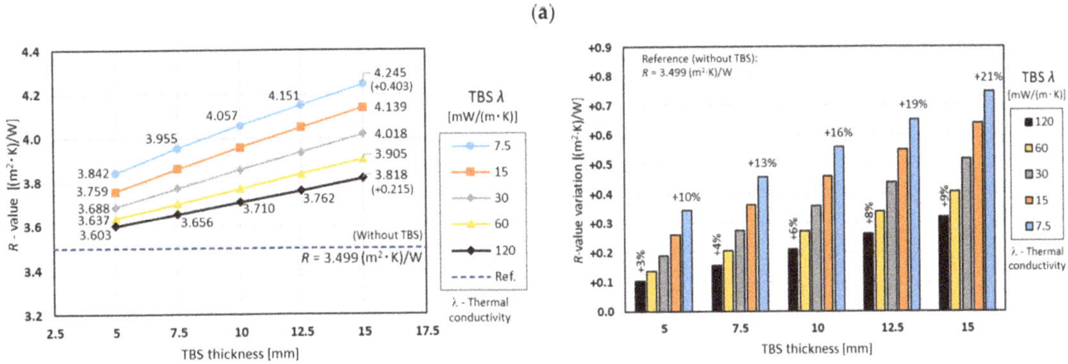

(b)

Figure 4. Surface-to-surface thermal resistances for facade LSF walls, having one thermal break strip with variable thickness and 50 mm wide, for two different steel stud spacings: (**a**) 400 mm; (**b**) 600 mm.

Nevertheless, with respect to the R-values presented in Figure 4a for the higher thermal conductivity (120 mW/m·K; black line), even the thinner TBS (5 mm) allowed increasing the thermal performance from 3.204 $m^2 \cdot K/W$ (reference value) to 3.293 $m^2 \cdot K/W$. Moreover, when there was an increase in the TBS thickness up to 15 mm, the thermal resistance also had a nearly linear variation increase, up to 3.486 $m^2 \cdot K/W$. In terms of percentages (see right plot), these R-value increments range from +3% to +9%.

Looking now to the other evaluated smaller TBS thermal conductivities, the slope of the corresponding R-values lines also increased with the decrease in TBS conductivity. This trend was predictable, and the major R-values were achieved for the smaller TBS thermal conductivity (7.5 mW/m·K, blue line), varying from 3.604 $m^2 \cdot K/W$ (5 mm thick) up to 4.045 $m^2 \cdot K/W$ (15 mm thick). As displayed in the right graph of Figure 4a, this thermal resistance variation, in relation to the reference facade LSF wall, increased from +12% (5 mm thick) to +26% (15 mm thick).

Two analogous plots are displayed in Figure 4b for 600 mm steel stud spacing, instead of 400 mm spacing. As predicted, all computed R-values were bigger than the preceding ones, including the new reference one (3.449 $m^2 \cdot K/W$). Moreover, for the same TBS thickness, the R-value increase originating from the TBS thermal conductivity decrease was now smaller relatively to the former studs' spacing (400 mm), illustrated in Figure 4a.

Additionally, in both plots (400 mm and 600 mm steel studs' spacing), increasing the TBS thickness always provided an improved thermal performance, regardless of the conductivity. This is due to the assumed mineral wool batt insulation's volumetric expansibility, fulfilling the increased wall cavity, whereby this wall thickness increment is equal to the TBS thickness.

3.2.2. The Influence of TBS Width and Thermal Conductivity

Figure 5a exhibits the surface-to-surface R-values obtained for the facade LSF walls having one TBS with variable width and constant thickness (10 mm), when the steel studs are spaced 400 mm. With respect to the R-values for the thermal conductivity equal to 120 mW/m·K (higher evaluated value; black line), even the smaller assessed TBS width (30 mm) enabled an R-value increase from 3.204 $m^2 \cdot K/W$ (reference value) to 3.447 $m^2 \cdot K/W$. However, when the TBS width increased to 70 mm, the R-value decreased to 3.361 $m^2 \cdot K/W$, i.e., a negative variation equal to −0.086 $m^2 \cdot K/W$. Looking now to the thermal conductivity of 30 mW/m·K (gray line), the R-value variation was very reduced (+0.018 $m^2 \cdot K/W$) when increasing the width of the TBS (3.564 to 3.582 $m^2 \cdot K/W$). Observing now the smaller evaluated TBS conductivity R-values (blue line), there was a major increment relative to the reference facade LSF wall (3.204 to 3.675 $m^2 \cdot K/W$), as well as when increasing the TBS width from 30 to 70 mm (+0.264 $m^2 \cdot K/W$). In terms of percentages (see blue vertical bars in the right plot of Figure 5a), this R-value increment ranged from +15% to +23%.

The computed R-values for a larger steel studs' spacing (600 mm) are displayed in Figure 5b. Comparing this new plot with the previous one (Figure 5a), the major differences are the higher predicted R-values and the reduced thermal performance variations due to the use of TBS. In fact, the R-value reduction for the higher TBS conductivity (120 mW/m·K; black line) became −0.068 $m^2 \cdot K/W$, instead of the previous decrease of −0.086 $m^2 \cdot K/W$. Observing the smaller TBS conductivity (7.5 mW/m·K; blue line), the thermal resistance increment, due to the TBS width increase, became (Figure 5b, 600 mm) only +0.197 $m^2 \cdot K/W$, when the previous value (Figure 5a, 400 mm) was +0.264 $m^2 \cdot K/W$.

Comparing Figures 4 and 5, it can be observed that a TBS having 10 mm thickness and 30 mm width (Figure 5) presented higher R-values in comparison to a TBS having 5 mm thickness and 50 mm width (Figure 4), regardless of the steel studs' spacing and the TBS conductivity. However, increasing the TBS thickness (Figure 4) was always more gainful (higher R-values increment) in comparison to a TBS width increase (Figure 5). Moreover, for the higher evaluated TBS conductivity values (60 and 120 mW/m·K), it was not adequate to increase the TBS width, since there was a consequent reduction in the facade LSF wall thermal performance (Figure 5). Thus, similarly to a previous study for partition

LSF walls [16], it can also be concluded for facade LSF walls that it is more effective to increase the thickness than the width of the TBS. However, in facade LSF walls, the thermal performance improvement due to the use of TBS was smaller, since the steel studs' thermal bridges were less relevant, due to the existence of a facade ETICS continuous insulation.

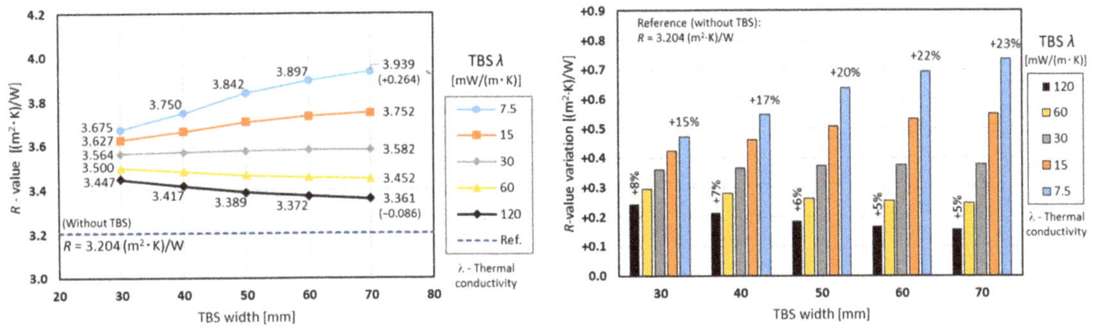

(**a**)

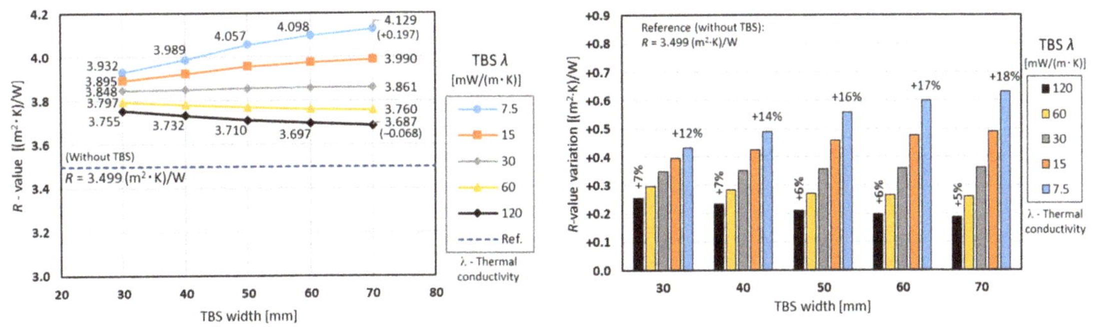

(**b**)

Figure 5. Surface-to-surface thermal resistances for facade LSF walls, having one thermal break strip with variable width and 10 mm thick, for two different steel stud spacings: (**a**) 400 mm; (**b**) 600 mm.

3.3. *Two Thermal Break Strips*

This subsection presents and discusses the computed *R*-values for facade LSF walls when using two TBS, instead of only one outer TBS.

3.3.1. The Influence of TBS Thickness and Thermal Conductivity

Figure 6a is similar to Figure 4a, but with the facade LSF walls having two TBSs instead of a single outer TBS. These new charts exhibit very similar features to the previous ones but with higher *R*-values. This was expected given the use of two TBS and their consequent thermal performance improvement. As mentioned before, this thermal performance enhancement was due not only to the use of the TBS itself, but also to the consequent increase in the wall air cavity and batt insulation thickness, which was equal to the thickness summation of both TBSs (10 + 10 mm).

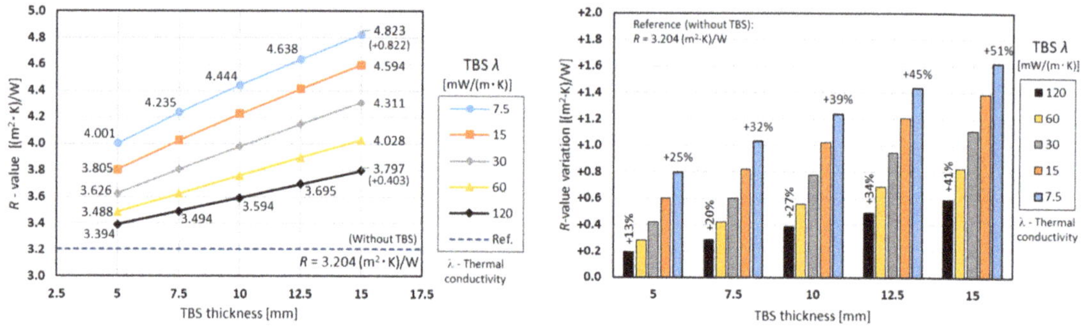

(a)

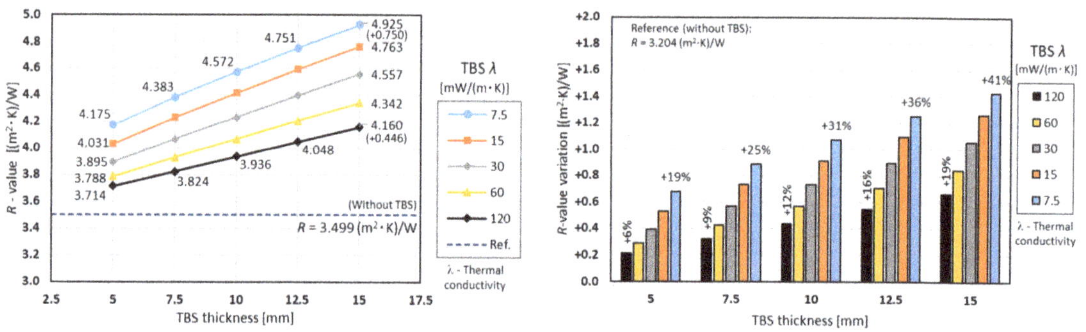

(b)

Figure 6. Surface-to-surface thermal resistances for facade LSF walls, having two thermal break strips with variable thickness and 50 mm wide, for two different steel stud spacings: (**a**) 400 mm; (**b**) 600 mm.

Other interesting findings could be obtained when comparing the case where the total thickness of TBSs was the same, but the number was different (one and two). For example, according to the blue line (λ = 7.5 mW/m·K) in the left graph of Figure 4a, the *R*-value was 3.842 [m^2·K)/W when the thickness was 10 mm (one TBS), while, according to the left graph of Figure 6a, the *R*-value was 4.001 [(m^2·K)/W] when the thickness of each TBS was 5 mm (the total thickness of two TBSs was 10 mm). This appears to indicate that it is more effective to split the TBSs (higher thermal resistance) instead of using a single one with the same total thickness.

The computed results for a 600 mm steel stud spacing is displayed in Figure 6b, exhibiting a similar trend to that for a 400 mm spacing (Figure 6a); however, all obtained *R*-values were higher than the previous ones, as anticipated given the smaller steel amount per wall square meter and consequent minor steel-related thermal bridge effect. Moreover, in Figure 6b, for the same TBS thickness, the thermal performance improvement due to the TBS conductivity decrease was now quite smaller when compared to the 400 mm steel stud spacing (Figure 6a). Additionally, the *R*-value increment due to the TBS thickness increase was reduced for smaller TBS conductivities (e.g., 7.5 mW/m·K; blue line), +0.750 instead of +0.822 m^2·K/W, and slightly increased for higher thermal conductivity values (e.g., 120 mW/m·K; black line), +0.446 instead of +0.403 m^2·K/W.

3.3.2. The Influence of TBS Width and Thermal Conductivity

Figure 7a is analogous to Figure 5a. However, instead of a single outer TBS, the facade LSF walls have two TBSs, i.e., one in the outer stud flange and another TBS in the inner flange. The trends exhibited in these two plots are identical, but with an improved thermal performance, i.e., increased *R*-values. In addition to this enhanced thermal performance, the relevance of increasing the TBS width was higher for the same TBS conductivity, which was expected since two TBSs were used (Figure 7a) instead of one (Figure 5a). In fact, for the smaller TBS conductivity (7.5 mW/m·K; blue line), the *R*-value increment was +0.430 $m^2 \cdot K/W$, instead of only +0.264 $m^2 \cdot K/W$. Moreover, for the higher evaluated TBS conductivity (120 mW/m·K; black line), the thermal resistance decrease was -0.169 $m^2 \cdot K/W$, instead of -0.086 $m^2 \cdot K/W$.

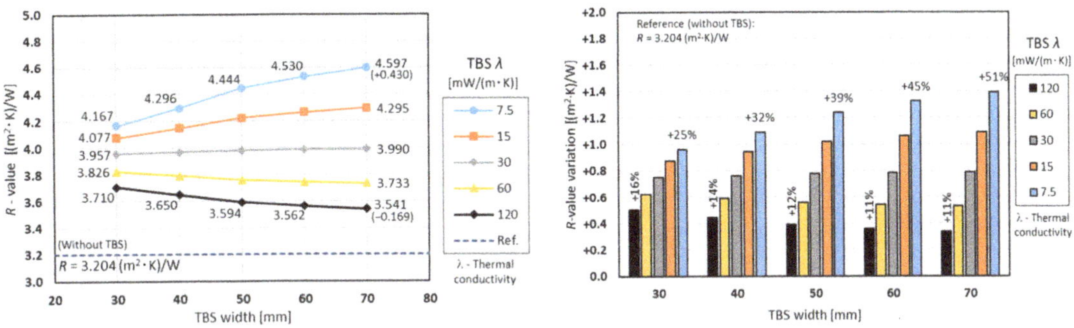

(a)

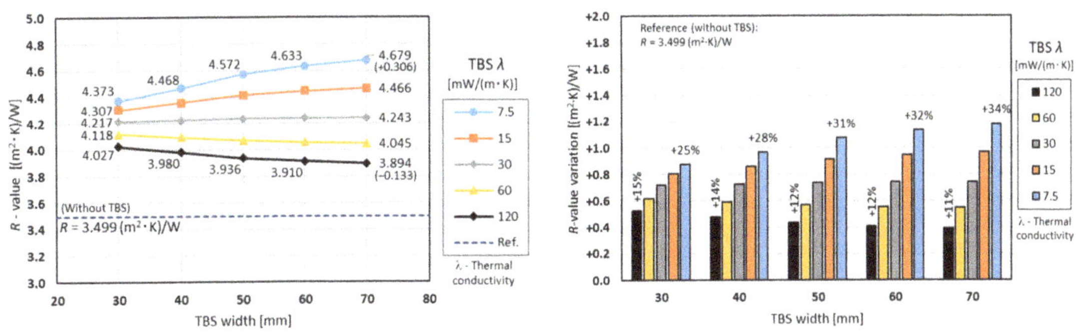

(b)

Figure 7. Surface-to-surface thermal resistances for facade LSF walls, having two thermal break strips with variable width and 10 mm thick, for two different steel stud spacings: (**a**) 400 mm; (**b**) 600 mm.

The predicted *R*-values for an increased steel stud spacing (600 mm, instead of 400 mm) are plotted in Figure 7b. As seen before for a single TBS (Figure 5), this chart had an identical tendency, but now all predicted *R*-values were bigger when compared to the previous ones (Figure 7a, 400 mm stud spacing), including the reference one (3.449 $m^2 \cdot K/W$), as mentioned before. Furthermore, the relevance of the TBS conductivity for the same TBS width, and the relevance of the TBS width for the same thermal conductivity were now smaller, when compared to the previous steel stud spacing (400 mm).

3.4. ETICS Insulation Thickness and Thermal Conductivity

In addition to the use of one (see Section 3.2) or two TBSs (see Section 3.3), this research assessed the relevance of changing the ETICS insulation thickness and thermal conductivity to improve the thermal performance of facade LSF walls. Figure 8a illustrates the computed surface-to-surface *R*-values, for facade LSF walls having ETICS insulation thickness changing from 30 to 70 mm and thermal conductivity values raging between 7.5 and 120 mW/m·K, with a steel stud spacing equal to 400 mm, without TBS.

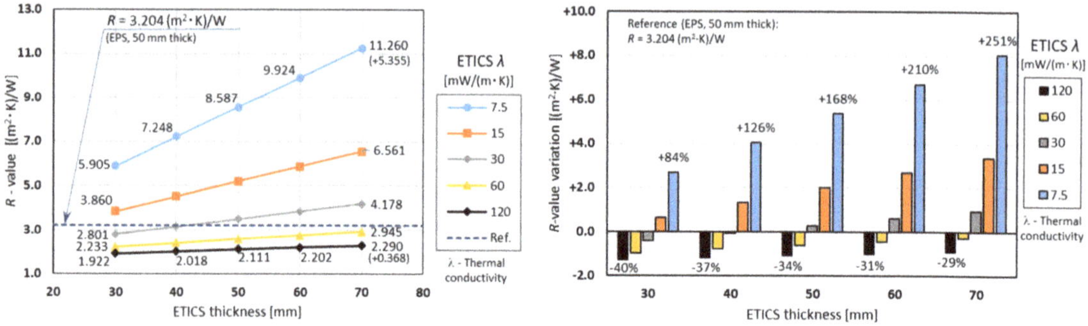

(a)

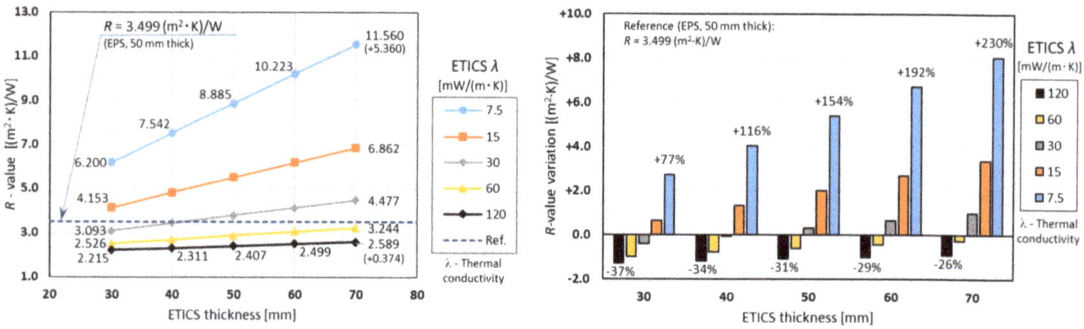

(b)

Figure 8. Surface-to-surface thermal resistances for facade LSF walls, without thermal break strips, having a variable ETICS insulation thickness and thermal conductivity, for two different steel stud spacings: (**a**) 400 mm; (**b**) 600 mm.

Compared to previous plots (one and two TBS), several main differences arise, such as (1) the ETICS insulation thermal conductivity lines for higher values (30, 60 and 120 mW/m·K) having smaller thermal resistances in relation to the reference facade LSF wall *R*-value (3.204 $m^2 \cdot K/W$), (2) the *R*-values variation range now being much larger, ranging (70 mm ETICS thickness) from 2.290 $m^2 \cdot K/W$ up to 11.260 $m^2 \cdot K/W$, for the 120 and 7.5 mW/m·K insulation thermal conductivities, respectively, and (3) the thermal resistance increment with the ETICS insulation thickness increase now also being larger, mainly for the smaller thermal conductivities (e.g., +5.360 $m^2 \cdot K/W$ for 7.5 mW/m·K, blue line).

Figure 8b shows the same parameters variation, but for LSF walls having a 600 mm steel stud spacing. In comparison to the previous steel stud spacing (400 mm) chart (Figure 8a), the main features are as follows: (1) as expected, due to the minor steel content per LSF wall area, all *R*-values were now increased; (2) however, the *R*-value increment due to ETICS insulation thickness increase, for each thermal conductivity value, was very similar (as expected, since this insulation layer was continuous). Note that, in terms of percentages, the *R*-value increment seemed larger for the 400 mm stud spacing (Figure 8a, left), but this was mainly due to a smaller reference *R*-value (3.204 instead of 3.499 $m^2 \cdot K/W$).

As illustrated in Figure 8b, insulation materials with smaller thermal conductivities, 7.5 and 15 mW/m·K, exhibited higher thermal performance, with maximum *R*-values (for 70 mm thickness and 600 mm steel stud spacing) of 11.560 and 6.862 $(m^2 \cdot K)/W$, respectively. Not surprisingly, for insulation materials with higher thermal conductivities, 60 and 120 mW/m·K, the *R*-value of the reference wall (having 50 mm of EPS thermal insulation, 36 mW/m·K) was not reached. Moreover, even for a smaller thermal conductivity

(30 mW/m·K; gray line), when the thickness was reduced (30 and 40 mm), the achieved *R*-values were also smaller than the reference one (3.499 m^2 K)/W).

4. Conclusions

In this article, a parametric study related to the thermal performance of load-bearing facade LSF walls was completed. This research is a continuation of a previous parametric study for partition LSF walls from the same authors [16]. In the present study, seven parameters were assessed: (1) the steel stud distance; (2) the TBS position and number along the steel stud's flanges; (3) the thermal conductivity of the TBS material; (4) the TBS thickness; (5) the TBS width; (6) the thickness of the ETICS thermal insulation layer; (7) the thermal conductivity.

The reliability of the obtained results was ensured by the experimental validation of the bidimensional THERM [23] models, which were used to simulate the thermal performance of the assessed facade LSF walls. Additionally, their accuracy was also successfully verified using three different approaches.

The main outcomes of this research are summarized as follows:

- The increase in the steel stud spacing from 400 mm to 600 mm allowed an *R*-value thermal performance improvement of +0.295 m^2·K/W, which is very similar to the result achieved previously for partition LSF walls: +0.292 m^2·K/W [16].
- Similarly to what was concluded before for load-bearing partition LSF walls [16], for load-bearing facade LSF walls, it is still more effective to increase the TBS thickness rather than their width.
- Nevertheless, the *R*-value increments are slightly smaller for facade LSF walls, due to the existence of an ETICS continuous thermal insulation layer, which decreases the steel studs' thermal bridges relevance, as expected.
- The previous features are valid for one or two TBSs placed along the vertical steel studs, but are more pronounced for the double TBS.
- The major thermal performance improvements were found when increasing the ETICS insulation thickness (from 30 to 70 mm) and decreasing their thermal conductivity (to 7.5 mW/m·K), for which it was found a relevant *R*-value increase of +5.360 m^2·K/W, for a 400 mm steel stud spacing.
- In fact, the abovementioned thermal performance improvement was significantly higher (around 6.5 times) than the most relevant one achieved when using two TBSs, having 15 mm thickness (increased from 5 mm), 50 mm width, and 7.5 mW/m·K thermal conductivity, which was only +0.822 m^2·K/W, for the reference steel stud spacing (400 mm).

Regarding the foremost limitations of this study, one can mention that all other steel profiles of the facade LSF wall were neglected, considering only the vertical load-bearing steel studs. Secondly, several batt insulation materials are available on the market, but only one was modeled (mineral wool). With respect to the first constraint, it can be mentioned that the modeled vertical load-bearing steel studs were the most frequent and relevant ones in facade LSF walls. Moreover, some other steel frame profiles (e.g., bottom and top wall trackers) are usually considered within the slab to contribute to the wall linear thermal bridge effect, which was outside of the scope of this study. Concerning the second restriction, mineral wool is perhaps the most used batt insulation material today. Moreover, it was supposed that this fibrous insulation material has enough expandability to fill the cavity of the facade LSF wall.

Through this work, it was possible to better comprehend, compare, and quantify the thermal performance improvement due to the use of TBS and ETICS in load-bearing facade LSF walls. Such a systematic parametric study did not previously exist in the literature. At the design stage, this knowledge could be advantageous for engineers and designers when there is a necessity to specify the TBS material, width, thickness, and number, as well as the ETICS insulation material and thickness.

Author Contributions: Conceptualization, P.S.; Methodology, P.S., P.L. and D.A.; Validation, P.S. and P.L.; Formal analysis, P.L. and D.A.; Investigation, P.S., P.L. and D.A.; Resources, P.S.; Writing—original draft, P.L. and D.A.; Writing—review & editing, P.S.; Visualization, P.S. and D.A.; Supervision, P.S.; Project administration, P.S.; Funding acquisition, P.S. All authors have read and agreed to the published version of the manuscript.

Funding: This research was funded by FEDER funds through the Competitivity Factors Operational Program—COMPETE, and by national funds through FCT, Foundation for Science and Technology, within the scope of the project POCI-01-0145-FEDER-032061.

Cofinanciado por: POCI-01-0145-FEDER-032061

Acknowledgments: The authors also want to thank the following companies: Pertecno, Gyptec Ibéria, Volcalis, Sotinco, Kronospan, Hulkseflux, Hilti, and Metabo.

Conflicts of Interest: The authors declare no conflict of interest.

References

1. Santos, P.; da Silva, L.S.; Ungureanu, V. *Energy Efficiency of Light-Weight Steel-Framed Buildings*, 1st ed.; Technical Committee 14-Sustainability & Eco-Efficiency of Steel Construction; European Convention for Constructional Steelwork (ECCS): Brussels, Belgium, 2012; ISBN 978-92-9147-105-8.
2. Erhorn-Kluttig, H.; Erhorn, H. *Impact of Thermal Bridges on the Energy Performance of Buildings*; European Commission: Brussels, Belgium, 2009; pp. 1–8.
3. Ge, H.; Baba, F. Dynamic effect of thermal bridges on the energy performance of a low-rise residential building. *Energy Build.* **2015**, *105*, 106–118.
4. Gomes, A.P.; de Souza, H.A.; Tribess, A. Impact of thermal bridging on the performance of buildings using Light Steel Framing in Brazil. *Appl. Therm. Eng.* **2013**, *52*, 84–89. [CrossRef]
5. Santos, P.; Mateus, D. Experimental assessment of thermal break strips performance in load-bearing and non-load-bearing LSF walls. *J. Build. Eng.* **2020**, *32*, 101693.
6. Santos, P.; Abrantes, D.; Lopes, P.; Mateus, D. Experimental and Numerical Performance Evaluation of Bio-Based and Recycled Thermal Break Strips in LSF Partition Walls. *Buildings* **2022**, *12*, 1237. [CrossRef]
7. Santos, P.; Mateus, D.; Ferrandez, D.; Verdu, A. Numerical Simulation and Experimental Validation of Thermal Break Strips' Improvement in Facade LSF Walls. *Energies* **2022**, *15*, 8169.
8. Kapoor, D.R.; Peterman, K.D. Quantification and prediction of the thermal performance of cold-formed steel wall assemblies. *Structures* **2021**, *30*, 305–315.
9. Martins, C.; Santos, P.; Simoesdasilva, L. Lightweight steel-framed thermal bridges mitigation strategies: A parametric study. *J. Build. Phys.* **2016**, *39*, 342–372. [CrossRef]
10. Santos, P.; Poologanathan, K. The Importance of Stud Flanges Size and Shape on the Thermal Performance of Lightweight Steel Framed Walls. *Sustainability* **2021**, *13*, 3970. [CrossRef]
11. Santos, P.; Lemes, G.; Mateus, D. Analytical Methods to Estimate the Thermal Transmittance of LSF Walls: Calculation Procedures Review and Accuracy Comparison. *Energies* **2020**, *13*, 840. [CrossRef]
12. Kosny, J.; Christian, J.E. Thermal evaluation of several configurations of insulation and structural materials for some metal stud walls. *Energy Build.* **1995**, *22*, 157–163.
13. Santos, P.; Gonçalves, M.; Martins, C.; Soares, N.; Costa, J.J. Thermal transmittance of lightweight steel framed walls: Experimental versus numerical and analytical approaches. *J. Build. Eng.* **2019**, *25*, 100776. [CrossRef]
14. Lupan, L.M.; Manea, D.L.; Moga, L.M. Improving Thermal Performance of the Wall Panels Using Slotted Steel Stud Framing. *Procedia Technol.* **2016**, *22*, 351–357.
15. Henriques, J.; Rosa, N.; Gervasio, H.; Santos, P.; da Silva, L.S. Structural performance of light steel framing panels using screw connections subjected to lateral loading. *Thin.-Walled Struct.* **2017**, *121*, 67–88.
16. Santos, P.; Lopes, P.; Abrantes, D. Thermal Performance of Load-Bearing, Lightweight, Steel-Framed Partition Walls Using Thermal Break Strips: A Parametric Study. *Energies* **2022**, *15*, 9271. [CrossRef]
17. Gyptec Ibérica. Technical Sheet: Standard Gypsum Plasterboard. 2022. Available online: https://www.gyptec.eu/documentos/Ficha_Tecnica_Gyptec_A.pdf (accessed on 10 February 2022).
18. Kronospan. *Technical Sheet: Kronobuild Materials*; KRONOSPAN GmbH: Steinheim-Sandebeck, Germany, 2022.
19. Volcalis. Technical Sheet: Alpha Mineral Wool. 2022. Available online: https://www.volcalis.pt/categoria_file_docs/fichatecnica_volcalis_alpharollo-386.pdf (accessed on 10 February 2022).
20. Santos, C.; Matias, L. *ITE50-Coeficientes de Transmissão Térmica de Elementos da Envolvente dos Edifícios (in Portuguese)*; LNEC-Laboratório Nacional de Engenharia Civil: Lisboa, Portugal, 2006.
21. LNEC. *Documento de Homologação: Tincoterm EPS-Sistema Compósito de Isolamento Térmico pelo Exterior*; LNEC: Maia, Portugal, 2015.

22. WEBERTHERM UNO. Technical Sheet: Weber Saint-Gobain ETICS Finish Mortar (in Portuguese). 2018. Available online: www.pt.weber/files/pt/2019-04/FichaTecnica_weberthermuno.pdf (accessed on 14 March 2019).
23. THERM. Software version 7.6.1. Lawrence Berkeley National Laboratory, United States Department of Energy. 2017. Available online: https://windows.lbl.gov/software/therm (accessed on 14 February 2019).
24. *ISO 6946*; Building Components and Building Elements—Thermal Resistance and Thermal Transmittance—Calculation Methods. International Organization for Standardization: Geneva, Switzerland, 2017.
25. *ISO 10211*; Thermal Bridges in Building Construction-Heat Flows and Surface Temperatures-Detailed Calculations Ponts Thermiques Dans les Bâtiments-Flux Thermiques et Températures Superficielles-Calculs Détaillés COPYRIGHT PROTECTED DOCUMENT. International Organization for Standardization: Geneva, Switzerland, 2017.
26. *ANSYS Workbench*, Software version 19.1; ANSYS, Inc.: Canonsburg, PA, USA, 2018. Available online: www.ansys.com/products/ (accessed on 8 June 2020).
27. Roque, E.; Santos, P. The Effectiveness of Thermal Insulation in Lightweight Steel-Framed Walls with Respect to Its Position. *Buildings* **2017**, *7*, 13.
28. Santos, P.; Lemes, G.; Mateus, D. Thermal Transmittance of Internal Partition and External Facade LSF Walls: A Parametric Study. *Energies* **2019**, *12*, 2671. [CrossRef]

energies

Article

Thermophysical Parameters and Hygrothermal Simulation of Aerogel-Based Fibre-Enhanced Thermal Insulating Renders Applied on Exterior Walls

Marco Pedroso [1], Maria da Glória Gomes [1], José Dinis Silvestre [1], Ahmed Hawreen [2,3,4,*] and Inês Flores-Colen [1,*]

[1] Civil Engineering Research and Innovation for Sustainability (CERIS), Departamento de Engenharia Civil, Arquitetura e Georecursos (DECivil), Instituto Superior Técnico (IST), Universidade de Lisboa, Av. Rovisco Pais, 1049-001 Lisbon, Portugal

[2] Department of Highway and Bridge Engineering, Technical Engineering College, Erbil Polytechnic University, Erbil 44001, Iraq

[3] Department of Civil Engineering, College of Engineering, Nawroz University, Duhok 42001, Iraq

[4] Civil Engineering, Architecture and Georesources Department, Instituto Superior Técnico, Universidade de Lisboa, Av. Rovisco Pais, 1049-001 Lisbon, Portugal

* Correspondence: hawreen.a@gmail.com (A.H.); ines.flores.colen@tecnico.ulisboa.pt (I.F.-C.)

Abstract: Aerogel-based renders have been the subject of research in the last few years due to their high thermal insulation characteristics and the need for buildings to become more energy-efficient. This study compares the hygrothermal behaviour of an aerogel-based render (reference) with the same base formulation, replacing the powder with three different fibres (aramid 0.5%, sisal 0.1%, and biomass 0.1%, by total volume) that can be used in buildings' envelopes. The experimental programme allowed us to characterise and compare the thermophysical properties of the different formulations and then simulate the hygrothermal performance of these solutions when applied to walls for different climatic conditions, considering additional parameters such as total water content, drying potential, water content levels, and thermal insulating performance. These thermophysical parameters were then included in hygrothermal numerical simulations. The results allowed us to verify that the incorporation of fibres improved the hygrothermal properties due to lower capillary absorption and higher water vapour permeability. These renderings showed a high potential for application to building envelopes in different climatic conditions, improving their energy efficiency by up to 20% when compared to other conventional solutions.

Keywords: energy-efficient building; fiber; hygrothermal performance; nanomaterial; silica aerogel; thermal insulation; thermal render

check for updates

Citation: Pedroso, M.; Gomes, M.d.G.; Silvestre, J.D.; Hawreen, A.; Flores-Colen, I. Thermophysical Parameters and Hygrothermal Simulation of Aerogel-Based Fibre-Enhanced Thermal Insulating Renders Applied on Exterior Walls. *Energies* **2023**, *16*, 3048. https://doi.org/10.3390/en16073048

Academic Editor: Paulo Santos

Received: 21 February 2023
Revised: 13 March 2023
Accepted: 13 March 2023
Published: 27 March 2023

1. Introduction

In the last few years, there have been several measures and methods to reduce energy consumption in buildings [1,2] through the study of innovative thermal insulating materials to be applied on their envelope, either in new construction or in retrofit scenarios [3,4], while also considering indoor thermal comfort [5].

The hygrothermal behaviour of the building envelope impacts energy needs, indoor comfort, and indoor air quality [6,7]. This behaviour is a function of the materials' thermal and hydrostatic performances, depending on several characteristics. The use of hygroscopic materials has been considered a good solution to reduce the internal condensation in the buildings' envelope while also impacting the reduction of heating and cooling energy consumption [8,9]. To understand and predict the hygrothermal behaviour of a building envelope, it is possible to use numerical models [10]. These models are based on experimental data that, to improve their prediction abilities, demands a complete set of experimental tests characterising the materials [11], e.g., through capillary water absorption, water vapour permeability, sorption curves, and thermal conductivity, among others.

To improve the façade's performance, silica aerogel, which is one of the most researched nanomaterials for application on buildings' envelopes for energy efficiency [12,13], has been used as a thermal super-insulating material [14,15]. Silica aerogel is classified as a 3-D nanomaterial (all dimensions above the nanoscale) [16], presenting a high quantity of pores with a reduced dimension (~10 to 100 nm) [17], supported by a silica (SiO_2) structure [14]. As a consequence of its nanostructure and highly tortuous paths that limit heat transport by radiation, convection, and conduction [18], it presents high thermal insulation properties with thermal conductivity values as low as 0.012 W m^{-1} K^{-1} [19]. Although its cost is usually high, it has been decreasing in the past years due to scale economy effects [20], promoting a higher spread in its application in building solutions.

Due to its properties, silica aerogel granules have been introduced in renders since the 2010s [21]. This shows a significant decrease in the render's thermal conductivity, with several researchers studying this solution and trying different formulations for new construction and retrofit scenarios [15,22,23]. However, these solutions still present some drawbacks, as their mechanical performance lowers and the thermal conductivity also lowers, as in several other lightweight materials [24]. They also tend to show high capillary water absorption [23], which could impair their outdoor use if no multi-layered coating systems are used [25].

Knowing the aerogel-based renders' main limitations, it was considered that fibres could be used to improve the overall mechanical and physical properties of these coating solutions [26,27]. Therefore, this study deals with the introduction of three fibres of synthetic (aramid) and natural origin (sisal and biomass-based) in a previously developed aerogel-based thermal render [23] that can potentially be used as coatings on buildings' envelopes. The used fibre quantities were a result of the improvement in the mechanical properties over the reference, using a substitution of render powder volumes of 0.5% aramid (vol.) and 0.1% (vol.) of sisal and biomass [27].

Although currently there are some research works [21,28,29] studying the hygrothermal properties of aerogel-based renders and plasters, there is a lack of study on the influence of fibres and how their use could potentiate the composites' multifunctional properties. Moreover, a complete set of hygrothermal results that allow others to simulate the material's performance under different climatic conditions is also needed. This fact can also be related to the non-existence of requirements in EN 998-1 [30] that could allow further characterisation of hygrothermal parameters needed to carry out hygrothermal simulations.

The present work aims to compare and analyse the hygrothermal behaviour of four different aerogel-based thermal render formulations, with and without fibres. This involved evaluating their influence through an extensive set of tests where their thermal conductivity for several moisture contents and parameters related to moisture behaviour and liquid water conduction were evaluated. Furthermore, other currently used thermal insulation materials were evaluated to benchmark their hygrothermal performance.

With this information, it was possible to carry out subsequent hygrothermal numerical models (using the software WUFI Pro®), which allowed evaluation of the performance of the analysed thermal render formulations under different application scenarios. These scenarios included: two different climatic areas: Lisbon and Zurich; constructive solutions: a new wall (Lisbon—NW—and Zurich—ZNW); and exterior and interior retrofit scenarios (only for Lisbon—RE and RI, respectively). Several hygrothermal analysis parameters were assessed, such as the total water content and drying potential, the mould growth potential, the water content in the thermal insulating layer as well as its performance over time, and finally the outdoor superficial temperatures and surface condensation potentials.

With such data gathered, it was also possible to evaluate how these materials are influenced by moisture over time and the repercussions that such moisture has on the potential increase in the buildings' energy consumption to keep their indoor conditions comfortable, due to higher thermal conductivities induced by moisture. Therefore, this study allows for further evaluation of these renders' performance and multifunctional capabilities.

Due to the extensive data presented, this work contributes to a better understanding of the thermophysical behaviour of renders incorporating aerogel and fibres and their implications in the future of energy savings associated with buildings, due to their performance in the hygrothermal simulations. With these findings, other researchers can support their study to improve current thermal insulating render formulations or to define new ones. The most original contribution is the connection between the thermophysical characterisation of the materials through an experimental programme (additional measurements) and the hygrothermal simulations under different application scenarios, for these innovative solutions.

2. Thermophysical Experimental Characterisation

2.1. Materials

To be able to carry out the hygrothermal simulations, a significant set of the materials' thermophysical parameters must be known. Therefore, in this research and due to its novelty, a laboratory campaign was performed to characterise the thermophysical properties of an aerogel-based thermal render (TR), previously developed [23], while also incorporating fibres. The aerogel-based render (TR reference) is mainly composed of a blend of mineral binders, rheological and hydrophobic agents, and resins, while also incorporating (≈70% (vol./vol.)) *Kwark* supercritical silica aerogel in the form of granules (diameter $\leq$ 3500 µm, bulk density $\leq$ 90 kg m^{-3}, and a thermal conductivity $\leq$ 0.020 W m^{-1} K^{-1} [31], Figure 1a). For those aerogel-based renders, three different fibre types were incorporated: aramid, sisal, and biomass fibres, whose origin, dimensions, and quantities were supported by previous research from the authors [27]. Therefore, the thermophysical characterization was carried out for: TR reference (0.00% vol./vol.), TR aramid (0.50% vol./vol.), TR sisal, and TR biomass (both 0.10% vol./vol.), all fibres with ≈5 mm in length, were produced, cured, and analysed in terms of their mechanical, microstructural, and some physical aspects [27].

Figure 1. Aerogel-based render's visual aspect during the mixing process (**a**) and the specimens (cylindric I with a linear probe; closer—cylindric II with a plane probe) and ISOMET 2114 equipment from Applied Precision supplier (**b**).

To obtain the additional thermophysical data needed for the hygrothermal simulations, some specimens had to be produced, as described in Table 1, where their shape, dimensions, and number of specimens and tests are shown for each formulation. Figure 1b) shows the specimens and the ISOMET equipment used for the thermal conductivity measurement, with two different probes. The specimens' dimensions were influenced by the minimum size requirements of the ISOMET to perform the measurements [32]. Since the hygrothermal simulations used several materials (e.g., EPS and XPS), the synthesis of their thermophysical properties, as they can be found in the literature, is presented in the Supplementary Data.

Table 1. Specimens produced, per formulation.

Specimens' Shape	Dimension [mm]	Number of Specimens
Cylindric I	$\phi \geq 70$; $h \geq 130$	3
Cylindric II	$\phi \geq 100$; $h \geq 40$	3

Note: ϕ—specimens' diameter; h—specimens' height; each test conducted at least three times for each specimen to obtain some statistical significance.

2.2. Experimental Methods

To characterise the hygrothermal properties of construction materials, when the objective is to use them in hygrothermal simulation models, as indicated by Fantucci et al. [11] and Barclay et al. [33], there is a need to systematise all the relevant data. As such, a comprehensive set of tests must be carried out to characterise and obtain the input for hygrothermal simulations [34,35].

The usually needed parameters are bulk density (ρ_{hard}), open porosity (P_0), specific heat capacity (c_p), thermal conductivity (λ) from a dry to a saturated state, capillary water absorption (A_w), liquid transport coefficient for suction (D_{ws}), water vapour diffusion resistance factor (μ), and the moisture content (w) for different conditions (hygroscopic and over-hygroscopic states). Table 2 presents the bulk density, open porosity, thermal conductivity in the dry state, capillary water absorption, and water vapour diffusion resistance coefficient for these formulations, as already presented in [27]. Therefore, it was also necessary to additionally characterise the specific heat capacity, the thermal conductivity for several moisture contents, the liquid transport coefficient for suction, and the moisture content for hygroscopic and over-hygroscopic conditions.

Table 2. Synthesis of some of the parameters previously characterised—average values adapted from [27].

Render Formulation	ρ_{hard} [kg m^{-3}]	P_0 [%]	$\lambda_{10^\circ C,dry}$ [W m^{-1} K^{-1}]	μ [-]	A_w [kg m^{-2} s$^{-1/2}$]
TR reference (0.0%)	158.7	86.3	0.0293	13.7	0.1090
TR aramid (0.50%)	164.5	85.1	0.0315	13.3	0.0286
TR sisal (0.10%)	160.3	86.9	0.0298	12.7	0.0325
TR biomass (0.10%)	161.6	87.0	0.0306	12.4	0.0310

Note: ρ_{hard}—bulk density in the hardened state [kg m^{-3}]; P_0—open porosity by mercury intrusion porosimeter [%]; $\lambda_{10^\circ C,dry}$—thermal conductivity at 10 °C and hardened dry state [W m^{-1} K^{-1}]; μ—water vapour diffusion resistance factor [-]; A_w—capillary water absorption [kg m^{-2} s$^{-1/2}$].

Moreover, it was also characterised by their thermal diffusivity (α), thermal inertia (I), thermal penetration depth at 10 min ($d_{p,10min}$), maximum water content (w_{max}), and saturation coefficient, adding a greater depth to the potential comparison with other existing materials already characterised. Another studied parameter was the practical moisture buffer value ($MBV_{practical}$) since this parameter is one way to quantify the material's capability to adsorb and desorb moisture [36] and thus evaluate the potential applications in indoor environments for moisture regulation.

2.2.1. Thermal Properties

To evaluate the formulations' thermal properties, an ISOMET 2114 device from Applied Precision supplier was used. This equipment's measurement principle is based on the analysis of the sample's temperature response to heat flow impulses. Heat flow is projected by the resistor's electrical heating, which is presented inside the probes and in direct heat contact with the sample. The evaluation of the thermal conductivity (λ) and volumetric heat capacity (c_ρ) (which then allows calculating the specific heat capacity, c_p, by dividing it by the bulk density) is based on periodically sampled temperature records as a function of time, where during the same measurement cycle, the thermal diffusivity (α) is also obtained [32]. Since different moisture contents had to be evaluated, which influence the thermal conductivity values, the two available probes (a linear probe with a measurement range between 0.015 and 0.050 W m^{-1} K^{-1}, and a plane probe with a measurement range between 0.040 and 0.300 W m^{-1} K^{-1}), were used, as shown in Figure 1b.

To carry out the measurements, and due to the individual probes' measurement ranges, the cylindric I specimens were selected to make the characterisation from the dry state until 93% RH (relative humidity). Additionally, the cylindric II specimens used in the saturated state were obtained through samples' immersion in liquid water until equilibrium was reached with the surrounding conditions.

The test procedures followed the indications of the ASTM-D-5334 [37], ASTM-D-5930 [38], EN 22007-2 [39], and Gomes et al. [40,41] studies. In this case, the tests were carried out at an average reference temperature of 20 ± 3 °C. The temperature conversion to 10 °C is in line with EN 10456 [42] for *IIa* conditions (reference temperature of 23 °C and dry state) and a temperature conversion coefficient of 0.003 K^{-1}. When the samples were evaluated, their weight was measured, and then they were wrapped in plastic film right away so they would not gain or lose moisture.

Since both the mass and thermal properties evaluations are not destructive tests, the same samples (cylindric I and II) were used again for the moisture storage function test. The samples' moisture variation procedure is presented in the following sub-section.

As the specific heat capacity corresponds to the material's ability to store heat relative to its weight, it is used to estimate the thermal inertia of the material (I), which can be calculated using Equation (1) [43], where I is the thermal inertia [J m^{-2} $s^{-1/2}$], λ is the thermal conductivity [W m^{-1} K^{-1}], ρ is the bulk density [kg m^{-3}], and c_p is the specific heat capacity [J kg^{-1} K^{-1}].

$$I = \sqrt{\lambda \times \rho \times c_p} \quad (1)$$

This study also assessed the thermal diffusivity (α), directly obtained by the equipment, and the thermal penetration depth ($d_{p,10\text{min}}$) at 10 min. The thermal diffusivity (α) indicates the heat transfer rate from the hot end to the cold end. As for the thermal penetration depth ($d_{p,10\text{min}}$), it allows us to directly compare different materials and how, in this case, in 10 min, the material will respond, quicker or slower, to temperature variations on their surface. This is a parameter used in the thermal characterisation of other materials, such as the one conducted by Palumbo et al. [44]. For calculating the penetration depth, Equation (2) [44] was used, where $d_{p,10\text{min}}$ is the penetration depth after 10 min [m], α is the thermal diffusivity [m^2 s^{-1}], and t is time [s].

$$d_{p,10\text{min}} = \sqrt{\pi \times \alpha \times t} \quad (2)$$

2.2.2. Moisture Storage Function

The moisture storage function relates the amount of equilibrium moisture content to the ambient RH for a given temperature for any given material. The samples used in this evaluation were the same as those used in the thermal properties tests: cylindric I and cylindric II. The cylindric I was for the air moisture (hygroscopic range), and the cylindric II was for the liquid water saturation (over-hygroscopic range).

The hygroscopic sorption properties (Figure 2a) were determined following the main indications and procedures presented in EN ISO 12571 [45], using a Heraeus Vötsch climatic chamber. Five different levels of RH were assessed under isothermal conditions (23 °C): 0, 35, 60, 85, and 93% RH. For each level of RH, when constant mass was achieved (the change in mass between three consecutive weightings, each made 24 h apart, differed by less than 0.10% of the total mass), the specimens' mass was evaluated.

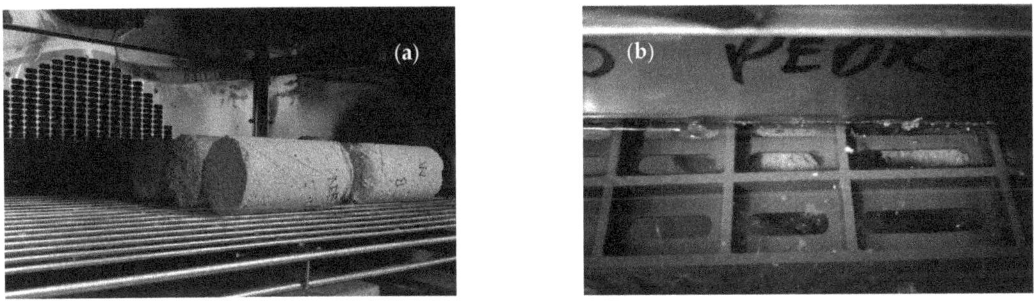

Figure 2. Aspects of the hygroscopic range evaluation (**a**) and the over-hygroscopic using a Heraeus Vötsch climatic chamber, with the samples immersed in water (**b**) in a sealed plastic box.

As for the over-hygroscopic range, i.e., the amount of water that a material absorbs at 100% RH, the samples were immersed in water at 20 °C (Figure 2b) in a sealed plastic box with the need to be restrained since otherwise they would float until achieving constant mass. The weighing occurred after carefully wiping off the sample's surface with a damp cloth. With this, it was possible to characterise their free water saturation (w_f).

For all the RH levels, when constant mass was achieved, the samples' mass was evaluated. Those different mass measurements allowed for the calculation of the moisture content mass by mass (u) and the moisture content mass by volume (w), in line with EN ISO 12571 [45] and EN ISO 12570 [46]. Using the previously presented parameters, two additional ones can be calculated: the maximum water content (w_{max}) and the saturation coefficient. The maximum water content (w_{max}) is the water content at full saturation when the porous structure is considered filled with water [47], in [kg m^{-3}]. This can be estimated by multiplying the material's open porosity (P_0, obtained by the mercury intrusion porosimeter technique, [%]) by the water density (ρ_w in [kg m^{-3}]) at 20 °C, according to Equation (3) [6]. The MIP equipment is an AutoPore IV from Micromeritics, presenting a maximum intrusion volume for low and high pressure of 1×10^5 mL g^{-1}, and a measuring range from 1×10^{-2} to 1×10^3 μm.

$$w_{max} = \varepsilon \times \rho_w \quad (3)$$

Considering both w_f and w_{max}, the saturation coefficient [-] can be calculated, which is a measure of the pore space quantity available after the free water saturation [48]. This parameter is calculated as the ratio between w_f in [kg m^{-3}] and w_{max} in [kg m^{-3}], Equation (4) [47], where the lower it is, the more space is available to accommodate the expansion of liquid water as it freezes [49].

$$Saturation\ coefficient = \frac{w_f}{w_{max}} \quad (4)$$

2.2.3. Liquid Water Transport Coefficient for Suction

Another important input parameter to consider in hygrothermal simulations is the liquid water transport coefficient for suction (D_{ws}) [35,50]. This parameter describes the situation where the material is applied to a surface in direct contact with liquid water or is exposed to wetting by rain action (fully wet) [51], while the capillary uptake of liquid water continues to occur [52]. Some techniques can be employed to characterize this parameter, such as nuclear magnetic resonance (NMR) or γ radiography [53,54]. However, since they were not available for this study, a previously used approach as a good estimation method by other researchers [55] was selected—Equation (5) [52,54], where D_{ws} is the liquid transport coefficient for suction [m^2 s^{-1}], A_w is the water absorption coefficient [kg m^{-2} s$^{-1/2}$]—previously characterised in [27], w is the moisture content [kg m^{-3}], and w_f is the free water saturation [kg m^{-3}].

$$D_{ws} = 3.8 \times \left(\frac{A_w}{w_f}\right)^2 \times 1000^{\left(\frac{w}{w_f}-1\right)} \tag{5}$$

2.2.4. Moisture Buffering Properties

The moisture buffering property of a given material is related to its capabilities to adsorb and desorb environmental moisture [36], as well as being useful to consider when the materials stay in contact with indoor environments [56]. Therefore, this property can help to reduce indoor superficial condensation in the buildings' envelope [8,9] and can also have an impact on reducing their heating and cooling energy consumption. Due to the novelty of aerogel-based, fibre-enhanced thermal renders (TRFs), this was an interesting characteristic to further evaluate to see the possibility of using these formulations for future indoor application scenarios, promoting their multifunctionality.

Although there are many test procedures, the one most broadly used is described by the NORDTEST project [57–59]. For this test procedure, a constant temperature of 23 °C must be attained, and then several cycles of exposure of 8 h with a 75% RH and 16 h with a 33% RH, simulating the usual indoor human occupancy of commercial buildings, are followed until stabilisation [58].

The common method to evaluate this concept is using the practical moisture buffer value, $MBV_{practical}$. This parameter is defined as the amount of moisture content that passes through or is captured by the material's open surface unit (moisture uptake and release) when exposed to variations in the RH of the surrounding air (daily cyclic variations) under the previously described test conditions. To calculate the $MBV_{practical}$, Equation (6) can be used [58], where $MBV_{practical}$ is the practical moisture buffer value [g m^{-2} %RH^{-1}], Δm is the mass variation of the moisture uptake/release [g], A is the open surface area [m^2]—here being considered 1 m^2, and the $RH_{high/low}$ are the higher and lower air RH limits of the test [%RH].

$$MBV_{practical} = \frac{\Delta m}{A \times \left(RH_{high} - RH_{low}\right)} \tag{6}$$

A direct comparison of the moisture buffer values of the different materials can be made, but at the same time, they can be classified in terms of their moisture buffer classes [57], as can be seen in Figure 3. With this classification, the materials can be classified as presenting negligible, limited, moderate, good, or excellent moisture buffering capabilities, considering the test conditions.

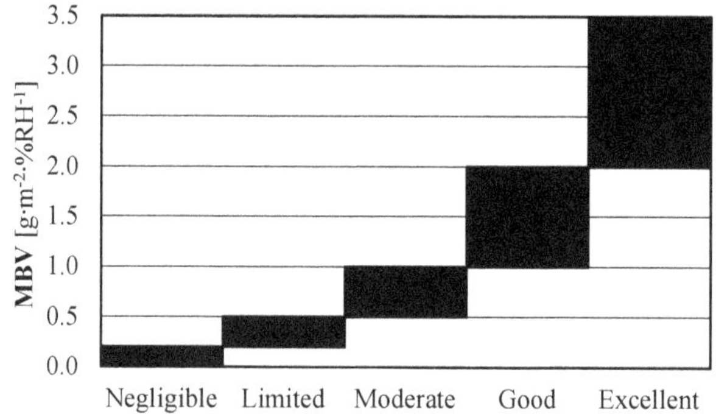

Figure 3. Ranges for practical moisture buffer value (MBVpractical) classes, adapted from [59].

Usually, this characterisation involves the application of specific test conditions (as previously described) using advanced climatic chambers and several samples. However, as Barclay et al. [33] and Wan et al. [60] showed, there is a good fit between the experimental and numerical simulation results using HAMT (combined heat and moisture transfer) models, such as WUFI software. Therefore, it was possible to simulate several application scenarios to evaluate the studied TRFs and their behaviour when subjected to water vapour cycles. However, for this to be possible, it was necessary to create an artificial weather file replicating the dynamic conditions of air temperature, RH, and time indicated by the NORDTEST project procedures (8 h at 75% RH and 16 h at 33% RH, at 23 °C) [57–59]. For the simulation, one square metre of exposed surface (1 × 1 m) with a 0.04 m thickness was considered, where all but one of the sample's faces were insulated by an aluminium foil, avoiding water vapour penetration while also defining its boundaries. Then, the virtual sample was subjected to the test cycle conditions until it was possible to obtain mass stabilisation of the samples, reached when the change in mass was the same between the last three cycles with a discrepancy of less than 5% [58]. For these simulations, the results from the previous tests and other previously performed tests and materials were used, with their characterisation presented in the Supplementary Data.

2.3. Experimental Results and Discussion

2.3.1. Thermal Properties

Table 3 summarises the results for the thermal properties. It can be seen that these TRF (Aerogel-based fibre-enhanced thermal renders) present very low thermal conductivity, being lower than other materials classified as classic thermal insulators (e.g., EPS $\lambda \approx 0.036$ W m^{-1} K^{-1} [61]), and much lower if already commercialised thermal renders are considered (e.g., thermal render with EPS granules $\lambda \approx 0.050$ W m^{-1} K^{-1} [62]).

Table 3. Thermal properties result in synthesis.

Render Formulation	$\lambda_{10°C,dry}$ [W m^{-1} K^{-1}] [41]	$\lambda_{10°C,sat}$ [W m^{-1} K^{-1}]	c_p [J kg^{-1} K^{-1}]	I [J m^{-2} K^{-1} s$^{-1/2}$]	α [m^2 s^{-1}]	$d_{p,10min}$ [mm]
TR reference (0.0%)	0.0293 ± 0.0031	0.1401 ± 0.0052	930.1 ± 5.1	65.8	$1.99 \times 10^{-7} \pm 1.91 \times 10^{-9}$	19
TR aramid (0.5%)	0.0315 ± 0.0019	0.1311 ± 0.0071	800.0 ± 7.9	64.4	$2.39 \times 10^{-7} \pm 2.63 \times 10^{-9}$	21
TR sisal (0.1%)	0.0298 ± 0.0027	0.1230 ± 0.0035	894.5 ± 10.2	65.4	$2.08 \times 10^{-7} \pm 2.70 \times 10^{-9}$	20
TR biomass (0.1%)	0.0306 ± 0.0032	0.1285 ± 0.0044	957.3 ± 12.8	68.8	$1.98 \times 10^{-7} \pm 2.77 \times 10^{-9}$	19

Note: ±SD—standard deviation. $\lambda_{10°C,dry}$—thermal conductivity at 10 °C and dry-state [W m^{-1} K^{-1}]; $\lambda_{10°C,sat}$—thermal conductivity at 10 °C and saturated-state [W m^{-1} K^{-1}]; c_p—specific heat capacity [J kg^{-1} K^{-1}]; I—material's thermal inertia [J m^{-2} K^{-1} s$^{-1/2}$]; α—thermal diffusivity [m^2 s^{-1}]; $d_{p,10min}$—heat penetration depth at 10 min after exposure [mm].

The fibre's incorporation leads to a thermal conductivity increase when compared to the reference (TR reference) of ≈7%, but with low significance since the results are within the standard deviation values. This can be related to the higher thermal conductivity of the fibres when compared with the silica aerogel, since cellulosic fibres present a $\lambda \approx 0.050$ W m^{-1} K^{-1} and aramid fibres a $\lambda \approx 0.040$ W m^{-1} K^{-1} [63], but also with the consequently lower aerogel quantity. The thermal conductivity difference between the formulations containing natural and aramid fibres can also be associated with the existence of voids around the fibres that result from the drying and shrinkage of the natural fibres after the hydration reactions. This is accompanied by forming closed pores and influencing their thermal performance (as seen in [27]), but also with the higher quantity of aramid fibres (0.50% vs. 0.10% (vol./vol.)). As a reference, air presents a $\lambda \approx 0.025$ W m^{-1} K^{-1} [64], where it can be seen that more pores can also influence the overall thermal conductivity performance.

When the thermal conductivity values in the dry-state ($\lambda_{10°C,dry}$) are compared with the ones obtained in the saturated state ($\lambda_{10°C,sat}$) it is seen that the highest increase was verified in the reference (TR reference), with the fibres' presence contributing to maintaining a lower thermal conductivity value for the saturated state. This can be related to the fact that the fibres' incorporation reduced the capillary pores [65], as observed in the MIP and

SEM analysis results presented in [27], reducing the liquid water contribution to increase the thermal conductivity.

The specific heat capacity presented by these aerogel-based thermal renders is much lower than that of expanded moulded polystyrene (EPS) ($c_p \approx 1568$ J kg^{-1} K^{-1} [43]) but very similar to glass wool ($c_p \approx 928$ J kg^{-1} K^{-1} [43]), while not differing by much from a conventional cement-based mortar ($c_p \approx 900$ J kg^{-1} K^{-1} [66]) or industrial thermal renders containing EPS ($c_p \approx 1000$ J kg^{-1} K^{-1} [62]). Considering the specific heat capacity, thermal conductivity, and bulk density, it is seen that the thermal inertia (I) presented by these renders, although with lower thermal conductivity than EPS, presents two times higher thermal inertia than this conventional thermal insulator ($I \approx 27.4$ J m^{-2} K^{-1} s$^{-1/2}$ [43]). Depending on the applied thickness, this can influence their potential applications.

Related to the previous characteristics, the thermal diffusivity (α) quantifies the rate of propagation of heat through the material, which relates to the penetration depth ($d_{p,10\text{min}}$), at a time of 10 min. The results show that if there is an increment of temperature applied to one side of a sample, the depth to which the temperature will change significantly after 10 min, being around 20 mm for all renders. Additionally, when compared with EPS ($d_{p,10\text{min}}$ = 63 mm) [44] it is significantly lower, being more similar to the behaviour of a wood fibreboard ($d_{p,10\text{min}}$ = 26 mm) [44]. Therefore, these renderings respond even slower to temperature variations than an EPS board.

When the thermal conductivity at 10 °C ($\lambda_{10°C}$) is evaluated against the moisture content (w), Figure 4, it can be observed that higher moisture content leads to higher thermal conductivity, as expected (since water presents a $\lambda_{10°C} \approx 0.600$ W m^{-1} K^{-1} [64]) and verified by other authors for other mortars [40,62,67]. The incorporation of fibres, as previously discussed, seems to reduce the free water saturation (w_f) of these renders [68]. Moreover, for the aramid (TR aramid), due to the porosity reduction (less than 1.2% of the TR reference—Table 2), and for the natural fibres (known to present more affinity to water [69]), due to the existence of air pockets trapped inside the matrices, they can maintain lower water contents (≈less than 5%) and consequently lower thermal conductivity (≈less than 6%). With the fibres' incorporation, it was possible to absorb less capillary water, but also lower thermal conductivity values were verified in the natural fibre-containing formulations (Table 2) [27].

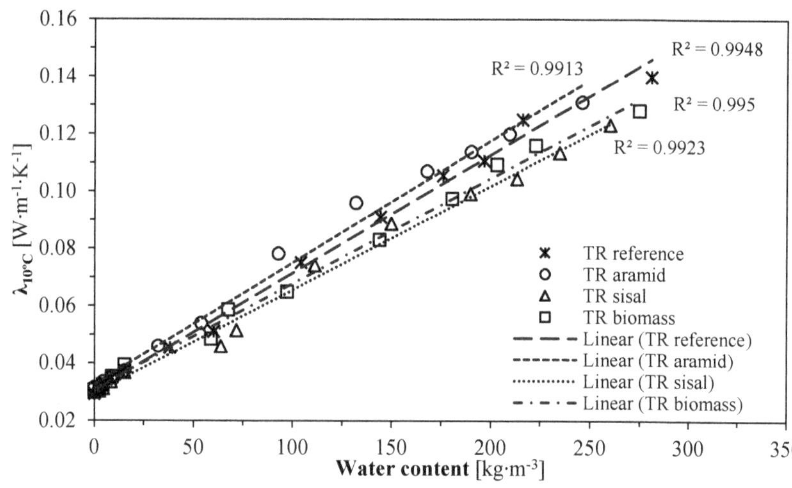

Figure 4. Thermal conductivity as a function of water content, at 10 °C.

2.3.2. Moisture Storage Properties

To determine the formulations' moisture content, they were evaluated in the dry state and then for 35, 60, 85, and 93% RH, at 23 °C. This way, it was possible to obtain the average sorption values for different moisture content indicators as shown in Figure 5, the hygroscopic regime of these TRFs is presented following EN ISO 12571 [45]. For comparison purposes, the data presented by Barclay et al. [33] for hemp-lime (Barclay et al.), by Maia et al. [62] for commercial thermal renders (Maia TR1 and Maia TR2), and by Fantucci et al. [11] for an aerogel-based internal plaster are also presented in Figure 5. It is possible to see that these TRFs showed a constant increase in moisture content from the dry state until 60% RH, and then, for higher RH, the moisture content strongly increased, reaching around 0.10 kg kg^{-1} at 93% RH. This highlights that these renders are hygroscopic materials, with the fibres contributing to decreasing the values of moisture content.

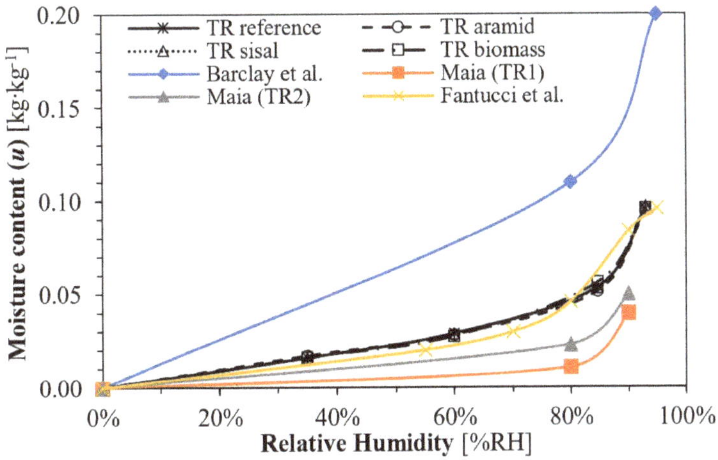

Figure 5. Sorption isotherms for the renders (black) at 23 °C, compared with other results [11,33,62].

In the same Figure 5, it is also seen that all aerogel-containing formulations (the ones herein studied and the one presented by Fantucci et al. [11]) show similar hygroscopic behaviour; therefore, it seems that the fibres did not significantly influence the hygroscopic behaviour already shown by the reference (TR reference) and other similar products. When compared with the thermal renders studied by Maia et al. [62], these aerogel-based thermal renders showed a higher degree of moisture uptake, but much lower than the hemp-lime studied by Barclay et al. [33]. These behaviours are probably linked to the microstructural aspects seen previously by the authors [27]. If other thermally insulating materials are considered, EPS also shows some significant moisture content from 80% RH onward, since at ≈ 90% RH it shows a water content of about 0.07 kg kg^{-1} [44], and even if wood wool is considered at ≈ 90% RH, it presents 0.15 kg kg^{-1} [44], not very far from these renders' performance.

For the over-hygroscopic range, the free water saturation (w_f) and the maximum water content (w_{max}) were also characterised (Table 4). Here, the formulations containing fibres lowered the w_f with the aramid fibres presenting the lowest value, followed by the natural fibres. Once again, the water affinity of the different fibres is the main aspect to consider. When comparing the w_f performance of these renders with other aerogel-enhanced insulating materials, it was seen that they showed a behaviour within the expected ranges, between 240 and 500 kg m^{-3} [11,28].

As for the saturation coefficient, the lower this coefficient, the more space is available to accommodate the expansion of liquid water as it freezes [49]. As a reference, for bricks, this coefficient is considered acceptable if below 0.78 [70], whereas, in this study (Table 4), the values were way below that value (<0.35), with the fibre-containing formulations presenting even lower values than the reference. This indicates that these aerogel-based renders have the potential to effectively accommodate the expansion of liquid water as it freezes.

Table 4. Moisture storage properties and porosity results in synthesis.

Render Formulation	$w_{80\%RH,eq}$ [kg m^{-3}]	w_f [kg m^{-3}]	w_{max} [kg m^{-3}]	Saturation Coefficient [-]
TR reference (0.0%)	7.80 ± 0.04	281.04 ± 5.12	863	0.33
TR aramid (0.5%)	7.12 ± 0.07	246.08 ± 7.88	851	0.29
TR sisal (0.1%)	7.28 ± 0.06	260.20 ± 8.29	869	0.30
TR biomass (0.1%)	7.56 ± 0.03	274.83 ± 9.87	870	0.32

Note: ±SD—standard deviation. Parameters: $w_{80\%RH,eq}$—moisture content at 80%RH [kg m^{-3}]; w_f—free liquid water saturation [kg m^{-3}]; w_{max}—maximum liquid water content [kg m^{-3}].

2.3.3. Liquid Water Transport Coefficient for Suction

To calculate the liquid water transport coefficient for the suction (D_{ws}) parameter, the results for capillary water absorption (A_w), previously obtained and presented in Table 2, were used. The fibre's incorporation by powder substitution led to lower water absorption coefficients (Supplementary Data), as a consequence of the fibres not only obstructing the capillaries but also forming air pockets inside the matrix (natural fibres)—as it could be observed in [27]—and hindering the capillary water progress [68]. This fact was expected to directly influence the liquid water transport coefficient for suction (D_{ws}) results due to the A_w contribution to its calculation.

Figure 6 shows the calculation results, where the fibres' use (TR aramid, TR sisal, and TR biomass) led to a lower liquid transport coefficient for suction (D_{ws}) when compared with the reference (TR reference), presenting better behaviour, as a construction material. When compared with other thermal renders, the reference (TR reference) shows a higher D_{ws} order of magnitude ($\approx 10^{-7}$ m^2 s^{-1}), but the fibre-containing formulations present the same order of magnitude as other thermal renders ($\approx 10^{-8}$ m^2 s^{-1}) [62]. Once again, the positive impact of the fibre's use is seen, due to its earlier verified impact on water absorption reduction.

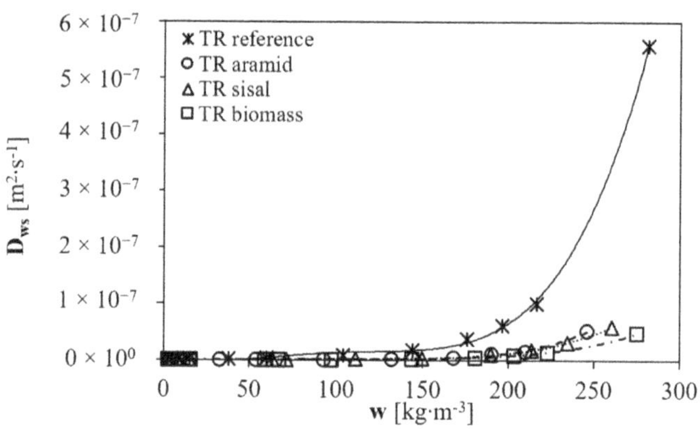

Figure 6. Liquid transport coefficient for suction (D_{ws}), as a function of moisture content (w).

2.3.4. Moisture Buffering Properties

The moisture buffering was another interesting property to be evaluated in these formulations since thermal renders can be applied in several scenarios, such as interior thermal insulation retrofits [28,71,72]. As this parameter is connected to the material's sorption and desorption properties [73], its evaluation allows for the classification of its potential to regulate indoor moisture [48], making it an interesting aspect to consider and analyse.

Following the work by Barclay et al. [33], three different potential indoor application scenarios were studied (Figure 7). Although the first scenario (Figure 7a) is not usually applied, the other two (Figure 7b,c) reflect real potential applications. The simulation and analysis of values were only carried out for the thermal insulation materials, with the other materials' influence being considered but not directly quantified. Therefore, a comparison was made between these TRFs and an industrial thermal render presented by Maia et al. [62] (designated as Ind Therm render), since other thermal insulation materials such as EPS and XPS do not present significant hygroscopic behaviours in these RH ranges [43,44].

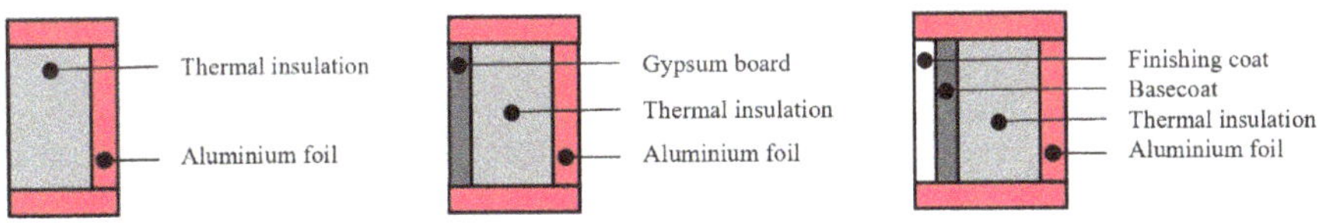

(**a**) (**b**) (**c**)

Figure 7. Section view of each analysed application: (**a**) thermal render; (**b**) thermal render with gypsum board; (**c**) multilayer system.

Figure 8 shows the results for the different simulations and solutions. (Figure 8a) shows that for the aerogel-based renders, the one containing aramid (TR aramid) showed a slightly lower change in mass. However, all presented significantly higher capabilities than the current thermal render (Ind Therm render).

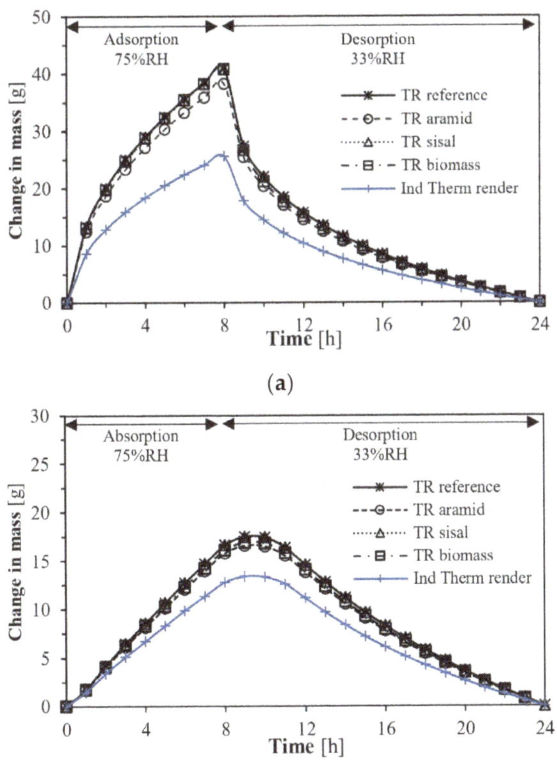

(**a**)

(**b**)

Figure 8. *Cont.*

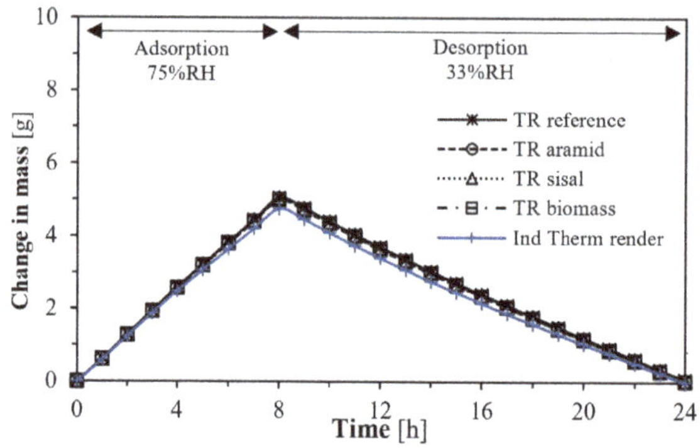

(c)

Figure 8. Moisture buffering performance for different applications: (**a**) thermal render; (**b**) thermal render + gypsum board; (**c**) thermal render + protective coating system—Ind thermal render from [62].

When a gypsum board (0.0125 m) was placed in front of these thermal insulation materials (Figures 7b and 8b), for mechanical protection, some started to emerge with interesting behaviour. The total change in mass due to moisture presence was reduced to half, which was expected since even one coating of paint can reduce their performance [74]. All the aerogel-based formulations presented similar behaviour; a lag appeared between the time where the test RH changed (at the 8 h mark) and where the maximum moisture content appeared (≈10 h mark), being this related to the moisture buffering properties also presented by the gypsum boards [75].

Finally, it was considered the application of a commercially available multi-layered coating system composed of a basecoat (3 mm) and a finishing coat (2.5 mm), as presented in [25]. Figure 8c shows that the maximum change in mass due to moisture was reduced by almost ten times when compared with Figure 8a, with the materials being now limited by the presence of the multilayer coating system. This behaviour was expected since these coating materials show low water vapour permeability due to their moisture-protective function [74,76].

When calculating their $MBV_{practical}$ and comparing with other materials [59,77], it can be seen in Figure 9 that the TRFs present a moderate-to-good classification, with higher values than the industrial thermal render (*Ind Therm render*) and much higher than the other materials. When the gypsum board is placed over their surface, the differences between the aerogel-based renders and the current thermal render almost disappear, as their behaviour is similar to that of concrete (however, in a solution like this, the gypsum board contribution should also be considered) and they are classified as presenting limited moisture buffering capabilities. When the multilayer system is applied on the surface, the moisture buffering behaviour of the renders is compromised by the performance of the coating system, with their classification now being negligible ($MBV_{practical} < 0.2$ g m^{-2} %RH^{-1} [58]) relative to their moisture buffering capabilities.

These results show that these TRFs can effectively improve the indoor environment, given that their performances are not impaired by surface coatings that degrade this property. In terms of differences between the fibre-containing formulations and the reference, the only variation worthy of note was when the render was directly exposed to the environment, with the aramid-containing render (TR aramid) showing a smaller change in mass (≈less than 3 g m^{-2}), with any difference eliminated in the other cases. The differences between the TRFs and the industrial thermal render can be related to their higher hygroscopicity, as previously seen.

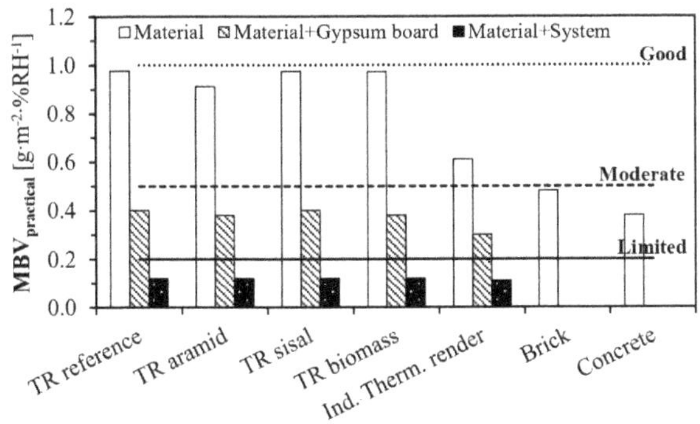

Figure 9. $MBV_{practical}$ when compared to other materials—Ind therm render [62], brick and concrete [59,77].

3. Hygrothermal Simulation

3.1. General Considerations

The hygrothermal simulations were carried out considering two approaches: new construction and retrofitting of building walls. Two geographical places were chosen for the new construction scenario: Lisbon, Portugal, and Zurich, Switzerland. Lisbon presents, according to the Köppen-Geiger climate classification [78,79], a hot summer Mediterranean climate (*Csa*), while Zurich presents, according to the same classification, a marine west coast climate (*Cfb*). These two areas present quite distinctive characteristics and are representative of generic European climates. At the same time, it is known that Switzerland already has a large construction market using aerogel-based thermal renders [71,80]. Therefore, it was important to see how these formulations would behave in such distinct conditions.

For the retrofit approach, since Lisbon has a significant number of buildings from the 1990s lacking thermal insulation [81,82], which leads to high energy consumption [83], two different retrofit approaches were selected: (*i*) thermal insulation material on the walls' inside surface and (*ii*) thermal insulation material on the walls' outside surface.

This study also compares the main thermal insulating materials used in this type of solution: EPS and XPS, and an industrial thermal render currently commercialised and applied in Portugal [62]. Due to their high capillary water absorption [23], these thermal renders need to be used in multilayer coating systems, like EPS and XPS in ETICS (External Thermal Insulating Composite System) or even in other thermal insulating renders [25]. Therefore, those coating layers were also considered when the solutions were applied outside the wall. When considered for application on the inside, a gypsum board was added as a protective and finishing element.

3.2. Simulations of Climatic and Boundary Conditions

With the formulations' hygrothermal characterisation, it was possible to study their performance in different solutions and geographic and climatic conditions. This was achieved using numerical models that predict the coupled heat, air, and moisture transport (HAMT) of building materials [33,84].

In the present study, WUFI Pro 6.4 (Wärme Und Feuchte Instationär—heat and moisture transiency) was the software used, allowing for realistic hygrothermal simulations [34]. This software needs as input: the composition and properties of the individual materials used; the solution's physical orientation and slopes; the initial simulation conditions (indoor and outdoor climates); and the duration of the simulation. Then, the numerical model, based on EN 15026 [10], evaluates, for the simulation period and chosen time step, the heat and moisture fluxes and the transient profiles of temperature, RH, and moisture content of the solution [85]. The model uses Equations (7) and (8) [85,86], where the storage terms are located on the equations' left side and the fluxes on the right side are affected by heat and moisture.

$$\frac{\partial w}{\partial \varphi} \times \frac{\partial \varphi}{\partial t} = \nabla \times \left(D_{\varphi} \nabla_{\varphi} + \delta_p \nabla(\varphi p_{sat})\right) \tag{7}$$

$$\frac{\partial H}{\partial T} \times \frac{\partial T}{\partial t} = \nabla \times (\lambda \nabla T) + h_0 \nabla\left(\delta_p \nabla(\varphi p_{sat})\right) \tag{8}$$

where $\partial w/\partial \varphi$ is the moisture storage capacity [kg m^{-3} %RH^{-1}]; $\partial H/\partial T$ is the moist heat storage capacity [J kg^{-1}]; w is the moisture content [kg m^{-3}]; λ is the thermal conductivity [W m^{-1} K^{-1}]; D_{φ} is the liquid conduction coefficient [kg m^{-1} s^{-1}]; δ_p is the water vapour permeability [kg m^{-1} s^{-1} Pa^{-1}]; h_v is the water evaporation enthalpy [J kg^{-1}]; p_{sat} is the water vapour saturation pressure [Pa]; T is the temperature [K]; φ is the RH [%]; and t is the time [s].

This software also considers the effects of short- and long-wave radiation on the surfaces, combined into a collective heat source [34,87], and the effect of wind-driven rain (WDR), where factors such as the horizontal rainfall, wind speed, orientation, and type of constructive solution are considered [34,88]. However, the model has some limitations, as it neglects [89]: the convective transport of heat and moisture; some of the liquid transport mechanisms; salts and water transport; the interface resistance between two capillary-active materials; and enthalpy flows resulting from the transport of liquid water due to temperature differentials. With knowledge of its main limitations, this software is widely used in research, with its validity extensively validated through comparisons between in situ and simulation results [88,90,91].

3.2.1. Climatic Conditions

To simulate the solutions' performance, the climatic conditions must be known, such as the hourly mean values of temperature and RH for the indoor and outdoor climates, solar radiation, wind speed and direction, and precipitation for the outdoor climate conditions to test [85]. For the Lisbon and Zurich outdoor conditions, values available in the WUFI database were used. The average conditions are shown in Table 5. The indoor climatic conditions, and although occupants' comfort greatly varies from individual to individual [5], are usually established as legal requirements, it is accepted that the comfort zone has temperature threshold values between 19 and 26 °C and RH between 30 and 70% RH [92], being in Portugal usually considered between 18 and 25 °C [93]. Therefore, for the indoor conditions, the high moisture load climate from the EN-15026 was used (temperatures between 20 and 25 °C and RH between 40 and 70% of the RH-standard B curve) [10], simulating an intensive indoor usage, as these are the most commonly used conditions when considering numerical simulations and accepted as a good benchmark for comparison [94,95].

Table 5. Outdoor and indoor average climatic conditions are available in the WUFI database.

Climatic Parameter	Lisbon	Zurich
	Outdoor climate average conditions	
Location	Lat. 38.736946; Long. −9.142685	Lat. 47.373878; Long. 8.545094
Temperature [°C]	15.6	8.3
Relative humidity [%]	74.6	78.2
Global radiation emitted by the sun [kW h m^{-2} year^{-1}]	-	2751.7
Mean cloud index [-]	-	0.7
Driving rain direction	South-west	West southwest
Rain (accumulated) [mm year^{-1}]	674.9	1007.9
	Indoor climate average conditions	
According to EN 15026—high moisture load [10]		

3.2.2. Boundary Conditions

Five years were chosen as the duration of the simulations to check the possibility of attaining some kind of dynamic equilibrium for the solutions [47], starting at the beginning of the rainy season (October) and, therefore, limiting the drying of the solutions at the initial stage. As for the facades' orientation, the north orientation was selected, as the northern hemisphere's orientation has less solar radiation, influencing the drying process, and the orientation would lead to the most WDR exposure. Additionally, 1% of the WDR value [96] was considered to penetrate the outer layers of the solutions, simulating construction defects [95,97].

Another parameter influencing the solution's performance is the short-wave radiation absorptivity (α_s) [88]. A lower value is associated with brighter colours while a higher value is associated with darker colours. From other studies, a lower α_s reduces the absorbed radiation, leading to lower surface temperatures, slowing the drying fluxes, and contributing to the material's increased water content. When comparing $\alpha_s = 0.2$ with $\alpha_s = 0.6$ or $\alpha_s = 0.8$ [62], significant differences in the drying of the construction elements were verified, leading to the choice of $\alpha_s = 0.25$ as a conservative approach. The boundary conditions can be seen in Table 6.

Table 6. Simulation boundary conditions.

Property	Adopted Values
Analysis period	Five years, starting and ending on 1 October
Timestep	1 h
Initial relative humidity	80%RH
Initial temperature	20 °C
Orientations	North and orientation with higher WDR: Lisbon—North and South West Zurich—North and West Southwest
Wind-driven rain	ASHRAE 160 [96] medium exposure, building with height ≤ 10 m and below a sloped roof; 1% of defects
Solar absorption coefficients	White smooth surface: 0.25
Long-wave radiation emissivity	0.9 (non-metallic component surface)
Rain absorption factor	0.70 (vertical wall)

3.3. *Wall Solutions and Simulations Evaluation Criteria*

3.3.1. Materials and Wall Assemblies

To model and simulate the behaviour of different wall assemblies, it was necessary to create new material entries in the WUFI database. With the experimental campaign, these TRFs were hygrothermally characterised, but other materials' properties had to be gathered from the literature. In the Supplementary Data, the detailed properties of the used materials can be seen.

Although it was considered a commercial protective coating system, composed of a basecoat, keycoat, fiberglass, mesh, and finishing coat, in these hygrothermal simulations only the contributions of the basecoat and finishing coat were considered. The fibreglass mesh, due to its open mesh, was considered to present a negligible influence. The keycoat, with a thickness of a few micrometres and a similar formulation to the finishing coat (acrylic co-polymers dispersion [25,98]), was considered a part of such a layer's properties and thickness.

With the individual materials characterised, to simulate the hygrothermal performance of a new wall scenario, one of the most common solutions was chosen: a multilayer coating protective system with a thermal insulation material, a lightweight concrete block, and an interior plaster (Figure 10a). Then, depending if it was applied in Lisbon or Zurich, due to the thermal insulating requirements (Lisbon: $U \leq 0.50$ W m^{-2} °C^{-1} [99]; Zurich: $U \leq 0.25$ W m^{-2} °C^{-1} [100]), the thermal insulating layer and the lightweight concrete block thicknesses varied to fulfil the requirements. For Lisbon, a thermal insulation of 0.025

m and a lightweight concrete block of 0.25 m were used, and in Zurich, the same materials but with 0.06 m and 0.38 m, respectively.

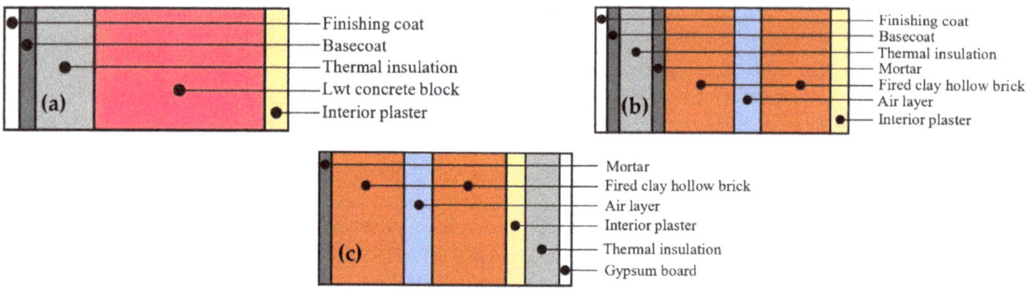

Figure 10. Section view of the studied wall assemblies: (**a**) new construction (NW and ZNW); (**b**) outdoor retrofit (RE); (**c**) indoor retrofit (RI).

For the retrofit scenario, two different thermal insulation applications were considered: one applied on the exterior (Figure 10b) and one applied on the interior surface (Figure 10c). The base wall was one of the most common types existent in buildings constructed in Lisbon in the 1980s [101]: double-leaf fired clay hollow bricks with an air layer in-between. This construction typology presents severe thermal insulation deficits due to the lack of thermal insulation materials [102]. The thickness of the thermal insulation was chosen to comply with the new buildings' requirements: $U \leq 0.50$ W m^{-2} °C^{-1} [99]. This way, it would be possible to evaluate the impact that moisture has on the constructive solutions' thermal insulating performance, since their starting point, from a hygrothermal point of view, was the same. The layers and respective thicknesses of all the studied solutions are presented in Supplementary Data.

The new wall solutions, both for Lisbon (NW) and Zurich (ZNW), had a similar constitution, only differing in the lightweight concrete block thickness. Solutions 1 to 4 represent the TRFs (TR reference, TR aramid, TR sisal, and TR biomass, respectively); solution 5 contains an industrial thermal render; solution 6 uses EPS; and solution 7 uses XPS. The full description of the solutions can be seen in the Supplementary Data.

The retrofit solutions carried out on the exterior surface of the wall (RE) had a reference solution without thermal insulating materials—solution RE1—and then RE2 to RE5 corresponded to TR reference, TR aramid, TR sisal, and TR biomass, respectively. The industrial thermal render was the solution RE6, and finally, the solutions RE7 and RE8 correspond to EPS and XPS, respectively. The retrofit carried out on the interior surface of the wall (RI) followed the same thermal insulator distribution as in the RE scenario. Once again, the full description of the solutions used in these scenarios can be seen in the Supplementary Data.

3.3.2. Hygrothermal Simulations Evaluation Criteria

This type of research returns several results, but they need to be contextualised to be valuable. As a result, different aspects of hygrothermal performance were considered in this study in order to evaluate and compare the solutions.

(i) Total water content and drying potential

One of the most critical factors in a hygrothermal analysis is the progression of a solution's total water content over time [4]. Hence, it is important for the simulation period to be long enough to see if a given solution can reach a dynamic equilibrium, or if it continues to accumulate moisture over the years. As a result, the total water content and the drying potential were analysed, where the first represents the progression of moisture over time and the second helps to verify if the solutions are drying, using Equation (9) [72].

A positive value is related to moisture drying (pass) and a negative value to moisture increase (fail) over that period.

$$DP = \frac{WC_{yearn-1} - WC_{yearn}}{WC_{yearn-1}} \times 100 \quad (9)$$

where DP is the drying potential [%]; $WC_{year\ n-1}$ is the solution's water content at year n − 1, and $WC_{year\ n}$ is the solution's water content at year n, both in [kg m^{-3}].

(ii) Mould growth potential

The appearance of mould in construction materials can lead to hygrothermal and mechanical performance deterioration [5]. This is usually only assessed for the indoor environment due to air quality and material integrity; for the exterior surfaces, it is harder to evaluate as temperatures, UV exposure, and washing of the spores due to rain tend to limit their growth [103]. For indoor conditions, the factors to avoid mould growth are a specific temperature, RH ranges [104], and substrate quality [105]. Mould appearance and growth can be evaluated in the simulations through two different tools. A graph (temperature vs. RH) that shows the limiting isopleths for building materials, *LIM B I* (for situations where there is a bio-utilisable substrate) and *LIM B II* (for substrates with porous structures), below which no mould growth is expected [106]. The other tool is the WUFI Bio postprocessor, which allows conducting a biohygrothermal evaluation by comparing the simulated transient ambient conditions with the needed growth conditions [106]. The simulated moisture content of the mould spores is compared with the critical water content, which allows a spore to germinate, and once germination occurs, it estimates its growth (fail) [104]—the objective being not to have any growth at all (pass). Figure S3 from Supplementary Material shows an example of an isopleth whose behaviour was shared by all solutions using the WUFI Bio postprocessor: no significant risk of mould growth since its results were kept below the identified LIM limits (the darkest area corresponds to the most common conditions).

(iii) Water content in the thermal insulation layers

Since this study involves highly porous thermal renders, one important assessment to be made was related to their water content over time. This directly influences the durability due to the potential running off of water inside the layer and on the frost resistance and thermal performance. Therefore, it is essential to know how the water content of the thermal insulating material is behaving relative to its maximum moisture capacity (w_f), where values of the ratio w/w_f below one indicate a low probability of water running off [85].

As for the frost damage risk, following the WTA guideline 6-5-12/D [107] referred to in [34], the contact area between the insulation and the most external layer (basecoat) should present a water content ratio (w/w_{max}) below 0.30, minimising the probability of frost damage. Since w_{max} for all solutions was much higher than w_f (more than three times higher), a ratio of w/w_f below 0.30 was considered a conservative approach in this study (pass).

(iv) Temperature variations

Temperature is a well-known factor that influences construction materials' expansion and retraction characteristics [108]. As such, the temperature on the solution's surface is an important parameter to evaluate since it can directly impact its durability [62]. If a given solution has a higher surface temperature for the same insulating performance, its durability can be decreased due to localised tensions on the material's surface [62,88]. Another important factor to be evaluated is the temperature gradients within the thermal insulating material since they can lead to internal tensions, degrading the material's structure and durability [109]. Therefore, the exterior surface temperature (99th percentile and maximum) and the temperature gradient (99th percentile) between the opposing surfaces of the thermal insulating materials were assessed.

(v) Surface condensation

One of the main pathologies that degrade buildings' coating systems is related to the defacement of some areas due to biological growth (e.g., mould and algae), usually related to surface condensation [90]. As water condensation occurs when the water vapour partial pressure in the air exceeds the water vapour saturation pressure on the surface [103], this risk for the construction solutions can be evaluated by two indicators: the risk of surface condensation [110] and the number of hours that the superficial RH φ = 100% RH [62], both evaluated for the 5th year of the simulation. As described by Zheng et al. [110], the condensation potential (*CP*) [Pa], Equation (10), is a result of the difference between the water vapour partial pressure in the air ($P_v(air)$) [Pa] and the water vapour saturation pressure on the surface ($P_{sat}(surface)$) [Pa]. If CP > 0, it is considered that there is condensation, and its value is accumulated for the evaluation period; otherwise, it is discarded. The other method considers the number of hours that superficial RH is equal to 100% (φ = 100%RH); the result is given in the simulation results and it is only necessary to accumulate the number of hours for the evaluation period.

$$CP = P_v(air) - P_{sat}(surface) \tag{10}$$

(vi) Thermal insulating performance—potential heat loss

The previous points are linked to the solution's hygrothermal integrity and durability, while this factor is more geared towards its thermal insulating performance. This was evaluated using two indicators: the ratio between the solution's monthly average U-value (U_{avg}) and its reference design's initial U-value (U_{ref}), and the heat loss (q) through the wall during a full winter season.

For the evaluation of the U_{avg}/U_{ref} ratio, the WUFI Thermal Transmission post-processor [34] was used, which calculates the solution's U_{avg} for each month, making it possible to evaluate if the solution is losing its thermal insulation performance. Then, knowing the U_{ref}, it is possible to see, for the different months, how the solution is behaving, with a ratio above 1 indicating loss of thermal insulation capacity. The post-processor uses Equation (11) [34] for its calculations.

Since during summer months, the direction of the interior heat flux usually changes twice a day, it can lead to a meaningless ratio—because more heat flows from the wall to indoors during the day than heat flows from indoors to the wall during the night. Therefore, some care must be taken when executing the analysis in these months. This fact contributed to the analysis period defined for the element's heat loss.

$$U_{avg} = \frac{-Q}{\Delta T_a} \tag{11}$$

where U_{avg} is the monthly average U-value [W m^{-2} °C^{-1}]; Q is the monthly mean heat flux per area through the interior surface [W m^{-2}]; and ΔT_a is the monthly mean value of the temperature difference (indoor and outdoor) [°C].

For the calculation of the solution's potential heat loss, the heat loss equation (Equation (12) [72]) is used. This was evaluated as a sum of the heat losses during the period from 1 October of the 4th year until 1 May of the 5th year, representing a full winter.

$$q = \frac{\sum U_i \times A_i \times (T_{in} - T_{out}) \times \Delta t}{1000} \tag{12}$$

where q is the element's heat loss [kW h]; U_i is the thermal transmittance of the assembly [W m^{-2} °C^{-1}]; A_i is the surface area [m^2]; T_{in} is the indoor temperature [°C]; T_{out} the outdoor temperature [°C] and Δt the time [h].

3.4. *Hygrothermal Simulations and Discussion*

This analysis was divided into three main scenarios: (i) a new wall with the thermal insulation material applied on the exterior surface (NW for Lisbon and ZNW for Zurich); (ii) a retrofit with the thermal insulation applied on the exterior surface (RE); and (iii) a retrofit with the thermal insulation applied on the interior surface (RI). The Supplementary Data presents graphics for all the scenarios and analyses of evaluation criteria that can further support the reading of these results.

3.4.1. New Wall Scenario (NW and ZNW)

Table 7 presents the most noteworthy results related to the solutions' total water content, both for Lisbon (NW) and Zurich (ZNW)—in the Supplementary Data, the graphics of these simulations can be seen. There, it is possible to see that the application under different conditions influences the materials' performance. In Lisbon, only solution NW5 (wall composition in the Supplementary Data) showed some water accumulation in year five, with the other solutions showing the capability to dry during the year. However, when the climate of Zurich (ZNW) is considered, with more accumulated rain, lower temperatures, and less solar radiation, it is seen that solutions ZNW5, ZNW6, and ZNW7 (wall compositions in the Supplementary Data) for the WDR exposed façade showed a lack of drying abilities, significantly impacting their potential integrity and durability. This lack of drying is also related to the penetrating 1% of driving rain due to construction defects. As for the aerogel-containing renders (NW1 to NW4, and ZNW1 to ZNW4), they showed the ability, in both climates, to attain a quicker dynamic equilibrium (around 3 years), whereas the natural fibre-containing formulations (NW3 and NW4, and ZNW3 and ZNW4) presented even better performance.

Table 7. Synthesis of the new wall simulations for Lisbon (NW) and Zurich (ZNW).

Solution *SW–N*	**TWC and DP**	**Mould**	$w/w_f \leq 0.30$	**Ext Surface Temp (max) [°C]**	**Temp Gradient Insulator [°C]**	$CP > 0$ **[Pa]**	φ = **100%RH [h]**	U_{avg}/U_{ref} ≤ 1 **[month]**	q **[kWh m^{-2}]**
NW1	pass–pass	pass–pass	pass–pass	39.1–35.5	7.0–3.9	1608–1711	164–168	12–3	11.92–13.84
NW2	pass–pass	pass–pass	pass–pass	39.1–35.5	6.9–3.9	1618–1704	166–167	12–3	11.84–13.94
NW3	pass–pass	pass–pass	pass–pass	39.1–35.5	7.0–4.0	1629–1729	164–168	12–3	11.53–13.55
NW4	pass–pass	pass–pass	pass–pass	39.1–35.5	6.9–3.9	1603–1700	164–167	12–3	11.97–14.00
NW5	fail–fail	pass–pass	pass–pass	38.9–35.5	7.3–4.2	1510–1670	155–166	12–1	13.79–15.04
NW6	pass–pass	pass–pass	pass–pass	39.1–35.5	7.3–4.1	1599–1658	163–167	12–1	11.67–14.20
NW7	pass–pass	pass–pass	pass–pass	39.2–35.5	4.3–4.3	1685–1746	167–171	12–0	10.64–12.91
ZNW1	pass–pass	pass–pass	pass–pass	33.0–27.5	6.9–4.0	27141–26263	1998–2134	6–2	12.34–13.01
ZNW2	pass–pass	pass–pass	pass–pass	33.0–27.5	6.8–4.0	26914–23303	2056–2250	6–2	11.84–13.26
ZNW3	pass–pass	pass–pass	pass–pass	33.0–27.5	6.8–3.9	27107–26392	2020–2212	6–2	12.48–12.67
ZNW4	pass–pass	pass–pass	pass–pass	33.0–27.5	6.8–4.0	26605–26083	2136–1988	6–2	12.48–13.20
ZNW5	fail–fail	pass–pass	pass–pass	32.8–27.4	7.2–4.4	26799–25010	2100–2000	1–1	15.24–16.09
ZNW6	fail–pass	pass–pass	pass–pass	33.0–27.6	7.2–4.2	26980–27172	2119–1980	3–2	12.94–13.0
ZNW7	fail–fail	pass–pass	pass–pass	33.0–27.6	4.2–4.2	27002–26992	2111–2136	2–2	13.05–13.10

Note: TWC and DP—total water content during the five years and respective drying potential; Mould—potential for mould appearance; $w/w_f \leq 0.30$—low probability of water running off and frost damage; Ext surface temp (max)—99th percentile of the maximum verified temperature at the surface [°C]; Temp gradient insulator—temperature gradient between the opposite surfaces of the thermal insulator [°C]; $CP > 0$—condensation potential [Pa]; φ = 100% RH—number of hours where it could be attained an air relative humidity of 100% RH [h]; $U_{avg}/U_{ref} \leq 1$—number of months in which the thermal transmittance of the solution was similar to or below the reference conditions [month]; q—element's heat loss during the full winter of the 5th [kWh], for a square meter.

The industrial thermal render (NW5 and ZNW5) was the solution that presented the lower performance in both locations, probably related to a higher w_f and the coating system limiting its drying. For the aerogel-based solutions (NW1 to NW4 and ZNW1 to ZNW4), since they present a lower w_f than the industrial thermal render (NW5 and ZNW5) but at the same time lower μ than EPS (NW6 and ZNW6) and XPS (NW7 and ZNW7), which can contribute to quicker drying, they could present better performance. The natural fibre formulations (NW3 and 4 and ZNW3 and 4) showed improved performance on this indicator due to their higher hygroscopic behaviour [27].

As for the mould growth, all solutions presented values of temperature and humidity that would not lead to mould development, although considering a high moisture indoor climate. These results were also verified, case by case, using the WUFI Bio postprocessor.

In Lisbon (NW), the aerogel-based renders share a similar behaviour with a higher baseline than the industrial thermal render (NW5). However, when the influence of the WDR is accounted for, the industrial thermal render (NW5) starts to show an increasing water accumulation, surpassing the aerogel-based thermal renders (NW1 to NW4 and ZNW1 to ZNW4). During most of the rainy season, for the North façade, all renders showed similar behaviours. In Lisbon, the higher number of days with solar radiation has a considerable influence on the solutions' drying [86]. From October to April, there is an increase in the water content, then drying until reaching the minimum water content in August.

For Zurich, all the aerogel-containing renders (NW1 to 4 and ZNW1 to 4) maintained similar behaviours throughout the year, being able to dry from one year to the next. The industrial thermal render (ZNW5) that was presenting good performance in Lisbon started to show signs of not being able to dry during the year. Where in Lisbon the industrial thermal render (NW5) can dry out relatively quickly due to higher solar radiation and a lower moisture presence, in Zurich (ZNW5), this was greatly reduced due to climate characteristics, also seen in the very similar values between the SW and the N façades. The industrial thermal render's (ZNW5) performance in Zurich seems limited by its higher w_f than the aerogel-containing renders (ZNW1 to 4).

In both Lisbon and Zurich, it can be seen the influence that the more exposed façade to the WDR has on the solutions. However, the ratio w/w_f was less than 30%, indicating negligible risk of frost damage. In both locations, the natural fibre-containing renders (TR sisal—NW3 and ZNW3—and TR biomass—NW4 and ZNW4) showed less water content year-round than the reference, once again certainly linked to their lower μ. The TR aramid renders (NW2 and ZNW2) with slightly higher μ and decreased porosity showed marginally lower performance than the TR reference (NW1 and ZNW1).

As can be seen, the maximum temperatures present in Lisbon are approximately 10 °C higher than in Zurich, as expected. Those higher temperatures can generate durability issues due to the materials' expansion and retraction [88,108]. In terms of drying, it is once again understood why in Lisbon, the solutions present a quicker drying process: the overall rainfall quantity is lower, and the temperatures are higher. Here, the coating system also limits the overall performance, since it is the same in all solutions.

As for the temperature gradients inside the thermal insulating material, in both places they are between 4 and 7 °C, indicating that the majority of times they are below those values, which are not very significant [109]. With the solutions containing XPS (NW7 and ZNW7) showing a lower gradient, this behaviour is potentially linked with a marked difference in this material's thermal diffusivity.

As expected, it is seen that the risk of surface condensation significantly increases in Zurich. This fact is related to the most demanding outdoor climatic conditions in Zurich and increased thermal insulation thicknesses. The higher thermal insulation needs lead to less heat migrating to the exterior surface, which increases the solutions' condensation potential [110].

When all solutions are compared, in Lisbon, the North façade is the one with the highest risk of exterior surface condensation, but in Zurich, the behaviours are similar between the SW and the N façades, and both are higher than Lisbon. The aerogel-containing solutions (NW1 to 4 and ZNW1 to 4) showed similar behaviours to the other materials, whereas the TR biomass solutions (NW4 and ZNW4) showed a slightly better performance. The industrial thermal renders (NW5 and ZNW5) showed lower values, which can be related to higher water content and lower thermal insulation, letting indoor heat pass and heat the exterior surface.

The climatic conditions differ between Lisbon (SW different from N) and Zurich (SW similar to N). In Lisbon's SW façade, due to the high number of non-raining days and solar radiation, there are a substantial number of months where the average thermal transmittance is below the initial values (20 °C and 80% RH conditions). For the north façade, due to the sun's low influence (without direct solar radiation), the values are maintained above the initial conditions, with the values increasing in late summer due to inversion, as previously seen. The industrial thermal renders (NW5 and ZNW5), due to their higher water absorption, have more months with lower performance (above 1.0) than the aerogel-containing renders (NW1 to 4 and ZNW1 to 4).

In Zurich, the differences between the SW and the N façades are much smaller, with all solutions lowering their thermal transmittance performances. Here, the most significant increase was related to the industrial thermal render (ZNW5), due to its higher water content, which influences its thermal conductivity and increases the solution's overall thermal transmittance.

Although all solutions were analysed for their 4th to 5th year, it should be noted that solutions ZNW5 (industrial thermal render), ZNW6 (EPS), and ZNW7 (XPS) in the subsequent years would start to exhibit lower performance due to water content accumulation. Therefore, from all solutions, the ones that could maintain better performance throughout the year were the ones containing aerogel and fibres: TR aramid (NW2 and ZNW2), TR sisal (NW3 and ZNW3), and TR biomass (NW4 and ZNW4).

Finally, for the solutions' heat loss, it can be seen that the aerogel-containing solutions (NW1 to 4 and ZNW1 to 4), although presenting a higher initial hygroscopicity, can contribute to higher energy savings than the other solutions containing industrial thermal render (NW5 and ZNW5), EPS (NW6 and ZNW6), and XPS (NW7 and ZNW7). This is more noticeable in Zurich than in Lisbon since, as could be seen in this analysis, the climate has a significant impact on the type of materials to be chosen and applied. This can be related to the fact that although EPS (NW6 and ZNW6) and XPS (NW7 and ZNW7) do not present significant hygroscopicity, from 80% RH on, they start to suffer the impact of moisture [44]. Therefore, when applied in a climate where moisture is always present, it impairs their performance since when moisture enters the solution, due to their very low water vapour permeability [43,44] it is retained due to the low number of drying hours in Zurich. The industrial thermal render (NW5 and ZNW5), even with an 8 cm thickness (the maximum allowed by the manufacturer), did not comply with Zurich's legal requirements.

The aerogel and fibre-containing renders (NW2 to 4 and ZNW2 to 4) showed good performances in both climates. The TR sisal (NW3) showed better performance than the TR reference (NW1) (less 3% of kWh), the industrial thermal render (NW5) (less 10%), and EPS (NW6) (less 5%) in Lisbon, being only surpassed by XPS (NW7) (more 5%); however, this latter solution tends to accumulate water during the service life, significantly decreasing its performance over time.

For Zurich, considering the SW orientation, the best solutions were the aerogel-based fibre-containing formulations (ZNW2 to 4), with the TR aramid (ZNW2) presenting the best performance (less than 4% kWh of the TR reference (ZNW1) and less than 9% of the XPS (ZNW7)). When considering the N orientation, the TR sisal (ZNW3) showed less than 3% of the TR reference (ZNW1) and less than 4% of the XPS (ZNW7). Where in Zurich industrial thermal render (ZNW5), EPS (ZNW6), and XPS (ZNW7) tend to degrade their performances due to an increase in water content. The difference in the industrial thermal render (ZNW5) is related to its being considered the same construction solution for all cases, but due to its higher thermal conductivity and maximum acceptable application thickness (0.08 m), it could not perform adequately.

Therefore, these aerogels and fibre-containing renders' hygrothermal properties could be evaluated against other currently used materials and showed potential to be applied while excelling in more demanding climates with a higher presence of rain and moisture, such as Zurich.

3.4.2. Wall Retrofit over the Exterior Surface (RE)

The presence of a superficial acrylic paint with two layers was considered for the existing reference wall (RE1) (see composition in the Supplementary Data), reducing the WDR water absorption (by 20% due to the usual low maintenance) while increasing the water vapour diffusion-equivalent air layer thickness (s_d = 0.56 m) [111]. This layer was considered to be removed when the new exterior multilayer coating systems were applied. Here, the defects were considered to be 1% of the WDR in the exterior mortar, even for the new systems.

In Table 8, it can be seen that the introduction of these thermal insulation systems drastically reduced the total water content, improving the integrity and durability of the construction. The reference solution (RE1), as well as the aerogel-based renders (RE2 to 5), quickly showed dynamic equilibrium. However, the solutions with EPS (RE7) and XPS (RE8) showed a limited capability of drying, while the solutions with TR sisal (RE4) and TR biomass (RE5) showed quicker drying. This, as previously seen, can be related to the water vapour permeability shown by these materials (lower in the EPS-RE7- and XPS-RE8-solutions) [43], associated with the high μ of the coating system, slowing their drying process. The solutions that reached a quasi-steady state took around 3 years.

Table 8. Synthesis of the exterior retrofit simulation for Lisbon (RE).

Solution SW–N	TWC and DP	Mould	$w/w_f \leq 0.30$	Ext Surface Temp (max) [°C]	Temp Gradient Insulator [°C]	$CP > 0$ [Pa]	φ = 100%RH [h]	$U_{avg}/U_{ref} \leq 1$ [month]	q [kWh m^{-2}]
RE1	pass–pass	fail–fail	fail–fail	35.6–33.3	-	5036–3040	282–39	4–0	81.45–96.82
RE2	pass–pass	pass–pass	pass–pass	39.3–35.6	9.2–5.1	1704–1793	165–171	12–5	11.85–13.79
RE3	pass–pass	pass–pass	pass–pass	39.2–35.6	9.2–5.1	1606–1766	159–171	10–2	11.84–14.23
RE4	pass–pass	pass–pass	pass–pass	39.2–35.6	9.2–5.1	1642–1807	159–171	9–3	12.90–13.57
RE5	pass–pass	pass–pass	pass–pass	39.2–35.6	9.2–5.1	1592–1764	158–170	6–3	13.64–14.27
RE6	pass–pass	pass–pass	pass–pass	39.1–35.6	9.4–5.3	1586–1736	164–164	8–1	14.21–15.14
RE7	fail–pass	pass–pass	pass–pass	39.2–35.6	9.4–5.2	1597–1675	160–167	11–0	13.72–15.66
RE8	fail–pass	pass–pass	pass–pass	39.4–35.7	9.5–5.3	1747–1848	169–175	10–0	11.47–12.64

Note: TWC and DP—total water content during the five years and respective drying potential; Mould—potential for mould appearance; $w/w_f \leq 0.30$—low probability of water running off and frost damage; Ext surface temp (max)—99th percentile of the maximum verified temperature at the surface [°C]; Temp gradient insulator—temperature gradient between the opposite surfaces of the thermal insulator [°C]; $CP > 0$—condensation potential [Pa]; φ = 100%RH—number of hours where it could be attained an air relative humidity of 100%RH [h]; $U_{avg}/U_{ref} \leq 1$—number of months in which the thermal transmittance of the solution was similar or below than the reference conditions [month]; q—element's heat loss during full Winter of the 5th [kWh], for a square meter.

Due to the low thermal insulation performance, it can be seen in Table 8 that the existing reference wall (RE1) showed a significant probability for mould to develop, but when the thermal insulation solutions were applied, that probability became negligible, being similar for all solutions. This shows the importance of adequate thermal insulation for improving indoor air quality.

The results are shown in Table 8, which also shows that the whole solution absorbed less water when the thermal insulation systems were used. This could be because the coating system is better at keeping liquid water from getting in, which lowers the water content. The TR sisal (RE4) and TR biomass (RE5) formulations were able to maintain lower water contents for SW and N orientations. The increase in total water content (ratio < 20%) when compared with the new wall solution for Lisbon can be related to the high water absorption properties presented by the existing mortar and fired clay hollow brick. If at the same time, the high of the coating system is considered, it can be explained that the rise in water content is due to the water vapour's difficulty exiting and, therefore, does not contribute to the solutions' drying.

The inclusion of the thermal insulating systems led to an increase in the exterior surface temperature over the reference (RE1). This seems linked to the coating system present in all solutions and the presence of the thermal insulators behind it that do not disperse heat as efficiently as other materials with higher thermal conductivity. The same type of behaviour was seen in the temperature gradient inside the thermal insulating materials, where all solutions presented similar performances.

When the thermal insulating systems were applied to the old walls, the risk of surface condensation was drastically reduced. The only situation worse than the reference (RE1) was related to the number of hours, over a year, with 100% RH. This fact can be associated with the increase in thermal insulation, in which, due to the lower influence of the WDR on the north façade, there was a loss of superficial heating, from indoor losses, which could maintain higher surface temperatures and, therefore, a lower risk of existing 100% RH. Nonetheless, all materials, for the analysis period, showed similar performance.

With the analysis of the thermal insulation loss over the year and the relationship with the heat losses during the winter season, it can be readily seen that all new retrofit solutions presented significant improvements in the building's thermal insulation efficiency.

These thermal insulation systems could maintain a better thermal performance over the year, with the TR sisal (RE4) showing a more equilibrated performance on both façades. The EPS (RE7) and XPS (RE8) tend to degrade their performance over the years due to their difficulty to dry. The industrial thermal render (RE6) solution, in this case, showed similar performance.

Table 8 shows that the inclusion of the thermal insulation systems reduced heat loss by 10 times, drastically increasing energy efficiency. Since EPS (RE7) and XPS (RE8) solutions suffer an increase in their water content, their performance will degrade over time. As for the solutions with lower heat loss, for the SW façade, the TR aramid formulation (RE3) improved over the TR reference (RE2) by around 2%, while for the N façade, TR sisal (RE4) showed a 3% improvement over the TR reference (RE2). TR aramid (RE3) showed the best overall performance (for SW and N façades conjugated), reducing heat loss by 1.5% kWh over TR reference (RE2) and ≈10-fold when compared to the original solution (RE1). With these results, the TRFs once again showed multifunctional application potential.

3.4.3. Wall Retrofit over the Interior Surface (RI)

Another application of thermal insulation on buildings' façades is usually considered an interior application [101]. In Table 9, it can be seen that in this retrofit scenario, all solutions except the aerogel and fibre-containing renders (RI2 to 5) showed high affinity to keep the water content increases. When compared with the exterior retrofit, all solutions present a higher baseline water content, linked with the reduction of heating of the walls in winter, which limits their drying, as well as the indoor environment presenting a high moisture content [112], also impairing such a drying process.

Table 9. Synthesis of the interior retrofit simulation for Lisbon (RI).

Solution SW–N	TWC and DP	Mould	$w/w_f \leq 0.30$	Ext Surface Temp (max) [°C]	Temp Gradient Insulator [°C]	$CP > 0$ [Pa]	φ = 100%RH [h]	$U_{avg}/U_{ref} \leq 1$ [month]	q [kWh m^{-2}]
RI1	pass–pass	fail–fail	pass–pass	35.6–33.3	-	5036–3040	282–39	3–0	81.45–96.82
RI2	pass–pass	pass–pass	pass–pass	35.5–33.2	1.0–0.2	11358–7887	311–42	9–0	14.91–18.40
RI3	pass–pass	pass–pass	pass–pass	35.5–33.2	1.0–0.2	11303–7839	292–42	8–0	11.84–18.70
RI4	pass–pass	pass–pass	pass–pass	35.5–33.2	1.0–0.2	11457–7839	281–39	9–0	14.41–17.80
RI5	pass–pass	pass–pass	pass–pass	35.5–33.2	1.0–0.2	11304–7970	282–39	8–0	15.17–18.74
RI6	pass–fail	pass–pass	pass–pass	35.5–33.2	0.6–0.2	11286–7832	304–55	6–0	15.42–18.99
RI7	fail–fail	pass–pass	pass–pass	35.5–33.2	1.1–0.3	10968–7589	290–38	4–0	16.93–20.79
RI8	fail–fail	pass–pass	pass–pass	35.5–33.2	1.2–0.3	12033–8552	270–60	4–0	11.94–14.54

Note: TWC and DP—total water content during the five years and respective drying potential; Mould—potential for mould appearance; $w/w_f \leq 0.30$—low probability of water running off and frost damage; Ext surface temp (max)—99th percentile of the maximum verified temperature at the surface [°C]; Temp gradient insulator—temperature gradient between the opposite surfaces of the thermal insulator [°C]; $CP > 0$—condensation potential [Pa]; φ = 100%RH—number of hours where it could be attained an air relative humidity of 100%RH [h]; $U_{avg}/U_{ref} \leq 1$—number of months in which the thermal transmittance of the solution was similar or below than the reference conditions [month]; q—element's heat loss during full Winter of the 5th [kWh], for a square meter.

Only the reference solution (RI1) presented a significant probability for mould development, as previously observed in the exterior retrofit study, where RE1 is the same as RI1 (see Supplementary Data). As in the other retrofit scenario, it took about 3 years for the solutions to reach a quasi-steady state.

Since the thermal insulation in this retrofit scenario was placed on the interior side of the wall, but with a gypsum board protecting it, it can be seen in Table 9 that in this case, the solutions' behaviour is very similar between the SW and the N façades.

On the SW façade, there is a more significant influence of the solutions' drying due to higher solar exposure. For the industrial thermal render (RI6), it is seen that it absorbs more water during the winter, although it loses it in the summer, being slower in this process than when applied on the exterior. This fact can be related to the influence that the outer climatic elements have on the drying of this material, and it can also be related to its lower MBV (moisture buffer value), showing more difficulty eliminating moisture in a short time.

However, for the aerogel-fiber-containing renders (RI3–5), there was a higher degree of stability. This could also have something to do with the fact that these aerogel renders had higher MBV values, which can help to keep their performance even throughout the year. In this specific case, the TR biomass formulation (RI5) showed better performance for both façades over the year, closely followed by TR sisal (RI4).

With the interior retrofit, the maximum exterior temperatures were not significantly affected (Table 9), since the exterior surface was similar. Regarding the temperature gradient for the thermal insulating layer, it is almost negligible since the indoor environment has a direct influence over it.

With this retrofit solution, the condensation risk increased significantly (Table 9), as expected. This is related to the indoor heat not being able to heat the exterior surface during the winter months, increasing that risk.

Both EPS (RI7) and XPS (RI8) solutions showed worse performance, not only for accumulating water but also by seeing their thermal performance degrade over time, probably related to their low water vapour permeability that impairs their drying.

Due to the much lower drying influence of the solar radiation in these solutions, they tended to present higher thermal transmittance values (Table 9) than their respective reference values over the year. Once again, the thermal inversion influence during the summer can be verified. Here, the TR sisal formulation (RI5) showed the best overall performance.

In Table 9, it can be seen the potential heat losses, per m^2, during the winter season for all solutions. As in the exterior retrofit scenario, all solutions decreased the heat losses over the reference wall (RI1) by almost 10 times. Once again, solutions EPS (RI7) and XPS (RI8) accumulate water, with their performance further decreasing over time. The other solutions show that the TR aramid (RI3) applied in the SW façade lowers the heat losses by almost 20% over the TR reference (RI2), while for the N façade, the best performance is from the TR sisal formulation (RI4), with less than 4% of the heat losses of the TR reference formulation (RI2).

When considering the overall performance (SW and N façades), the TR aramid (RI3) shows less than 8% heat loss compared to the TR reference (RI2). In this case, since the thermal insulation is inside, it seems that the higher moisture affinity of the natural fibre-containing solutions had some influence on their overall performance, as already described in other research works [9]. Additionally, in this case, the aerogel-based fibre-containing formulations (RI3–5) showed better performance, confirming, once again, their multi-functionality.

4. Critical Analysis of the Results

4.1. Thermophysical Properties of the Materials

When considering the thermal properties, the fibres' incorporation, due to their higher thermal conductivities and reduction of aerogel quantity, increased the overall thermal conductivity in the dry state, but by less than 7% relative to the reference, and still below EPS ($\lambda_{10°C} \approx 0.036$ W m^{-1} K^{-1} [61]) or thermal renders with EPS ($\lambda_{10°C} \approx 0.050$ W m^{-1} K^{-1} [62]).

This was an interesting aspect since the incorporation of fibres in cement-based composites is usually linked with lowering thermal conductivity [113]. In this case, as the silica aerogel has a low thermal conductivity, such behaviour was not verified. However, the formulations with fibres could lower the water saturation, contributing to lower thermal conductivity (at least by 7%) in the presence of moisture. This fact was attributed to the presence of air pockets trapped in the pore structure and to the capillaries' interruption due to the fibres' presence, seen in [27], which impaired water progression. Here, the TR aramid showed less water absorption and could maintain a more stable thermal conductivity increase. As for thermal inertia and penetration depth, significant differences between the distinct formulations were not found.

For the moisture storage, it seems that the fibres did not significantly influence the TR reference's hygroscopic behaviour. As seen through the dynamic vapour sorption (DVS) results [27], the paste already presented considerable influence in the hygroscopic range; therefore, the similar behaviour between all the formulations seems related to the paste. However, when tested in the over-hygroscopic range, it started to show differences, with less water absorbed by the fibre-containing formulations, either by reduction of capillaries or reduction of pores linked to the exterior [65,68]. Additionally, the water vapour permeability increased with the use of natural fibres.

Here, it was also found that the fibres' inclusion led to a lower liquid transport coefficient for suction (D_{ws}), since its calculation is dependent on the A_W coefficient. As seen in [27], the fibres' use led to a significant reduction of the capillary water absorption coefficient (A_W), therefore the difference between the D_{ws} parameters between the reference (TR reference) and the fibre-containing formulations was already expected, representing better behaviour as a construction material.

Finally, for the analysis of moisture buffering properties, it was seen that when only considering the TRFs, the TR aramid formulation showed a slightly lower performance but still presented higher capabilities than the studied industrial thermal render. Nonetheless, the TR sisal and TR biomass showed very promising performances. When a gypsum board was placed in front of these materials, the total moisture-absorbing capability was reduced to half, with all the aerogel-based formulations presenting similar behaviours. When the application of a commercial multilayer coating system was considered, the maximum moisture absorbing capability for all thermal insulating materials was reduced almost ten times due to its intrinsic properties [25].

With this experimental campaign, it was seen that the most significant influence of the fibres' use was related to the reduction of capillary water absorption, either by reducing capillary paths or by changing the render's pore structure [27]. This factor greatly influenced the thermal conductivity of the fibre-containing formulation, since less water uptake led to overall lower thermal conductivities.

4.2. Hygrothermal Simulation Performance

In this study, several scenarios were studied, from new walls placed in different climates (Lisbon and Zurich) to thermal insulation retrofit scenarios on the exterior and interior. From the results, it could be seen that the climate and the legal requirements should dictate the choice of the materials since their hygrothermal behaviour influences the performance of the construction solutions where they are applied.

For the new wall solution, in both Lisbon and Zurich, the TRFs presented better performance than the other solutions (industrial thermal render with EPS granules, EPS, and XPS). The best performance was attained by the natural fibres (TR sisal and TR biomass), as it seems that their higher water vapour permeability conjugated with their low capillary water absorption contributed to a quicker drying of the respective solutions.

In the exterior thermal insulation retrofit scenario, the TR sisal and TR biomass formulations continued to present better performance than the other solutions (industrial thermal render, EPS, and XPS). This continued in the same line as in the new wall scenario; however, EPS and XPS solutions showed higher water contents, probably related to the constructive solution's differences. Nonetheless, these innovative TRFs showed their multifunctionality potential, presenting better performances than the other solutions, and even better than the TR reference formulation, achieving dynamic equilibrium more quickly.

When the interior thermal insulation retrofit scenario was considered, although the fibre-containing formulations (TR aramid, TR sisal, and TR biomass) continued to show better performance, the best formulation of all was the TR aramid. This can be related to this fibre's lower moisture affinity (as seen in the DVS analysis in [27]) in an indoor environment, where the solutions have more difficulties drying due to the presence of the gypsum board. Nonetheless, the TR sisal could present some reliable performance. In this type of solution, since the indoor temperature is higher than the outdoor temperature (during Winter), moisture is transferred from the inside to the outside. Therefore, the materials' water-vapor permeability has a significant influence on the behaviour of the whole constructive system, contributing to or not increasing its water content. With indoor thermal insulation, the wall becomes in contact with the cold environment outside (during winter), lowering the constructive system's temperature, which significantly increased the risk of surface condensation [114].

As expected, the application of thermal insulation on the exterior side of the walls presented more advantages than on the inside [34,88,90]: reduction of global heat losses and energy consumption; a higher temperature of the wall, improving drying and lowering condensation risk; reduction of thermal bridges; more mass in contact with the interior, contributing to better indoor temperature regulation.

Another aspect that must be considered is that the protective coating system significantly limits the water vapour permeability of the solutions [25], lowering it by more than 90%. Nonetheless, due to the higher water vapour permeability of the TR sisal and TR biomass when compared with, for example, EPS or XPS, they contribute to a higher overall water vapour permeability of the multilayer coating systems, improving their drying capabilities.

4.3. Study Considerations

Nonetheless, for all the simulations made, several limiting factors should be considered:

- The hygrothermal simulations of Lisbon and Zurich, which were based on this analysis, were run with validated files made available by the WUFI software. However, those files represent one year (1998) of hourly data for several climatic parameters. This is one important factor to consider, as in different years the conditions can be quite different, which can impact the performance of constructive solutions;
- It is currently known that climate change is happening [115]. This can further impact the performance of constructive solutions since climate variations can lead to more moisture and rainfall, influencing the solutions' performance and needs. This fact can also be limited by microclimatic phenomena, which should be considered when evaluating the applicability of any given material [116];
- Another aspect that was not considered, even in the simulations, was the presence of rising damp and salts in buildings, which is one of the significant building defects verified in Portugal [117]. Although this factor was not considered, it should be accounted for, mainly in retrofit scenarios, since it can lead to several anomalies;
- The indoor occupancy can also greatly influence the overall performance (users' occupancy 100% of the time vs. 40%, for example), as well as the type of use (e.g., kitchen vs. bedroom);
- As these simulations represented broad application scenarios, each singular case should be analysed, since the geographic location, utilisation conditions, used materials, and climate, among other factors, will influence the obtained results.

5. Conclusions

In this study, the influence of a synthetic fibre (TR aramid) and two natural fibres (TR sisal and TR biomass) on the hygrothermal properties of an aerogel-based thermal render (TR reference—without fibers) was evaluated. For hygrothermal evaluation, the water content with time and its effect on the thermal insulation of the solution were the primary considerations. Both of these factors affected a number of other variables. Moreover, the behaviour of these renders was evaluated when applied in new construction and retrofit scenarios under different conditions compared with other current thermal insulating solutions. From the gathered results and respective analyses, it is possible to draw the following conclusions:

- Although low amounts of fibres were used (0.5% (vol./vol.) for the TR aramid and 0.1% (vol./vol.) for the TR sisal and TR biomass), they significantly improved the hygrothermal properties over the TR reference.
- These aerogel and fibre-containing formulations showed adequate moisture buffering potential; therefore, their future use indoors can be considered as presenting an air moisture regulation capacity, contributing to their application multifunctionality;
- The aerogel and fibre-containing formulations, especially the ones containing natural fibres (TR sisal and TR biomass), showed the most balanced hygrothermal performance of all the studied solutions, either in quickly attaining dynamic equilibrium or dealing with moisture throughout the year. However, depending on the application scenario, the choice of materials must be a careful process. Nonetheless, these aerogel-based thermal renders showed that they could be applied in several distinct geographical and climatic conditions while maintaining good hygrothermal performance;
- In new construction, the aerogel and fibre-containing solutions presented better characteristics than the other studied solutions (industrial thermal render, EPS, and XPS), mainly due to lower capillary water absorption and higher water vapour permeability. Moreover, when a colder and moister climate was chosen (Zurich), these TRFs excelled over the other solutions, since after two years they showed dynamic equilibrium, whereas other solutions (e.g., with EPS and XPS) did not, showing their application potential;
- It was found that the used finishing coating (with an acrylic base) and basecoat have a big effect on the multilayer systems that were looked at. Other materials might have different effects. Although such a protective coating reduced the liquid water penetration from the outside, it also presented low water vapour permeability, hindering the elimination of water that penetrated the interior layers; therefore, they usually contributed to the increase in the solutions' interior water content;
- In a retrofit scenario, it was seen that the exterior application of the thermal insulation is more effective in reducing energy losses than the interior application. Additionally, in both cases (interior and exterior retrofit), the aerogel and fibre-containing solutions (TR aramid, TR sisal, and TR biomass) presented better performances than the other solutions. In addition, when applied to the interior retrofit, those renderings can improve the indoor environment due to their moisture buffering capabilities;

As it has been seen, the incorporation of fibres (aramid, sisal, and biomass) in the aerogel-based formulation of thermal render has shown improvements over the reference physical properties, which led to an improvement in their hygrothermal performance. With such findings, it can be attributed to some multifunctionality in these formulations, since they can show benefits for different uses and applications while also contributing to energy savings in buildings when compared with other solutions, with a reduction of heat transfer of more than 20%.

In future research, these formulations should be studied in terms of their environmental impacts and how their demonstrated capabilities of saving energy consumption would influence the overall performance over the life cycle of a building. These formulations could also benefit from their simulation with FEM models and, above all, be applied in situ to validate the results of these simulations. Climate change scenarios and occupancy internal rates could also be investigated

Supplementary Materials: The following supporting information can be downloaded at: https://www.mdpi.com/article/10.3390/en16073048/s1, Table S1. Thermal insulation materials' properties. Table S2. Properties of other materials used in the simulations. Table S3. New wall solutions for Lisbon (NW). Table S4. New wall solutions for Zurich (ZNW). Table S5. Retrofit solutions are applied on the wall's external surface (RE). Table S6. Retrofit solutions are applied on the wall's internal surface (RI). Figure S1. Flowchart of the conducted hygrothermal simulations. Figure S2. Total water content and drying rates for Lisbon (NW) and Zurich (ZNW). Figure S3. Isopleth example for the ZNW1 solution. Figure S4. Ratio of water content to w_f for the renders. Figure S5. Temperatures were verified for the different solutions. Figure S6. Superficial condensation potential. Figure S7. Thermal insulation losses over the reference performance during the year. Figure S8. Heat loss during the winter per square metre of wall. Figure S9. Total water content and drying rates of the solutions (RE scenario). Figure S10. Mould growth potential. Figure S11. Ratio of water content to w_f for the renders. Figure S12. Temperatures were verified in the simulation. Figure S13. Conditions for the formation of superficial condensation. Figure S14. Thermal insulation loss over the year. Figure S15. Heat loss during the winter, per square metre of wall. Figure S16. Total water content and drying potential rate. Figure S17. Ratio of water content to w_f for the renders. Figure S18. Temperatures were verified in the simulation. Figure S19. Verification of the conditions for the formation of superficial condensation. Figure S20. Thermal insulation loss over the year. Figure S21. Heat loss during the winter, per square metre of wall.

Author Contributions: M.P.: Conceptualization, Methodology, Investigation, Writing—original draft. M.d.G.G.: Supervision, Writing—review and editing. J.D.S.: Supervision, Writing—review and editing. A.H.: Writing—review and editing. I.F.-C.: Supervision, Writing—review and editing. All authors have read and agreed to the published version of the manuscript.

Funding: This research was funded by Fundação para a Ciência e a Tecnologia (F.C.T.), Ph.D. grant SFRH/BD/132239/2017, and CERIS Research Unit (UIDB/04625/2020).

Acknowledgments: The authors wish to acknowledge Saint-Gobain Weber Portugal for making available the aerogel-based thermal render used in this study. The authors gratefully acknowledge the support of the CERIS Research Centre (UIDB/04625/2020), IST—University of Lisbon, and the FCT—Foundation for Science and Technology. The first author also wants to thank FCT for the Ph.D. grant SFRH/BD/132239/2017.

Conflicts of Interest: The authors declare no conflict of interest.

Abbreviations

RH	Relative humidity
TR	Thermal render
TR aramid	Aerogel-based render with 0.5% (volume) of aramid fibre
TR biomass	Aerogel-based render with 0.1% (volume) of biomass fibre
TR reference	Aerogel-based render without fibres
TR sisal	Aerogel-based render with 0.1% (volume) of sisal fibre
TRF	Aerogel-based fibre-enhanced thermal renders with either aramid or biomass or sisal
WDR	wind-driven rain

References

1. Berardi, U. A cross-country comparison of the building energy consumptions and their trends. *Resour. Conserv. Recycl.* **2017**, *123*, 230–241. [CrossRef]
2. Marwan, M. The effect of wall material on energy cost reduction in building. *Case Stud. Therm. Eng.* **2020**, *17*, 100573. [CrossRef]
3. Berardi, U. Aerogel-enhanced systems for building energy retrofits: Insights from a case study. *Energy Build.* **2018**, *159*, 370–381. [CrossRef]
4. Fenoglio, E.; Fantucci, S.; Serra, V.; Carbonaro, C.; Pollo, R. Hygrothermal and environmental performance of a perlite-based insulating plaster for the energy retrofit of buildings. *Energy Build.* **2018**, *179*, 26–38. [CrossRef]
5. Bui, R.; Labat, M.; Lorente, S. Impact of the occupancy scenario on the hygrothermal performance of a room. *Build. Environ.* **2019**, *160*, 106178. [CrossRef]
6. Reuge, N.; Collet, F.; Pretot, S.; Moisette, S.; Bart, M.; Style, O.; Shea, A.; Lanos, C. Hygrothermal effects and moisture kinetics in a bio-based multi-layered wall: Experimental and numerical studies. *Constr. Build. Mater.* **2019**, *240*, 117928. [CrossRef]

7. Buratti, C.; Belloni, E.; Merli, F. Water vapour permeability of innovative building materials from different waste. *Mater. Lett.* **2020**, *265*, 127459. [CrossRef]
8. Zhang, H.; Yoshino, H.; Hasegawa, K. Assessing the moisture buffering performance of hygroscopic material by using experimental method. *Build. Environ.* **2012**, *48*, 27–34. [CrossRef]
9. Latif, E.; Lawrence, M.; Shea, A.; Walker, P. Moisture buffer potential of experimental wall assemblies incorporating formulated hemp-lime. *Build. Environ.* **2015**, *93*, 199–209. [CrossRef]
10. *CEN EN 15026*; Hygrothermal Performance of Building Components and Building Elements—Assessment of Moisture Transfer by Numerical Simulation. Comité Européen de Normalisation: Brussels, Belgium, 2007.
11. Fantucci, S.; Fenoglio, E.; Serra, V.; Perino, M.; Dutto, M.; Marino, V. Hygrothermal Characterization of High-Performance Aerogel-Based Internal Plaster. In *Sustainability in Energy and Buildings. Smart Innovation, Systems and Technologies*, 1st ed.; Littlewood, J., Howlett, R., Capozzoli, A., Jain, L., Eds.; Springer: Singapore, 2020; pp. 259–268. [CrossRef]
12. European Commission. HOMES Key INsulating Material-HOMESKIN. 2019. Available online: http://cordis.europa.eu/project/rcn/193413_en.html (accessed on 27 December 2019).
13. Hipin, Hipin Project—High Performance Insulation Based on Nanostructure Encapsulation of Air. 2019. Available online: http://www.hipin.eu/ (accessed on 29 December 2019).
14. Baetens, R.; Jelle, B.P.; Gustavsen, A. Aerogel insulation for building applications: A state-of-the-art review. *Energy Build.* **2010**, *43*, 761–769. [CrossRef]
15. Westgate, P.; Paine, K.; Ball, R.J. Physical and mechanical properties of plasters incorporating aerogel granules and polypropylene monofilament fibres. *Constr. Build. Mater.* **2018**, *158*, 472–480. [CrossRef]
16. Vajtai, R. (Ed.) *Springer Handbook of Nanomaterials*; Springer: Berlin/Heidelberg, Germany, 2013. [CrossRef]
17. de Fátima Júlio, M.; Soares, A.; Ilharco, L.M.; Flores-Colen, I.; de Brito, J. Aerogel-based renders with lightweight aggregates: Correlation between molecular/pore structure and performance. *Constr. Build. Mater.* **2016**, *124*, 485–495. [CrossRef]
18. Ebert, H.-P. Thermal Properties of Aerogels. In *Aerogels Handb*, 1st ed.; Aegerter, M., Leventis, N., Koebel, M.M., Eds.; Springer: New York, NY, USA, 2011; pp. 537–564. [CrossRef]
19. Aegerter, M.A.; Leventis, N.; Koebel, M.M. (Eds.) *Aerogels Handbook*; Springer: New York, NY, USA, 2011. [CrossRef]
20. Garrido, R.; Silvestre, J.D.; Flores-Colen, I.; de Fátima Júlio, M.; Pedroso, M. Economic assessment of the production of subcritically dried silica-based aerogels. *J. Non-Cryst. Solids* **2019**, *516*, 26–34. [CrossRef]
21. Stahl, T.; Brunner, S.; Zimmermann, M.; Wakili, K.G. Thermo-hygric properties of a newly developed aerogel based insulation rendering for both exterior and interior applications. *Energy Build.* **2012**, *44*, 114–117. [CrossRef]
22. Fantucci, S.; Fenoglio, E.; Grosso, G.; Serra, V.; Perino, M.; Marino, V.; Dutto, M. Development of an aerogel-based thermal coating for the energy retrofit and the prevention of condensation risk in existing buildings. *Sci. Technol. Built Environ.* **2019**, *25*, 1178–1186. [CrossRef]
23. Pedroso, M.; Flores-Colen, I.; Silvestre, J.D.; Gomes, M.G.; Silva, L.; Ilharco, L. Physical, mechanical, and microstructural characterisation of an innovative thermal insulating render incorporating silica aerogel. *Energy Build. Energy Build.* **2020**, *211*, 109793. [CrossRef]
24. Pedroso, M.; Flores-Colen, I.; Silvestre, J.D.; da Glória Gomes, M. Nanomaterials' Influence on the Performance of Thermal Insulating Mortars—A Statistical Analysis. *Appl. Sci.* **2020**, *10*, 2219. [CrossRef]
25. Pedroso, M.; Flores-Colen, I.; Silvestre, J.D.; Gomes, M.G.; Silva, L.; Sequeira, P.; de Brito, J. Characterisation of a multilayer external wall thermal insulation system. Application in a Mediterranean climate. *J. Build. Eng.* **2020**, *30*, 101265. [CrossRef]
26. Dams, B.; Amornrattanasereegul, N.; Shepherd, P.; Ball, R.J. Cement-fibre composites for additive building manufacturing. In *39th Cement and Concrete Science Conference*; Ball, R.J., Dams, B., Ferrandiz-Mas, V., Ke, X., Paine, K., Tyrer, M., Walker, P., Eds.; University of Bath: Bath, UK, 2019; pp. 14–18.
27. Pedroso, M. Eco-Efficient and Multifunctional Thermal Renders Based on Silica Aerogel and Fibres. Ph.D. Thesis, University of Lisbon, Lisbon, Portugal, 2021.
28. Stahl, T.; Wakili, K.G.; Hartmeier, S.; Franov, E.; Niederberger, W.; Zimmermann, M. Temperature and moisture evolution beneath an aerogel based rendering applied to a historic building. *J. Build. Eng.* **2017**, *12*, 140–146. [CrossRef]
29. Nosrati, R.H.; Berardi, U. Hygrothermal characteristics of aerogel-enhanced insulating materials under different humidity and temperature conditions. *Energy Build.* **2018**, *158*, 698–711. [CrossRef]
30. *CEN EN 998-1: 2010*; Specification for Mortar for Masonry–Part 1: Rendering and Plastering Mortar. Comité Européen de Normalisation: Brussels, Belgium, 2010.
31. Enersens, Product Data Sheet: Aerogel Kwark. 2019. Available online: http://enersens.fr/en/home/ (accessed on 7 January 2020).
32. Applied Precision Ltd. *Isomet 2114 Thermal Properties Analyzer User's Guide*; Version 120712; Applied Precision Ltd.: Bratislava, Slovakia, 2019.
33. Barclay, M.; Holcroft, N.; Shea, A.D. Methods to determine whole building hygrothermal performance of hemp–lime buildings. *Build. Environ.* **2014**, *80*, 204–212. [CrossRef]
34. Fraunhofer-IBP. *WUFI Pro 5.0*; Fraunhofer-IBP: Holzhirchen, Germany, 2009.
35. Kehrer, M. *Determination of Material Properties for Hygrothermal Calculation with WUFI*; Munich, Germany, 2019. Available online: https://wufi.de/img_wufiforum/WUFI-Materialmessung_EN.pdf (accessed on 12 March 2023).

36. Ansell, M.P.; Ball, R.J.; Lawrence, M.; Maskell, D.; Shea, A.; Walker, P. Green composites for the built environment. In *Green Composites*; Elsevier: Amsterdam, The Netherlands, 2017; pp. 123–148. [CrossRef]
37. *ASTM D5334-14*; Standard Test Method for Determination of Thermal Conductivity of Soil and Soft Rock by Thermal Needle Probe Procedure, USA. American Society for Testing and Materials: West Conshohocken, PA, USA, 2014. [CrossRef]
38. *ASTM D5930-17*; Standard Test Method for Thermal Conductivity of Plastics by Means of a Transient Line-Source Technique, USA. American Society for Testing and Materials: West Conshohocken, PA, USA, 2009.
39. *ISO/TC 61/SC 5, ISO 22007-2: 2015*; Plastics—Determination of Thermal Conductivity and Thermal Diffusivity—Part 2: Transient Plane Heat Source (Hot Disc) Method. ISO: Geneva, Switzerland, 2015.
40. Gomes, M.G.; Flores-Colen, I.; Manga, L.M.; Soares, A.; de Brito, J. The influence of moisture content on the thermal conductivity of external thermal mortars. *Constr. Build. Mater.* **2017**, *135*, 279–286. [CrossRef]
41. Gomes, M.G.; Flores-Colen, I.; da Silva, F.; Pedroso, M. Thermal conductivity measurement of thermal insulating mortars with EPS and silica aerogel by steady-state and transient methods. *Constr. Build. Mater.* **2018**, *172*, 696–705. [CrossRef]
42. *ISO/TC 163/SC 2, ISO 10456: 2007*; Building Materials and Products—Hygrothermal Properties—Tabulated Design Values and Procedures for Determining Declared and Design Thermal Values. ISO: Geneva, Switzerland, 2007.
43. Ducoulombier, L.; Lafhaj, Z. Comparative study of hygrothermal properties of five thermal insulation materials. *Case Stud. Therm. Eng.* **2017**, *10*, 628–640. [CrossRef]
44. Palumbo, M.; Lacasta, A.M.; Giraldo, M.P.; Haurie, L.; Correal, E. Bio-based insulation materials and their hygrothermal performance in a building envelope system (ETICS). *Energy Build.* **2018**, *174*, 147–155. [CrossRef]
45. *CEN EN ISO 12571*; Hygrothermal Performance of Building Materials and Products—Determination of Hygroscopic Sorption Properties. Comité Européen de Normalisation: Brussels, Belgium, 2013.
46. *CEN EN ISO 12570:2000+A1*; Hygrothermal Performance of Building Materials and Products–Determination of Moisture Content by Drying at Elevated Temperature. Comité Européen de Normalisation: Brussels, Belgium, 2013.
47. Dhakal, U.; Berardi, U.; Gorgolewski, M.; Richman, R. Hygrothermal performance of hempcrete for Ontario (Canada) buildings. *J. Clean. Prod.* **2017**, *142*, 3655–3664. [CrossRef]
48. Evrard, A.; De Herde, A. Hygrothermal Performance of Lime-Hemp Wall Assemblies. *J. Build. Phys.* **2010**, *34*, 5–25. [CrossRef]
49. Mensinga, P. Determining the Critical Degree of Saturation of Brick Using Frost Dilatometry. Master's thesis, University of Waterloo, Waterloo, ON, Canada, 2009.
50. Abadie, M.O.; Mendonça, K.C. Moisture performance of building materials: From material characterization to building simulation using the Moisture Buffer Value concept. *Build. Environ.* **2009**, *44*, 388–401. [CrossRef]
51. *CEN EN 1015-18*; Methods of test for mortar for masonry—Part 18: Determination of Water Absorption Coefficient Due to Capillary Action of Hardened Mortar. Comité Européen de Normalisation: Brussels, Belgium, 2002.
52. Krus, M. Moisture transport and storage coefficients of porous mineral building materials. In *Theoretical Principles and New Test Methods*; IRB-Verlag Stuttgart: Stuttgart, Germany, 1996.
53. de Freitas, V.P.; Krus, M.; Kunzel, H.; Quenard, D. Determination of the water diffusivity of porous materials by gamma-ray attenuations and nmr. In Proceedings of the International Symposium On Moisture Problems In Building Walls, Porto, Portugal, 11–13 September 1995; pp. 445–460.
54. Krus, M.; Holm, A. Simple methods to approximate the liquid transport coefficients describing the absorption and drying process. In Proceedings of the Symposium on Building Physics in the Nordic Countries 1999, Göteborg, Sweden, 24–26 August 1999; pp. 241–248. [CrossRef]
55. Maia, J.; Ramos, N.M.M.; Veiga, R. A new durability assessment methodology of thermal mortars applied in multilayer rendering systems. *Constr. Build. Mater.* **2019**, *222*, 654–663. [CrossRef]
56. Zhang, M.; Qin, M.; Chen, Z. Moisture Buffer Effect and its Impact on Indoor Environment. *Procedia Eng.* **2017**, *205*, 1123–1129. [CrossRef]
57. Rode, C. *Moisture Buffering of Building Materials—DTU BYG-Rapporter*; No. R-126; Technical University of Denmark: Lyngby, Denmark, 2006.
58. Rode, C.; Peuhkuri, R.; Time, B.; Svennberg, K.; Ojanen, T. Moisture Buffer Value of Building Materials. In *Heat-Air-Moisture Transport: Measurements on Building Materials*; ASTM International: West Conshohocken, PA, USA, 2009; p. 33-33-12. [CrossRef]
59. Rode, C.; Peuhkuri, R.; Time, B.; Svennberg, K.; Ojanen, T.; Mukhopadhyaya, P.; Kumaran, M.; Dean, S.W. Moisture Buffer Value of Building Materials. *J. ASTM Int.* **2007**, *4*, 100369. [CrossRef]
60. Wan, H.; Sun, Z.; Huang, G.; Xu, X.; Yu, J. Calculation of the maximum moisture buffering thickness of building wall layer of hygroscopic material. *Build. Environ.* **2019**, *160*, 106173. [CrossRef]
61. Abu-Jdayil, B.; Mourad, A.-H.; Hittini, W.; Hassan, M.; Hameedi, S. Traditional, state-of-the-art and renewable thermal building insulation materials: An overview. *Constr. Build. Mater.* **2019**, *214*, 709–735. [CrossRef]
62. Maia, J.F. Durability of Thermal Rendering and Plastering Systems. Ph.D. Thesis, University of Porto, Porto, Portugal, 2019.
63. Guyer, E.; Brownell, D. (Eds.) *Handbook of Applied Thermal Design*, 1st ed.; Taylor & Francis: Philadelphia, PA, USA, 1999.
64. Pina dos Santos, C.; Matias, L. *ITE50—Thermal Transmittance of Building Envelope Elements*, 1st ed.; LNEC: Lisbon, Portugal, 2006. (In Portuguese)
65. Broda, J. Application of Polypropylene Fibrillated Fibres for Reinforcement of Concrete and Cement Mortars. In *High Performance Concrete Technology and Applications*, 1st ed.; Yilmaz, S., Ozmen, H., Eds.; InTechOpen: London, UK, 2016; pp. 189–204. [CrossRef]

66. Kumaran, M.K. *Heat, Air, and Moisture Transfer in Insulated Envelope Parts: Final Report; Task 3: Material Properties*; National Research Council Canada: Ottawa, ON, Canada, 1996.
67. Mňahončáková, E.; Jiřičková, M.; Pavlík, Z.; Fiala, L.; Rovnaníková, P.; Bayer, P.; Černý, R. Effect of Moisture on the Thermal Conductivity of a Cementitious Composite. *Int. J. Thermophys.* **2006**, *27*, 1228–1240. [CrossRef]
68. Richardson, A.E. Freeze/thaw durability in concrete with fibre additions. *Struct. Surv.* **2003**, *21*, 225–233. [CrossRef]
69. Filho, R.D.T.; Ghavami, K.; England, G.L.; Scrivener, K. Development of vegetable fibre–mortar composites of improved durability. *Cem. Concr. Compos.* **2003**, *25*, 185–196. [CrossRef]
70. Crawford, C.B. Frost durability of clay bricks—Evaluation criteria and quality control. In *CBAC/DBR Manufacturers' Symposium*; NRC of Canada: Ottawa, ON, Canada, 1984.
71. Ganobjak, M.; Brunner, S.; Wernery, J. Aerogel materials for heritage buildings: Materials, properties and case studies. *J. Cult. Herit.* **2020**, *42*, 81–98. [CrossRef]
72. Ibrahim, M. Improving the Buildings Envelopes Energy Performance Using Aerogel-Based Insulating Mineral Rendering. Ph.D. Thesis, Ecole Nationale Supérieure des Mines de Paris, Paris, Frnace, 2014.
73. Collet, F.; Chamoin, J.; Pretot, S.; Lanos, C. Comparison of the hygric behaviour of three hemp concretes. *Energy Build.* **2013**, *62*, 294–303. [CrossRef]
74. Ramos, N.M.M.; Delgado, J.M.P.Q.; de Freitas, V.P. Influence of finishing coatings on hygroscopic moisture buffering in building elements. *Constr. Build. Mater.* **2010**, *24*, 2590–2597. [CrossRef]
75. Cerolini, S.; D'Orazio, M.; Di Perna, C.; Stazi, A. Moisture buffering capacity of highly absorbing materials. *Energy Build.* **2009**, *41*, 164–168. [CrossRef]
76. Yu, S.; Cui, Y.; Shao, Y.; Han, F. Research on the Comprehensive Performance of Hygroscopic Materials in an Office Building Based on EnergyPlus. *Energies* **2019**, *12*, 191. [CrossRef]
77. Wu, Y.; Gong, G.; Yu, C.W.; Huang, Z. Proposing ultimate moisture buffering value (UMBV) for characterization of composite porous mortars. *Constr. Build. Mater.* **2015**, *82*, 81–88. [CrossRef]
78. Beck, H.E.; Zimmermann, N.E.; McVicar, T.R.; Vergopolan, N.; Berg, A.; Wood, E.F. Present and future Köppen-Geiger climate classification maps at 1-km resolution. *Sci. Data* **2018**, *5*, 180214. [CrossRef] [PubMed]
79. Peel, M.C.; Finlayson, B.L.; McMahon, T.A. Updated world map of the Köppen-Geiger climate classification. *Hydrol. Earth Syst. Sci.* **2007**, *11*, 1633–1644. [CrossRef]
80. Koebel, M.M.; Wernery, J.; Malfait, W.J. Energy in buildings—Policy, materials and solutions. *MRS Energy Sustain.* **2017**, *4*, E12. [CrossRef]
81. Alves, S.; Ferreira, C.; de Freitas, V.P.; Guimarães, A. *Existing Buildings—Improvement Measures of Energy Performance and Indoor Air Quality*; 2011.
82. CCE. *Energy Rehabilitation Methodology for Buildings Located in Urban Areas—A SAVE II Programme Action*; CCE: Amadora, Portugal, 2000.
83. Uidhir, T.M.; Rogan, F.; Collins, M.; Curtis, J.; Gallachóir, B.P.Ó. Improving energy savings from a residential retrofit policy: A new model to inform better retrofit decisions. *Energy Build.* **2020**, *209*, 109656. [CrossRef]
84. Yang, J.; Fu, H.; Qin, M. Evaluation of Different Thermal Models in EnergyPlus for Calculating Moisture Effects on Building Energy Consumption in Different Climate Conditions. *Procedia Eng.* **2015**, *121*, 1635–1641. [CrossRef]
85. Künzel, H.M. Simultaneous Heat and Moisture Transport in Building Components. In *One- and Two-Dimensional Calculation Using Simple Parameters*; IRB-Verlag Stuttgart: Stuttgart, Germany, 1995.
86. Künzel, H.M.; Kiessl, K. Calculation of heat and moisture transfer in exposed building components. *Int. J. Heat Mass Transf.* **1996**, *40*, 159–167. [CrossRef]
87. Kehrer, M.; Schmidt, T. Radiation effects on exterior surfaces. In Proceedings of the 8th Symposium on Building Physics in the Nordic Countries, Copenhagen, Denmark, 16–18 June 2008; DTU: Copenhagen, Denmark, 2008; pp. 207–212.
88. Barreira, E.; de Freitas, V.P. External Thermal Insulation Composite Systems: Critical Parameters for Surface Hygrothermal Behaviour. *Adv. Mater. Sci. Eng.* **2014**, *2014*, 650752. [CrossRef]
89. Mendes, N.; Chhay, M.; Berger, J.; Dutykh, D. *Numerical Methods for Diffusion Phenomena in Building Physics: A Practical Introduction*; PUCPress: Curitiba, Brazil, 2016. [CrossRef]
90. Barreira, E.; de Freitas, V.P. *External Thermal Insulation Composite Systems (ETICS)*; Springer International Publishing: Cham, Switzerland, 2016. [CrossRef]
91. Delgado, J.M.P.Q.; Barreira, E.; Ramos, N.M.M.; de Freitas, V.P. Inputs for Hygrothermal Simulation Tools. In *Hygrothermal Numerical Simulation Tools Applied to Building Physics*, 1st ed.; Delgado, J., Barreira, E., Ramos, N.M.M., de Freitas, V.P., Eds.; Springer: Dordrecht, The Netherlands, 2013; pp. 7–20. [CrossRef]
92. *ASHRAE ANSI/ASHRAE 55*; Thermal Environmental Conditions for Human Occupancy. ANSI/ASHRAE: Peachtree Corners, GA, USA, 2017.
93. *Decree-Law 118/2013*; Energy Certification of Buildings (SCE), Regulation of Energy Performance of Residential Buildings (REH) and Regulation of Energy Performance of Buildings Trade and Services (RECS). Diário da República 2013. Volume 159, pp. 4988–5005. (In Portuguese)
94. Cho, H.M.; Park, J.H.; Wi, S.; Chang, S.J.; Yun, G.Y.; Kim, S. Energy retrofit analysis of cross-laminated timber residential buildings in Seoul, Korea: Insights from a case study of packages. *Energy Build.* **2019**, *202*, 109329. [CrossRef]

95. Lee, J.; Wi, S.; Chang, S.J.; Choi, J.; Kim, S. Prediction evaluating of moisture problems in light-weight wood structure: Perspectives on regional climates and building materials. *Build. Environ.* **2020**, *168*, 106521. [CrossRef]
96. *ASHRAE ANSI/ASHRAE 160-2016*; Criteria for Moisture-Control Design Analysis in Buildings (ANSI Approved). ANSI/ASHRAE: Peachtree Corners, GA, USA, 2016.
97. Park, J.H.; Kang, Y.; Lee, J.; Wi, S.; Chang, J.D.; Kim, S. Analysis of walls of functional gypsum board added with porous material and phase change material to improve hygrothermal performance. *Energy Build.* **2019**, *183*, 803–816. [CrossRef]
98. Weber, S.-G. *Technical Datasheet: Weberplast Decor Plus*; 2019. (In Portuguese)
99. *Assembleia da República, Statute 379-A/2015—1st Alteration to Statute 349-B/2013, Presenting the Thermal Insulation Requisites for Buildings*; Diário da República: Portugal, 2015; Volume 207. (In Portuguese)
100. *SIA 180*; Protection Thermique, Protection Contre L'humidité et Climat Intérieur Dans les Bâtiments. SIA: Zürich, Switzerland, 2014.
101. de Freitas, V.P. (Ed.) *Support Manual for the Rehabilitation Project of Old Buildings*, 1st ed.; Ordem dos Engenheiros da Região Norte: Porto, Portugal, 2012. (In Portuguese)
102. Appleton, J. *Old Buildings Rehabilitation (In Portuguese)*, 2nd ed.; Editora ORION: Lisboa, Portugal, 2011.
103. Trechsel, H.R.; Bomberg, M.T. *Moisture Control in Buildings: The Key Factor in Mold Prevention*, 2nd ed.; ASTM International: Baltimore, MD, USA, 2009.
104. Sedlbauer, K.; Krus, M. A new model for mould prediction and its application in practice. In *2nd International Building Physics Conference*; Carmeliet, J., Hens, H., Vermeir, G., Eds.; CRC Press: Leuven, Belgium, 2003; pp. 276–282.
105. Ojanen, T.; Viitanen, H.; Peuhkuri, R. Modelling of Mould Growth in Building Envelopes—Existing Models, Discussion on Improvement Aspects, Sensibility Analysis. In Proceedings of the Annex 41 ST4, Porto, Portugal, 22–24 October 2007.
106. Sedlbauer, K. *Prediction of Mould Fungus Formation on the Surface of/and Inside Building Components*; Universität Stuttgart: Stuttgart, Germany, 2001.
107. WTA. *WTA-Merkblatt E 6-5-12/D: Innendämmung nach WTA II—Nachweis von Innendämmsystemen Mittels Numerischer Berechnungsverfahren, Germany*; WTA: London, UK, 2012.
108. Ashby, M.; Jones, D. *Engineering Materials: An Introduction to Their Properties and Applications*, 2nd ed.; Pergamon Press: Cambridge, UK, 1982.
109. Le, Q.X.; Dao, V.T.; Torero, J.L.; Maluk, C.; Bisby, L. Effects of temperature and temperature gradient on concrete performance at elevated temperatures. *Adv. Struct. Eng.* **2018**, *21*, 1223–1233. [CrossRef]
110. Zheng, R.; Janssens, A.; Carmeliet, J.; Bogaerts, W.; Hens, H. An evaluation of highly insulated cold zinc roofs in a moderate humid region—Part I: Hygrothermal performance. *Constr. Build. Mater.* **2004**, *18*, 49–59. [CrossRef]
111. Šadauskiene, J.; Monstvilas, E.; Stankevicius, V. The impact of exterior finish vapour resistance on the moisture state of building walls. *Technol. Econ. Dev. Econ.* **2007**, *13*, 73–82. [CrossRef]
112. Hamid, A.A.; Wallentén, P. Hygrothermal assessment of internally added thermal insulation on external brick walls in Swedish multifamily buildings. *Build. Environ.* **2017**, *123*, 351–362. [CrossRef]
113. Pedroso, M.; Flores-Colen, I.; Silvestre, J.D.; Gomes, M.G. Study on the thermal conductivity performance of cement-based composites incorporating natural organic fibres. In *39 th Cement and Concrete Science Conference 2019*; Ball, R., Dams, B., Ferrandiz-Mas, V., Ke, X., Paine, K., Tyrer, M., Walker, P., Eds.; University of Bath: Bath, UK, 2019; pp. 112–115.
114. Hansen, E. RIBuild—Internal Insulation in Historic Buildings—Horizon 2020 Project. 2020. Available online: https://www.ribuild.eu/ (accessed on 28 September 2020).
115. Rubel, F.; Kottek, M. Observed and projected climate shifts 1901-2100 depicted by world maps of the Köppen-Geiger climate classification. *Meteorol. Zeitschrift.* **2010**, *19*, 135–141. [CrossRef]
116. Graham, J.; Berardi, U.; Turnbull, G.; McKaye, R. Microclimate Analysis as a Design Driver of Architecture. *Climate* **2020**, *8*, 72. [CrossRef]
117. Freitas, A.M.; de Freitas, V.P. *Miranda, Main Pathologies Associated with External Thermal Insulation Composite Systems (ETICS)—LFC-IC-282A-2014*; APFAC: Porto, Portugal, 2014. (In Portuguese)

energies

MDPI

Article

Simplified Guidelines for Retrofitting Scenarios in the European Countries

André Furtado [1,*], Hugo Rodrigues [2] and Humberto Varum [3]

1 CERIS, Instituto Superior Técnico, 1049-001 Lisboa, Portugal
2 RISCO, Universidade de Aveiro, 3810-193 Aveiro, Portugal
3 CONSTRUCT, Faculdade de Engenharia da Universidade do Porto, 4200-465 Porto, Portugal
* Correspondence: andre.furtado@tecnico.ulisboa.pt

Abstract: A large part of the European building stock was built before implementing the recent energy and structural codes, resulting in buildings characterized by deficiencies in terms of comfort, energy savings and structural safety. The retrofitting and rehabilitation of the existing building stock need to be adequately performed, aiming to improve the seismic and energy performance simultaneously. The work summarized here is dedicated to defining priority scenarios for buildings' retrofitting to improve the seismic safety and energy efficiency of the European Union (EU) building stock. First, the state of the EU building stock is analysed in terms of buildings' age, types of structures, energy efficiency, energy consumption and energy poverty. Then, the EU climate demands are presented, namely the regions with higher temperature variations, i.e., heating or cooling degree days. The EU seismic risk is also presented and discussed in terms of average annual losses, average annual economic losses and average annual life losses. Based on these input parameters, nine seismic–climate regions in the EU are proposed using a simplified approach. Finally, retrofitting scenarios are proposed for two types of buildings (i.e., masonry and reinforced concrete) based on their seismic–climate region.

Keywords: building stock; climate indicators; seismic retrofitting; energy retrofitting

Citation: Furtado, A.; Rodrigues, H.; Varum, H. Simplified Guidelines for Retrofitting Scenarios in the European Countries. *Energies* **2023**, *16*, 2408. https://doi.org/10.3390/en16052408

Academic Editor: Paulo Santos

Received: 14 February 2023
Revised: 26 February 2023
Accepted: 1 March 2023
Published: 2 March 2023

1. Introduction

A large part of the European building stock was built before implementing the recent energy and structural codes, resulting in buildings characterized by deficiencies in terms of comfort, energy savings and structural safety [1]. Regarding the last aspect, it is particularly relevant that most of the buildings located in seismic-prone regions were designed before the enforcement of current seismic regulations and with no modern concepts of antiseismic detailing and philosophies, such as capacity design. Thus, the structural safety of these buildings may not be satisfactory when subjected to seismic actions, but it could also hamper any refurbishment investment in the case of an earthquake [2]. Therefore, the transition toward a neutral carbon society needs to go through the structural strengthening of the existing building stock regarding seismic actions.

According to a recent report by the European Commission, the construction industry is accountable for 36% of carbon dioxide emissions, 40% of energy usage and 55% of electricity consumption in the European Union (EU) [3]. The majority of energy consumption and carbon dioxide emissions stem directly from the heating and cooling of buildings. The inadequate energy performance of existing buildings in the EU can be attributed to the belated implementation of the first energy codes for buildings, which only became official in the 1970s, by which point approximately 66% of the current EU building stock had already been constructed. Improving the energy efficiency of buildings can play a pivotal role in realizing the objectives outlined by the United Nations and the New European Green Deal, namely, reducing emissions of greenhouse gases that contribute to climate change to zero by 2050.

Roughly 40% of buildings in the EU are situated in seismic zones and constructed with inadequate safety measures, with 65% of these structures requiring both seismic and energy retrofitting. While independent retrofitting approaches for seismic [4] or energy purposes [5,6] are accessible and frequently utilized, comprehensive SpEU techniques that integrate both aspects are yet to be developed, validated and implemented. The most common insulation building materials used to improve a building's energy efficiency can be classified into three categories: (a) conventional insulation materials [7]; (b) insulation materials including nano-insulation materials [8]; and (c) smart insulation materials such as phase-change materials [9]. The economic issues and limitations lead to the need to use low-cost and conventional materials, such as external thermal insulation composite systems for infilled reinforced concrete (IRC) buildings' envelopes [10]. However, this technique only provides energy upgrading, neglecting the seismic safety of the building. An adequate upgrade of this technique would have great potential due to the low unitary cost of the intervention.

Concerning the strategies available to reduce the IRC buildings' seismic vulnerability, three different approaches are available: (a) global structure retrofitting (i.e., introducing steel braces or reinforced concrete (RC) shear walls or energy dissipation devices) [11]; (b) local retrofitting [12]; and (c) combining global and individual retrofitting measurements. The choice of the most suitable techniques depends on the desired performance standards as well as economic considerations, and potentially other non-technical factors. Some research works were recently carried out on strengthening solutions to reduce the seismic vulnerability and collapse probability of masonry infill walls (MIW) [13], using techniques such as fibre-reinforced polymers [14], engineered cementitious composites and textile-reinforced mortar (TRM) [15,16].

Various performance assessment methodologies are only available for a specific field, i.e., seismic safety and energy performance, and most of them were developed for new buildings. Seismic vulnerability assessment tools are based on a quantitative four-step evaluation consisting of hazard, structural, damage and loss analysis [17]. Several detailed [18] and simplified [19] numerical tools are available to simulate the behaviour of non-strengthened IRC building structures. However, they need proper calibration and validation to assess their capabilities to simulate the behaviour of retrofitted structures with SpEU solutions. The energy/environmental efficiency assessment is carried out through hygrothermal, life cycle (LC) analysis [20,21] and LC energy assessment [22]. Building sustainability assessment is based on European (DGNB) [23] or non-European rating systems (BREEAM) [24]. A holistic performance assessment methodology that combines sustainability, energy and structural performance indicators for existing IRC buildings is missing. It will allow decision-making bodies to identify buildings requiring renovation and to optimise those interventions.

Consequently, enhancing the current building inventory to become more sustainable, energy-efficient and resilient is crucial. The substantial investments earmarked over the next ten years for achieving a climate-neutral society by 2050 (EUR 1 bn in the EU; EUR 145 M in Portugal) under the European Green Deal demonstrate the paramount significance of this issue for the future of our communities. These strategies must also be consistent with the Sendai Framework Action Plan for Disaster Risk Reduction 2015–2030 and the United Nations' 2030 Agenda to fulfil all of the outlined objectives.

Based on this motivation, this research aims to define priority scenarios for buildings' retrofitting to simultaneously improve the seismic safety and energy efficiency of the EU building stock. First, the state of the EU building stock is analysed in terms of buildings' age, types of structures, energy efficiency, energy consumption and energy poverty. Then, the EU climate demands are presented, namely, the regions with higher temperature variations, i.e., heating or cooling degree days. The EU seismic risk is also presented and discussed in terms of average annual losses, average annual economic losses, and average annual life losses. Based on these input parameters, nine seismic–climate regions in the EU are proposed using a simplified approach. Finally, retrofitting scenarios are proposed for two types of buildings (i.e., masonry and reinforced concrete) based on their seismic–climate region.

The definition of the priority regions for which the combined seismic plus energy retrofitting has more impact on the building's performance and on the society as well requires the correlation of three different data inputs: (i) EU building stock characteristics (i.e., buildings age, type of constructions (i.e., masonry or reinforced concrete, buildings' distribution, energy poverty and energy consumption); (ii) EU climate (i.e., heating and cooling degree days); and (iii) EU seismic risk (average annual losses, average annual economic losses, average annual life losses), as shown in Figure 1. The data concerning these three inputs are presented and discussed in Section 2. This information will support the definition of seismic–climate scenarios in the EU using a simplified approach. Finally, retrofitting scenarios will be proposed based on the seismic–climate location in the EU.

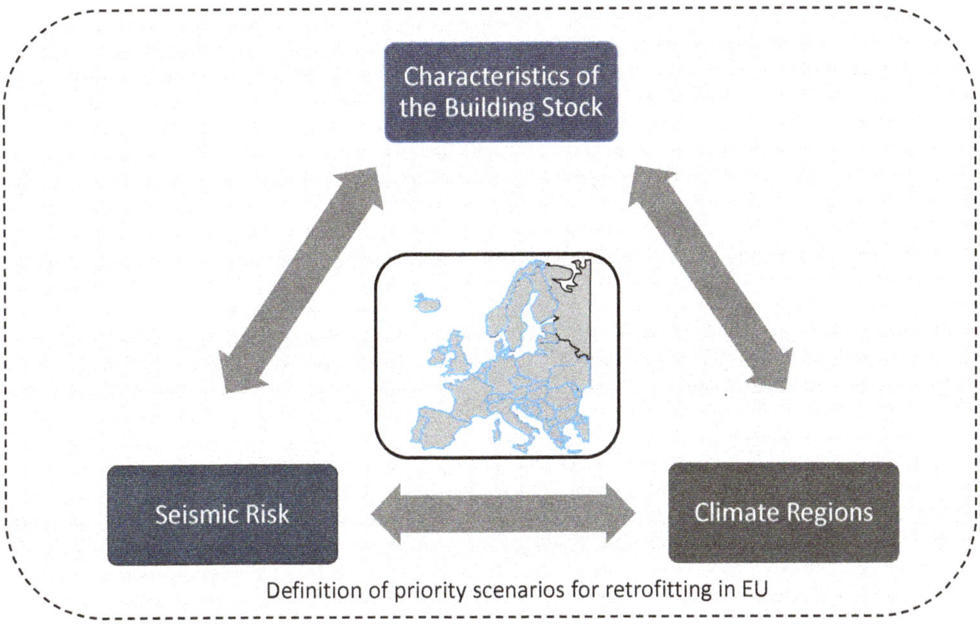

Figure 1. Flowchart of the identification process of scenarios for retrofitting interventions in the EU.

2. Overview of the EU Building Stock

2.1. General Characteristics

According to public data from the EU Buildings Observatory, a majority of buildings in the EU are residential and were erected prior to 1969 [25]. A significant portion of Europe's building inventory is over 50 years old, with many structures still in use that are over a century old. Over 40% of residential buildings were constructed before the 1960s, when energy regulations for buildings were limited. Several factors impact the energy efficiency of buildings, including the performance of heating systems and building envelopes, climatic conditions, building attributes and societal conditions. Information regarding typical heating consumption levels for the existing stock by age demonstrates that the most substantial potential for energy savings is associated with older buildings.

In certain instances, buildings constructed during the 1960s exhibit inferior performance compared to those from earlier periods. The historical U-value data reinforces the insufficiency of insulation in older buildings, which is unsurprising given the limited insulation standards during that era. Countries with the highest proportion of older buildings include Denmark, Sweden, France, the Czech Republic and Bulgaria, with a significant construction boom between 1961 and 1990.

About half of the EU building stock is older than 50 years, which means that it has already completed its conventional life and may need retrofitting or a replacement of building components due to durability-related issues.

The first edition of the European seismic codes (e.g., Eurocodes) in the EU was published in 1990, meaning about 77% were built before [1], and in several countries the implementation was done over the years after. Based on that, it may be concluded that the existing EU building stock may need both seismic and energy retrofitting since they may be inefficient and/or seismically vulnerable.

The age distribution of existing buildings and the proportion of new buildings within the total stock serve as reliable indicators of the overall efficiency of the building stock. A higher proportion of newly constructed dwellings built with more efficient standards generally corresponds to a better overall energy performance of the building stock. Figure 2 depicts the distribution of buildings' construction in the years up to 2014, revealing that the majority of buildings in most EU countries are over fifty years old. However, in some countries such as Spain, Cyprus and Ireland, there is a noticeable percentage of recent buildings (i.e., less than 20 years old).

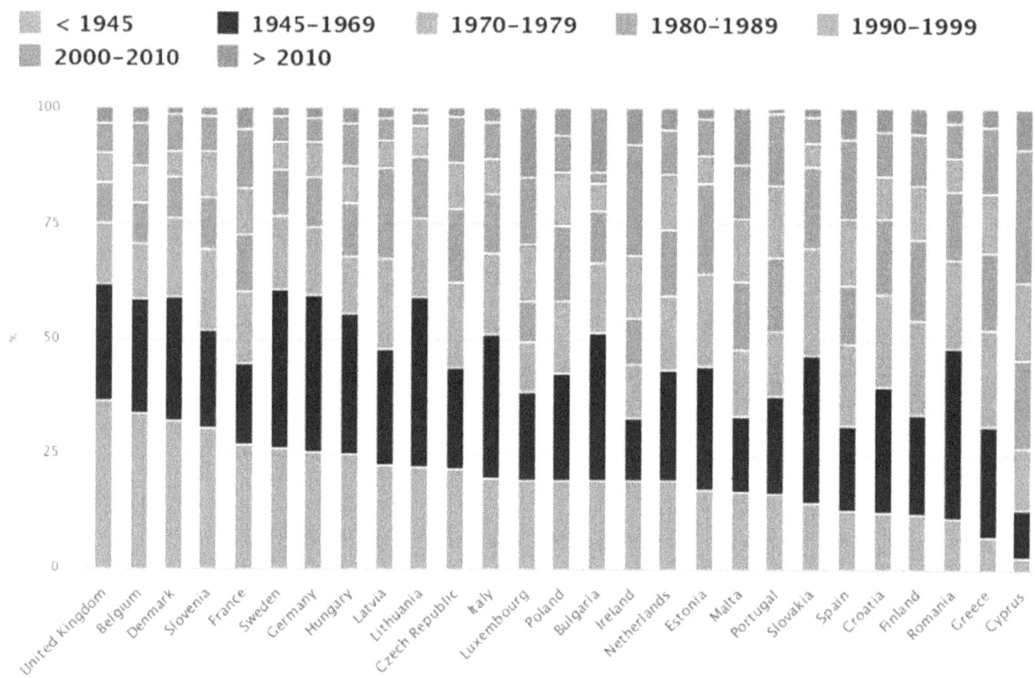

Figure 2. Distribution of building construction by year in EU countries up to 2014. (Sources: EU Buildings Factsheets).

Concerning the buildings constructed before 1945, Belgium (34%), Denmark (32%), Slovenia (30%) and France (27%) have the most significant percentage. Romania (37%), Lithuania (36.9%), Sweden (34.3%) and Germany (34.1%) are the countries with a larger percentage of buildings built between 1945 and 1969. Between 1970 and 1979, all the countries presented similar rates, but the highest one was achieved by Slovakia (23.3%). Between 1980 and 1989, the percentage of buildings in the EU countries varies between 9 and 20%. The highest rate is observed in Cyprus. Between 1990 and 1999, the same observation can be drawn. Cyprus (17%), Malta (14%) and Ireland (13.3%) are the countries with large percentages of buildings. Finally, Cyprus (>29%) and Ireland (>15%) have the highest rates of buildings built after 2000.

According to the European Union Commission, multi or single dwellings (residential) distribution differs from country to country. On average, about 52.7% are single-family dwellings and 47.3 are multi-family dwellings. Countries such as Ireland, Belgium and the Netherlands have a high percentage of single-family dwellings (87%, 73% and 70%, respectively). On the other hand, Estonia, Italy and Spain have the most multi-family dwellings (75%, 74% and 71%, respectively). The type of dwelling directly impacts the space heating energy performance since different insulation characteristics imply different specific space heating consumption (due to other wall areas in contact with the outdoors).

2.2. Seismic Vulnerability of the Building Stock

Masonry buildings are prevalent in Europe and are, on average, older than RC ones. For example, the Italian building stock comprises about 3.7 million RC buildings and 7.2 million masonry buildings. Moreover, by 1981, about 86% of masonry buildings were already constructed. The year 1981 is a year of a massive seismic classification due to the consequences of the 1980 Irpinia-Basilicata destructive earthquake. Additionally, the first seismic design code for masonry structures was issued only in 1987, which means that almost the totality of masonry buildings were designed without any seismic criteria. Recent earthquakes such as the one in 2016 demonstrated the problem dimension at the Italian scale [26]. The same approach can be taken at the European scale, even though northern European countries have lower seismic hazards than the European Mediterranean ones [27].

The low strength of the masonry units (both stone or brick) and mortar affects the wall capacity under vertical and seismic loads. Additionally, the irregular disposition of the masonry units and the excessive mortar beds' thickness reduces the wall's vertical and seismic capacity [28]. On the other hand, the irregular wall arrangement due to the dimension of stone/brick elements and the weak connection among the different wall leaves are responsible for the masonry disintegration when subjected to lateral loads. It is pretty standard in older buildings.

The poor connection between transverse walls (i.e., corner walls), the connection between walls and the floors/slabs (diaphragm effect) and the roofs contribute to avoiding the possibility of developing a "box behaviour". The retrofitting of a masonry structure must ensure that the "box behaviour" effect is guaranteed.

All these deficiencies are responsible for activating local mechanisms, such as wall overturning, inhibiting global seismic behaviour, which is usually more efficient [29].

Structural failures due to earthquakes do not affect environmental measures, such as CO_2 emissions or energy consumption, in a direct way as they are rather abrupt events. It might not be intuitive to include them when examining energy reduction policies. However, their impact on the local economies can be devastating. When quantified with absolute economic measures, it can be seen that the economic losses they can bring about can be comparable to those due to energy deficiencies.

One of the most relevant problems was the lack of connection between transverse walls, roofs or floors. The "box behaviour" is fundamental to achieving good seismic performance of a masonry structure.

Undoubtedly, structural deficiencies play a much more important role in the seismic-prone areas of the European south. Buildings are frequently subjected to seismic events in countries that are located in seismic-prone regions. In such places, structural deficiencies are often brought to the surface by major seismic events. Moreover, seismic standards have improved considerably during the last 50 years, both in terms of the prescribed loads and detailing measures, as opposed to concrete and structural steel codes, for example, which have received much fewer significant updates.

The Eurocode 8 [30] provides a classification for structural elements of RC structures, differentiating them as either structural or non-structural. Structural elements are further classified into primary members (SP) or secondary members (SS). Primary members (SP) are considered part of the structural system that can withstand seismic demands, and are designed and detailed for earthquake resistance. Secondary elements, however, are not considered part of the seismic-resisting system, and their strength and stiffness against seismic actions are neglected. They are designed and detailed to only support gravity loads when subjected to seismic actions, and are not required to meet all Eurocode 8 [30] requirements but are designed and detailed to maintain the support of gravity loads when subjected to the displacements caused by the seismic actions.

Non-structural elements (NS) are composed of various architectural, mechanical or electrical components and systems that are not load-bearing elements in the seismic design due to their lack of strength or the way they are connected to the structure. Among these elements, masonry infill walls play a significant role in the seismic performance of RC buildings. Eurocode 8 [30] recognizes that infill panels are part of the non-structural-elements group.

Following the last major earthquakes in the EU, different types of damage have occurred and affected the RC structures [31]. During seismic events, reinforced concrete structures can experience different types of damage, including but not limited to, inadequate detailing of stirrups and hoops (Damage Type 1), issues with longitudinal reinforcement detailing (Damage Type 2), reduced shear and flexural capacity of elements (Damage Type 3), insufficient shear capacity of structural joints (Damage Type 4), strong-beam weak-column mechanism (Damage Type 5), short-column mechanism (Damage Type 6), structural irregularities (Damage Type 7), pounding (Damage Type 8), damage to secondary elements (Damage Type 9) and damage to nonstructural elements (Damage Type 10).

The structural primary elements are associated with the first eight damage types, while the ninth damage type is related to secondary structural members and the tenth damage type specifically pertains to infill walls and other non-structural elements. Based on the post-earthquake damage survey assessment, it can be concluded that there is an interdependence among the last five types of damage. The damage observed during the post-earthquake field trips indicates that masonry infill walls cannot be considered as non-structural or secondary elements and cannot be ignored from the expected building seismic behaviour [32,33]. Based on these observations, it is apparent that infill walls, especially those located in the building envelope, play a crucial role as they can alter the overall structural behaviour of the building. Proper retrofitting is necessary to prevent out-of-plane (OOP) collapse. Infill walls were responsible for over 30% of the rehabilitation costs of buildings damaged by earthquakes [34].

2.3. *Energy Deficiencies of the Building Stock*

The poor buildings' energy efficiency is due to the later development and implementation of regulations and functional requirements in European countries. The first building energy codes were published in the 1960s in the Scandinavian countries [35]. Later, other countries have progressively introduced some regularly updated standards, especially to match European directives. From the energy perspective, the first Energy Performance of Buildings European Directive was issued in 2002 [36], although most European countries already have developed their national standards since the beginning of the 1990s. Once again, the masonry building stock at the European level has been built mostly without any energy efficiency rule.

The most significant deficiencies concerning the energy efficiency are due to inadequate thermal insulation provided by the building envelope and due to the roof and windows characteristics. Additionally, inefficient mechanical services providing heating, cooling and domestic hot water needs are responsible for high energy consumption, especially in older buildings. Finally, old lighting systems can further increase the energy need and CO_2 emissions.

An essential aspect of masonry buildings is related to the fact that load-carrying components constitute the building envelope. Therefore, some limitations are present in the insulation interventions, contrarily to RC buildings for which the building envelope (infills) can even be replaced.

According to the latest EU report [25], the energy used for space heating seems to be the most important end-use in the residential sector (68%). Countries such as Italy, Poland, Belgium, Luxembourg and Finland have a high percentage of energy consumption for space heating (>69%). The average percentage of the energy consumption in the EU for space heating is 67.74%, which is quite significant. On the one hand, the climate conditions can be strongly linked with higher needs for space heating (such as Finland, Denmark and Poland). On the other hand, the poor energy efficiency of buildings plays a significant role, particularly in countries such as Italy or Croatia that have a more moderate climate. Space cooling emerges with high importance in Malta (15%), Bulgaria (5%) and Cyprus (5%). The other ways of consuming energy are water heating (second most relevant), lighting, appliances (third most relevant) and cooking (fourth most relevant).

The average annual specific consumption per m^2 for all types of buildings in the EU was approximately 180 kWh/m^2 in 2013. However, there are variations among countries, with Malta having the lowest consumption at 47 kWh/m^2, followed by Portugal and Cyprus at 70 kWh/m^2, while Romania has the highest consumption at 300 kWh/m^2 (or 290 kWh/m^2 in Latvia and Estonia), which is significantly higher than the EU average. It is worth noting that even countries with similar climates exhibit substantial differences in consumption, such as Sweden with 210 kWh/m^2, which is 18% lower than Finland. The differences can be attributed in part to climatic conditions and statistical definitions.

The final energy consumption in households in 2020 is shown in Figure 3. This parameter focuses on the energy spent in households for heating purposes and how the amelioration of buildings can contribute to energy-saving plans. The data were extracted by Eurostat, which collects information on total energy consumption in households split by fuel category. This parameter is highly related to the climate conditions and the country's population size. Above the average European line (7618), it is possible to find Germany (57,743), France (39,619), Italy (31,138), Poland (20,993), Spain (14,448) and the Netherlands (9460). The countries with low final energy consumption are Albania (504), Cyprus (362), Luxembourg (461), Malta (313) and Montenegro (243).

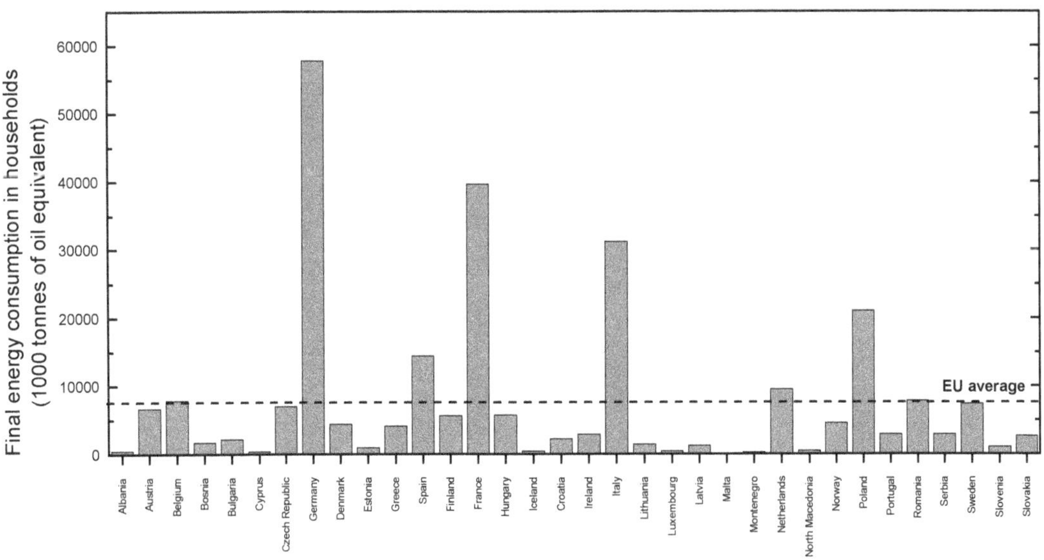

Figure 3. Final energy consumption in households in 2020 (Sources: EU Buildings Factsheets).

One of the most critical parameters under discussion is the percentage of the population unable to keep their home adequately warm (i.e., energy poverty). According to Eurostat, the average rate in Europe was 7.3% in 2018, which means that more than fifty million people were in this position. From the country-by-country analysis, it is possible to highlight the high percentages in Bulgaria (33.7%), Lithuania (27.9%), Greece (22.7%), Cyprus (21.9%), Portugal (19.4%) and Italy (14.1%).

Based on the above results, it can be concluded that addressing the above issues of older buildings via suitable seismic plus energy retrofitting solutions seems to be very promising in reducing the energy consumption and seismic vulnerability of the current building stock. That is why many states are already offering incentives to citizens to renovate their dwellings based on certain goals. An energy upgrade of a given building is an investment that can be achieved at reasonable costs and will have an immediate effect on its consumption. Nonetheless, as stated earlier and explained further below, an energy upgrade will not be effective when applied in a building of questionable structural integrity due to safety reasons.

2.4. *Brief Considerations on the Climate Conditions*

The Köppen–Geiger climate classification is the most widely used climate classification system. This classification divides the climate into five main climatic groups based on seasonal precipitation and temperature patterns [1]. The main five groups are tropical, arid, temperate, continental and polar.

The parameters used to understand the building energy requirements are the heating degree days (HDD) and cooling degree days (CDD). They are quantitative indices that reflect the demand for energy to heat or cool buildings. Both variables are derived from measurements of the outside temperature. According to Eurostat, the baseline temperatures for HDDs and CDDs are 15 °C and 24 °C, respectively. The outside temperature difference with the baseline temperature gives the HDD and CDD indexes. The sum of the index over the year results in the annual HDD and CDD. According to the Eurostat database (Eurostat), the available data for the HDD and CDD started in 1979 up to 2020. It should be underlined that these indicative indexes are related to average country values, which means that there are regions in each country that may have higher HDD and/or CDD than the average value estimated by Eurostat.

The correlation between the average national HDD and CDD indexes in 2020 is shown in Figure 4a. As expected, countries with high HDD such as Denmark or Sweden have at the same time a low CDD. The same can be said regarding countries with high CDD, such as Malta or Cyprus, that simultaneously have a short number of HDD. The most critical scenario can be described with countries with medium/high CDD and HDD at the same time, such as Hungary, Bulgaria, Italy, Spain and Greece. They are expected to consume more energy for space heating and cooling.

Another critical issue that can be highlighted is that the countries with higher HDD have, at the same time, the lowest percentage of the population unable to keep homes adequately warm (e.g., Finland and Sweden). Typically, this trend is observed in the north European countries. However, some countries present dangerous results concerning this relationship, i.e., medium/high HDD and a high percentage of the population unable to keep homes adequately warm. The highlighted countries are Lithuania, Romania, Bulgaria, Italy and Greece. Energy retrofitting measures would greatly benefit these countries, allowing them to find a balance between seasons without needing to spend much energy on space heating or cooling.

No direct relationship between the HDD and the building energy consumption can be found, as shown in Figure 4c. Nonetheless, countries such as Spain, Poland, Italy, France and Germany present a higher ratio between these two parameters, which means that energy retrofitting would help to reduce energy consumption for space heating due to the high HDD.

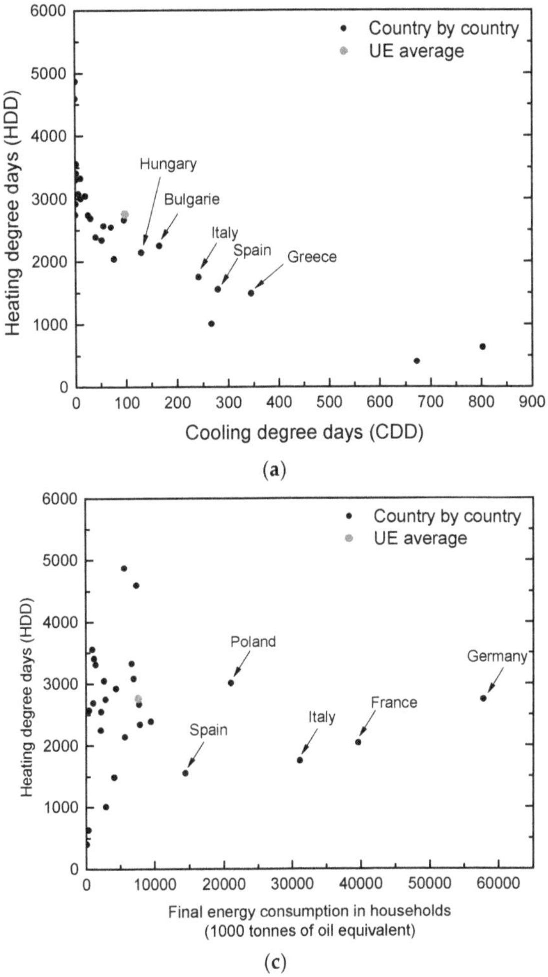

Figure 4. Climate data and building energy performance: (**a**) HDD vs. CDD; (**b**) HDD vs. energy poverty; (**c**) HDD vs. building energy consumption; and (**d**) CDD vs. building energy consumption (Sources: EU Buildings Factsheets).

In the same way, no clear trend is observed in the analysis of CDD versus the building energy consumption. Again, countries such as Spain, Italy and France presented a high ratio between these two parameters.

From the analysis of the HDD and CDD range distribution over the EU, a considerable variation can be observed for each country, which means that the analysis of the EU climate should be performed region by region. For example, the average HDD in Bulgaria is 2273, but the maximum value of 3036 is reached in the Smolyan region (BG424 in NUT 3 subdivision), about 34% higher. The same can be pointed out concerning the CDD since the average value found for Bulgaria is 160, but the maximum value of 379 was reached in the Yambol region (BG 343 in NUT 3 subdivision), about 136% higher. This observation is more notorious in the HDD when compared with the CDD ones.

In Italy, the average HDD found for the country was 1646, but the minimum and maximum values found were 800 (Cagliari) and 4119 (Valle d'Aosta/Vallée d'Aoste), about

−52% and +150%, respectively. The average CDD found was 246, and the minimum and maximum values found were 1.3 (Sondrio) and 457 (Napoli), which are −99% and +50%, respectively.

A box plot concerning the heating (HDD) and cooling degree days (CDD) according to NUTS 3 subdivision is presented in Figure 5a,b. The box plot, also called a box and whisker plot or box chart, represents key values from summary statistics. Each Y column of data is represented as a separate box in each of these plots. The countries' nomenclatures supply the X-axis tick labels. By default, the box is determined by the 25th and 75th percentiles. The whiskers are determined by the 5th and 95th percentiles. Additional values are represented, including the minimum, median, mean, maximum, the 1st and 5th percentiles and the 95th and 99th percentiles.

From this analysis, it can be seen that the countries with larger HDD variations are Greece, Spain, Italy, Austria, Portugal and Romania. Concerning the CDD variations, the countries with larger dissimilarities are Bulgaria, Greece, Spain, Italy, Portugal and Romania.

Figure 5c presents the ratio computed between the minimum and average HDD values, maximum and average HDD values, minimum and average CDD values and maximum and average CDD values. The most critical ones are the ratio between the maximum (HDD and CDD) and the average value found for each country since it is directly related to higher needs for energy consumption. The countries with higher variation in terms of HDD (i.e., the ratio between the maximum and average HDD) are Italy (+150%), Portugal (+90%), Greece (+78%) and Spain (+71%). Concerning the higher variation in terms of CDD (i.e., the ratio between the maximum and average HDD), it can be observed that the higher variation can be found in Austria (+700%), Czech Republic (+316%) and Denmark (+316%). However, the effect of this variation is more important for higher average CDD and/or HDD values.

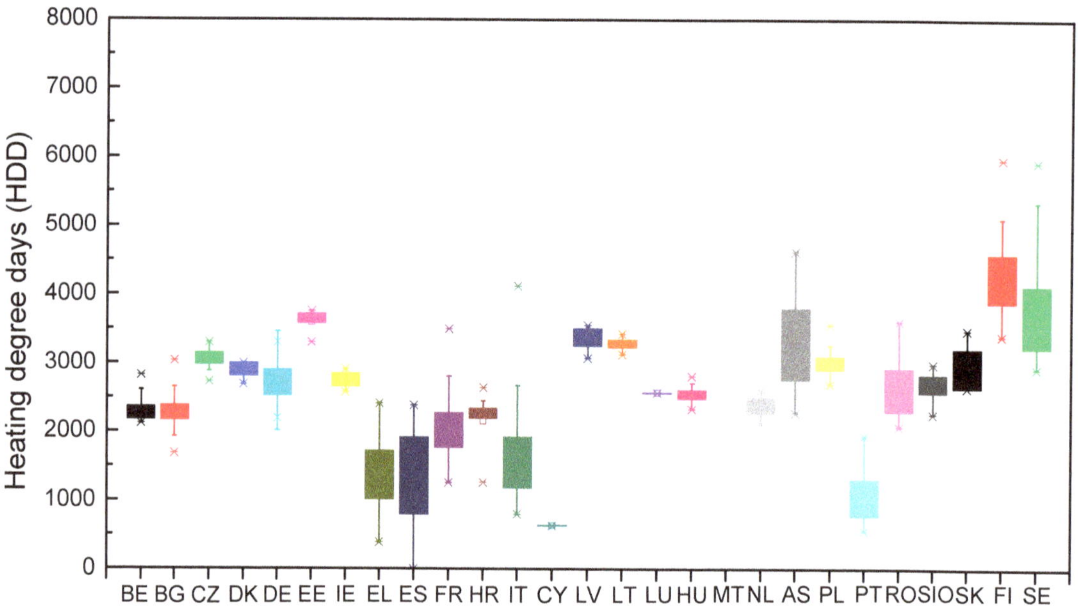

(a)

Figure 5. *Cont.*

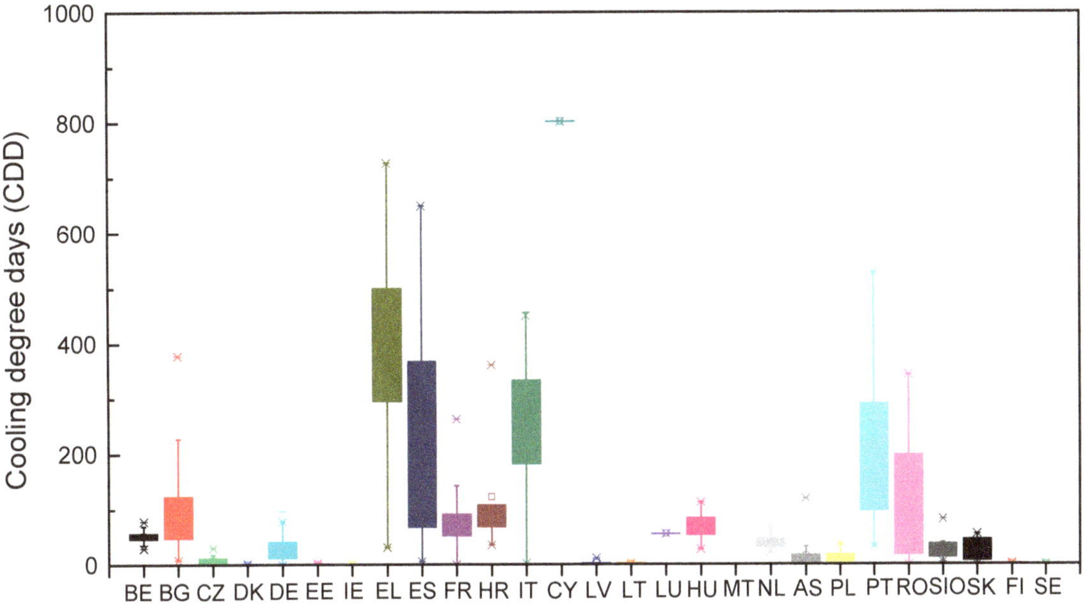

(**b**)

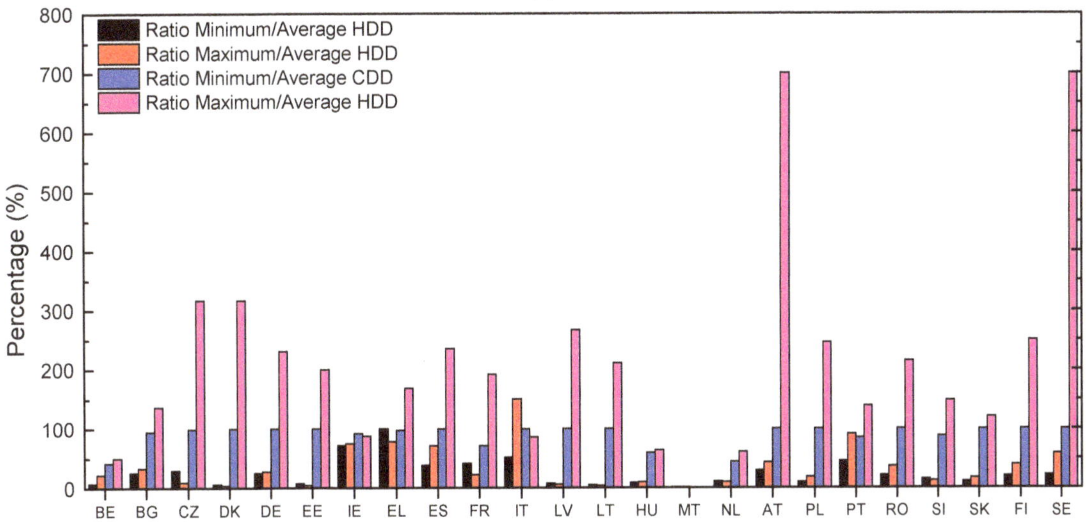

(**c**)

Figure 5. (**a**) Heating degree days; (**b**) cooling degree days; and (**c**) ratio between minimum/maximum HDD and CDD values and the average ones for each country (Sources: EU Buildings Factsheets).

The identification of the countries that most need energy retrofitting measures needs to be performed by interpolating the HDD, energy poverty (i.e., the percentage of the population unable to keep homes adequately warm) and buildings' energy consumption. The countries that present higher values in these three categories are the ones that should be the focus of energy retrofitting. According to Pohoryles et al. [37], HDD values are more significant since heating and hot water account for the high energy consumption in Europe.

Future investigations must be performed towards a European Map that presents the distribution of the buildings' energy efficiency based on the climates and characteristics of buildings. Then, it will be possible to compute the annual average losses due to the energy required for heating and cooling space.

2.5. Seismic Risk

The buildings that most need seismic retrofitting depend on the respective seismic risk. It is estimated that 30% of EU buildings are located in areas of moderate seismic hazard (where the design peak ground acceleration, PGA, is at least 0.1 g). The seismic risk can be computed as the product between seismic hazard, exposure and vulnerability.

Gkatzogias et al. [38] presented the EU seismic risk results in which average annual losses (AAL) in USD per country were computed. It is possible to observe that the country with the highest AAL is Italy by far. After that, Greece, Romania, France and Germany come in the following positions. The ranking of AAL is strongly influenced by the level of seismic hazard but also the country's size and the exposure value, hence the average annual loss ratio (which is the AAL divided by the replacement value, AALR).

The analysis of the AALR highlights countries where the losses are high relative to the exposure value. So, countries with lower construction costs are often higher. To look at the areas where absolute losses are expected to be high but not necessarily due to the higher replacement costs of the buildings, another risk metric has been considered, namely the average annual loss per building.

From the analysis of the AALR, it can be observed that the ten countries with higher AALR are Cyprus (1.32‰), Greece (1.11‰), Romania (0.93‰), Italy (0.76‰), Bulgaria (0.58‰), Slovenia (0.39‰), Croatia (0.35‰), Austria (0.12‰), Hungary (0.11‰), Portugal (0.11‰) and Slovakia (0.11‰). The countries with lower AALR are Finland, Ireland, Lithuania and Sweden.

Moreover, Gkatzogias et al. [38] computed the average annual economic losses ratio (AAELR). The countries with higher AAELR are Cyprus (1.16‰), Greece (0.90‰), Italy (0.58‰), Croatia (0.42‰) and Bulgaria (0.31‰). Conversely, the countries with low AAELR are Lithuania, Latvia, Poland, Sweden, Estonia, Denmark, Finland and Ireland. The comparison between the AAELR and AALR shows that the countries follow a linear trend (grey line). Romania is the country furthest from this trend, but not significantly.

In addition, Crowley [39] computed the average annual life loss (AALL) by the number of occupants and loss of life. This parameter was given per 100,000 occupants to avoid very low numbers. Because this metric is one of those proposed by the Sendai Framework for Disaster Risk Reduction, the countries with higher AALL are Greece (0.36), Cyprus (0.18), Croatia (0.17), Bulgaria (0.12), Italy (0.09), Austria (0.09), Slovenia (0.08) and Romania (0.07). From the relationship between the AALR and the AALL, shown in Figure 6a, it is possible to observe that Greece appears to be the country with the highest correlation (i.e., high AALR and high AALL). It can also be stated that Cyprus, Romania, Italy and Bulgaria present high AALR and medium–high AALL at the same time.

Finally, the last parameter that must be analysed is the average annual economic losses (AAEL) per building. Cyprus, Greece and Italy are clearly on the front line of this parameter, followed by Croatia, Austria, Slovenia and Spain. The countries with low AAEL per building are Estonia, Ireland and Finland.

A specific trend cannot be concluded from the analysis of the relationship between AALR and AAEL per building, shown in Figure 6b. For example, Cyprus and Italy present an increasing linear trend, i.e., the AALR increases with the increase in the AAEL per building. However, the same was not observed for other countries such as Romania, Bulgaria and Slovenia.

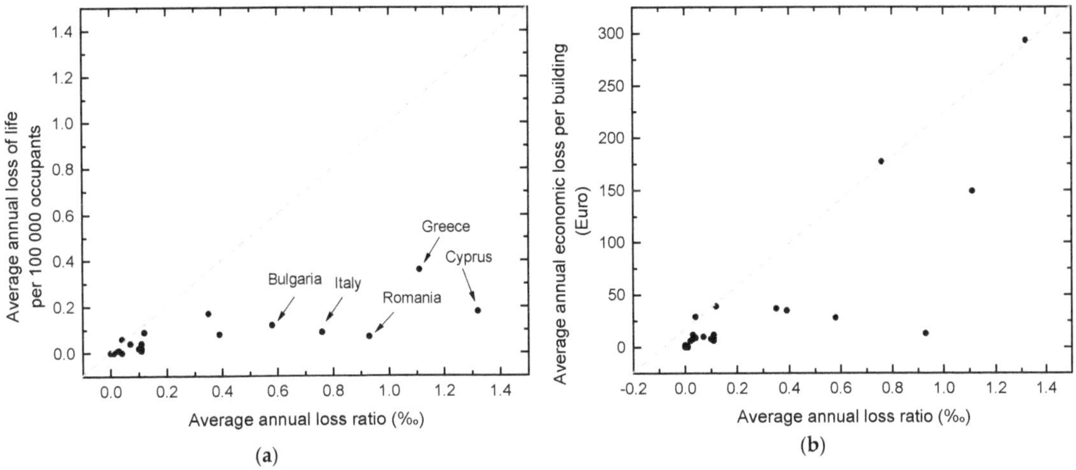

Figure 6. (**a**) Comparison between AALL and AALR; and (**b**) AAEL per building and AALR.

Furthermore, one example of regulation is the Greek Interventions Regulation, also known as KANEPE [40], which is a legal framework designed to regulate the provision of emergency liquidity assistance by the Bank of Greece to Greek banks. KANEPE was introduced in 2015 during a period of economic turbulence in Greece, and it establishes a set of criteria that must be met by banks in order to receive emergency funding. These criteria include the need for the bank to have sufficient collateral and the requirement for the bank to submit a business plan outlining how it will return to financial stability. The aim of KANEPE is to ensure that emergency funding is only provided to banks that have a viable long-term future and that the risks to the Greek taxpayer are minimized. This regulation is fundamental in a post-earthquake scenario.

3. Identification of Relevant Scenarios Based on Climate and Seismic Risk Maps

Combined seismic and energy upgrading is essential for countries exposed simultaneously to high seismic hazards and climatic exposures. Butenweg (2021) proposed a methodology to identify European countries and regions with a higher correlation between seismic and energy upgrading. A score-based approach was proposed to determine whether combined upgrading is required for a building in a specific region, depending on the seismic hazard level and the climatic conditions. The authors proposed a score-based approach to identify these countries. The score is calculated based on the seismic hazard level in terms of PGA and the indicators HDD and CDD for the climatic conditions.

The strategy adopted to identify the relevant scenarios for combined structural plus energy retrofitting herein proposed is based on the inputs from the EU seismic risk (i.e., average annual expected losses ratio), climate maps (heating and cooling) and characteristics of the EU building stock (total building energy consumption, energy poverty). First, the analysis is performed at the macro-level perspective, i.e., using country average values.

The relationship between the AALR and the HDD is shown in Figure 7a. It can be seen that several countries simultaneously have low AALR and high HDD (e.g., Finland and Sweden), which means that energy retrofitting would be adequate for this type of region. On the other hand, it is possible to observe countries with simultaneous low AALR and HDD (e.g., Malta and Portugal). Countries such as the Netherlands, Ireland and Denmark have medium–high HDD combined with low AALR. The most critical scenario is the combination of medium–high AALR with medium–high HDD, such as in Romania, Italy and Bulgaria.

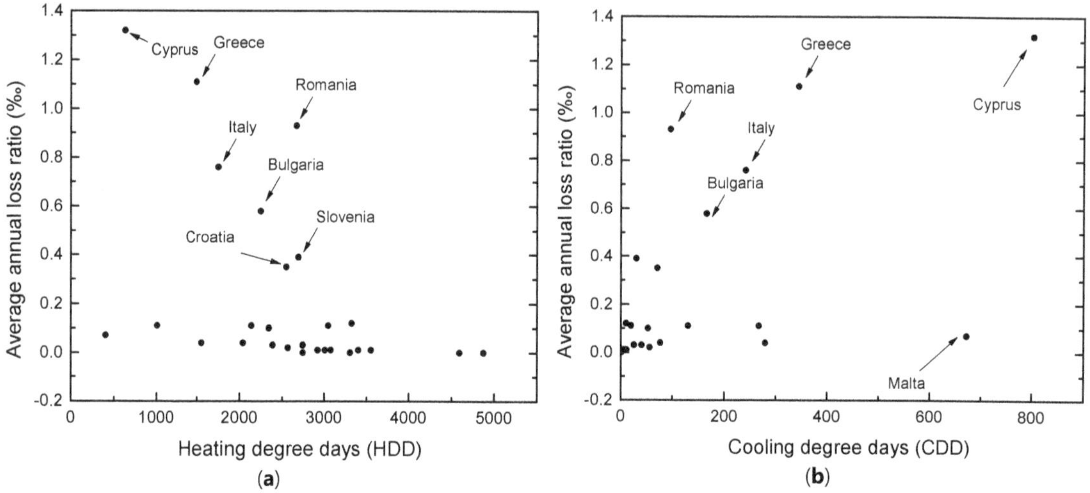

Figure 7. Definition of priority regions: (**a**) AALR vs. HDD; and (**b**) AALR vs. CDD.

The same observation can be made for the relationship between CDD and AALR, shown in Figure 7b. It can be observed that Cyprus is a country with high AALR and CDD simultaneously. After that, Greece, Italy, Bulgaria and Romania have higher AALR and CDD. The remaining countries, typically north and central European Countries, have low CDD and AALR. Malta is the second country with a higher CDD but has a low AALR.

The definition of climate plus seismic scenarios for each region of the EU is herein performed based on the two inputs: AALR and a climate indicator (CI). Nine different EU zones are herein proposed based on the combination of AALR and CI. The zones were defined based on the EU's AALR and CI range values. Three different levels of AALR are proposed: low (AALR < 0.30‰); medium (0.30‰ $\leq$ AALR < 0.60‰); and high (AALR $\geq$ 0.60‰). The CI was computed for each country according to Equation (1). Then, three different levels are proposed: low (CI < 1500); medium (1500 $\leq$ CI < 3000); and high (CI $\geq$ 3000). Thus, the seismic–climate matrix for the definition of the EU zone is presented in Figure 8.

$$CI = HDD + CDD \quad (1)$$

It should be stressed that CI greatly depends on the HDD and more minorly on CDD. The relationship between HDD and CDD justifies this, i.e., HDD is much higher in almost all countries than CDD. Since a significant part of the building energy consumption is due to heating spaces, highly dependent on the HDD, the CI herein proposed represents the climate severity representative of each country well. However, it is expected that the CDD will increase significantly in future years due to climate change. The characteristics of each combined seismic and climate scenario are presented in Table 1.

The results concerning the association of the EU countries to the respective seismic–climate zone (SCZ), using average country reference data, are presented in Table 2 and Figure 8b. Additionally, the seven countries with a higher percentage of the population unable to warm their homes were highlighted. One of the EU and United Nations' objectives is to reduce the population's portion under energy poverty conditions. For example, Portugal has a high percentage of energy poverty and is located in SCZ_A (i.e., low CI and low AALR). Usually, low-energy and seismic strengthening measures are recommended for zone A. However, in the case of Portugal, strong energy retrofitting measures are recommended to reduce the percentage of the population in energy poverty conditions.

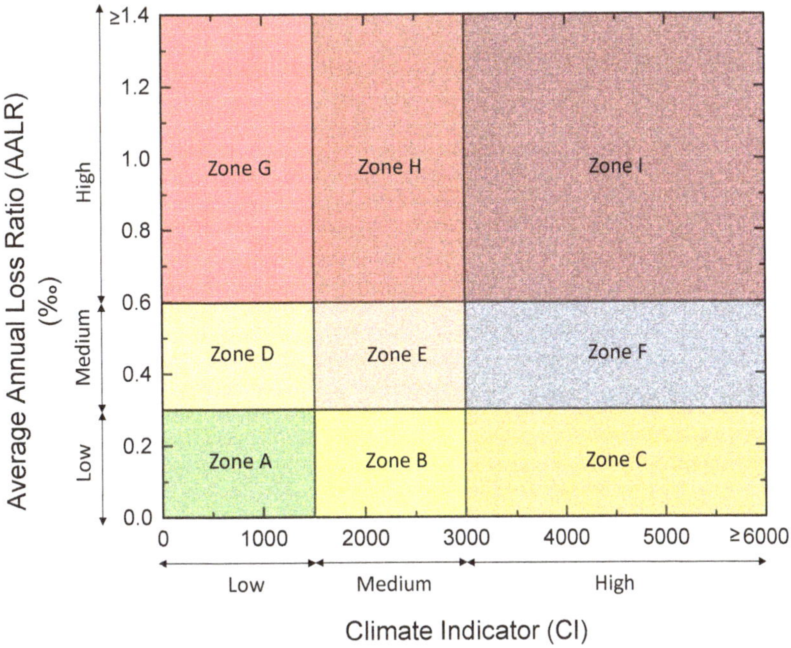

(a)

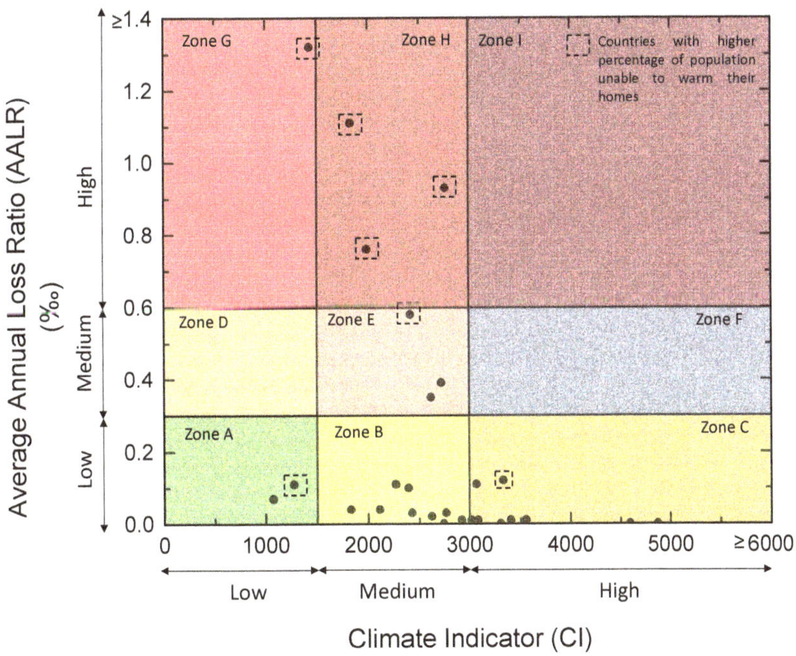

(b)

Figure 8. Seismic–climate zone matrix: (**a**) schematic layout; (**b**) average country values.

Table 1. Definition of seismic–climate zones (SCZ).

Seismic–Climate Zone	AALR (‰)	CI	General Recommendation on Retrofitting
A	<0.30	<1500	Low energy retrofitting Low seismic retrofitting
B	<0.30	1500 ≤ CI < 3000	Medium energy retrofitting Low seismic retrofitting
C	<0.30	≥3000	High energy retrofitting Low seismic retrofitting
D	0.30 ≤ AALR < 0.60	<1500	Low energy retrofitting Medium seismic retrofitting
E	0.30 ≤ AALR < 0.60	1500 ≤ CI < 3000	Medium energy retrofitting Medium seismic retrofitting
F	0.30 ≤ AALR < 0.60	≥3000	High energy retrofitting Medium seismic retrofitting
G	≥0.60	<1500	High energy retrofitting Low seismic retrofitting
H	≥0.60	1500 ≤ CI < 3000	High energy retrofitting Medium seismic retrofitting
I	≥0.60	≥3000	High energy retrofitting High seismic retrofitting

Table 2. Distribution of EU countries over the seismic–climate zones proposed.

Countries	AALR (‰)	HDD	CDD	CI	Percentage of the Population Unable to Warm Their Homes (%)	SCZ
Austria	0.12	3323	10	3333	1.6	C
Belgium	0.10	2340	52	2392	5.2	B
Bulgaria	0.58	2247	166	2413	33.7	E [A]
Cyprus	1.32	630	803	1433	21.9	G [A]
Czech Republic	0.01	3079	6	3085	2.7	C
Germany	0.03	2741	25	2766	2.7	B
Denmark	0.01	2921	0.6	2922	3	B
Estonia	0.01	3553	1.7	3555	2.3	C
Greece	1.11	1489	345	1834	22.7	H [A]
Spain	0.04	1553	279	1832	9.1	B
Finland	0	4871	0.4	4871	1.7	C
France	0.04	2038	76.4	2114	5	B
Hungary	0.11	2138	130	2268	6.1	B
Croatia	0.35	2547	70.5	2618	7.7	E
Ireland	0	2744	0	2744	4.4	B
Italy	0.76	1750	241	1991	14.1	H [A]
Lithuania	0	3305	1.06	3306	27.9	C [A]
Luxembourg	0.02	2567	55.6	2623	2.1	B
Latvia	0.01	3404	1.7	3406	7.5	C
Malta	0.07	402	672	1074	7.6	A
Netherlands	0.03	2386	40.1	2726	2.2	B
Poland	0.01	3006	11.2	3017	5.1	C
Portugal	0.11	1008	266	1274	19.4	A [A]
Romania	0.93	2666	96.4	2762	9.6	H [A]
Sweden	0	4593	0.08	4593	2.3	C
Slovenia	0.39	2691	29.8	2721	3.3	E
Slovakia	0.11	3047	20	3067	4.8	C

[A]–Country with a high percentage of population unable to warm their homes.

From the results, it can be seen that there are no countries located in zones D, F and I. Portugal and Malta are positioned in zone A. Zone B comprises Belgium, Germany, Denmark, Spain, France, Hungary, Ireland, Luxembourg and the Netherlands. Zone C includes Austria, Czech Republic, Finland, Estonia, Lithuania, Latvia, Poland and Slovakia. Zone E comprises Bulgaria, Croatia and Slovenia. Zone G includes Cyprus. Finally, Greece, Italy and Romania belong to zone H.

Once again, it should be underlined that the data used for selecting the seismic–climate scenario are based on average values found for each country. Different regions can be positioned in different SCZs in each country, depending on the seismic plus climate demands. The strategy herein recommended to perform this analysis involves interpolating this input data at the region level (i.e., using NUTS 1 or NUTS 3), as Gkatzogias et al. [38] indicated.

The identification of relevant seismic–climate zones herein proposed is based on the inputs from the seismic risk (i.e., average annual expected losses) and climate indicators (heating and cooling degree days). A detailed analysis was performed at a macro level (i.e., country level). Regional zonation must be a priority in the future for defining priority regions in the EU by correlating seismic and climate inputs for each admin level. Some regions are suggested for each seismic–climate zone based on seismic risk and climate indicators (Table 3). It should be noted that these are some examples and not the total number of regions in the EU.

Table 3. List of relevant regions for each seismic–climate zone.

CSZ	Country/Region
A	Porto (Portugal) Valletta (Malta)
B	Lagos (Portugal) Madrid (Spain) Barcelona (Spain) Montana (Bulgaria) Bratislava (Slovakia) Berlin (Germany)
C	Tyrolean Oberland (Austria) Norrbotten County (Sweden) Unterallgäu (Germany) Aosta (Italy) Krakow (Poland)
D	Lisboa (Portugal)
E	Primorsko-Goranska (Croatia) Olt County (Romania)
F	Kardzhali Province (Bulgaria) Suceava County (Romania)
G	Calabria (Italy)
H	Ljubljana (Slovenia) Galati (Romania)
I	Covasna (Romania)

Moreover, the lack of an EU energy vulnerability map was identified, which will allow the regions with a higher number of buildings with poor energy efficiency to be identified. Future studies must prioritize developing this new map and integrate it with the seismic risk map towards a future EU seismic plus energy risk map.

4. Definition of Scenarios for Retrofitting Strategies in the European Union

The present section proposes scenarios for retrofitting specific building typologies located in the different seismic–climate zones defined. The recommended retrofitting strategies depend highly on the building seismic design, energy efficiency and the seismic–climate demand. For example, if a building typology is located in zone A (i.e., low climate indicators and low expected seismic losses), the seismic retrofitting is only addressed if the seismic vulnerability assessment, according to Eurocode 8 [30], concludes that it is necessary. No seismic retrofitting is required if the building verifies the Eurocode 8 safety assessment methodology. The same exercise needs to be carried out regarding energy efficiency.

In the case of relevant seismic and climate indicators, the synergy between the seismic and energy retrofitting interventions needs to be prioritised in the rehabilitation of existing EU building stock. This new approach will reduce energy consumption and CO_2 emissions while buildings' seismic vulnerability is reduced. It will also be possible to take advantage of the existing policies proposed by the EU and use the incentives to complement the energy retrofitting and reduce buildings' seismic vulnerability.

For this purpose, nine different types of building typologies are considered concerning their seismic design and energy efficiency, namely, low seismic design (LSD) combined with low energy efficiency (LEE); high seismic design (HSD) combined with low energy efficiency (LEE); or low seismic design (LSD) combined with high energy efficiency (HEE). The building typologies matrix is presented in Figure 9.

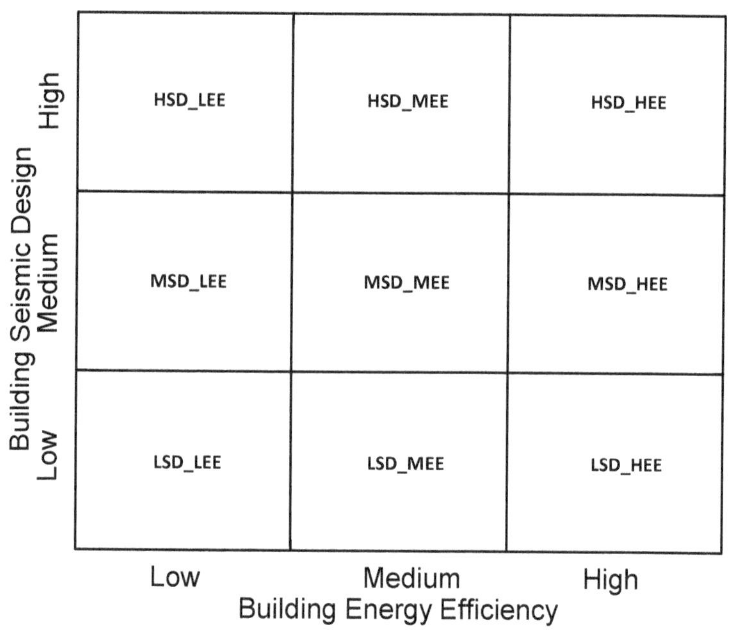

Figure 9. Building typologies matrix.

From the seismic vulnerability and energy efficiency point of view, these nine typologies represent the possible different building typologies existing in the EU building stock. The most vulnerable typologies are the LSD_LEE, LSD_HEE, and HSD_LEE for different reasons. The typology LSD_LEE is related to buildings with low (or no) seismic design and low/poor energy efficiency. The buildings constructed before 1970 can represent the LSD_LEE typology, considering the EU's implementation date of the first seismic and energy codes (around 1970).

The typology LSD_HEE is related to buildings with low or no seismic design and high energy efficiency. Buildings with a year of construction before 1970 that were recently subjected to energy strengthening can be representative of this typology.

The typology HSD_LEE is dedicated to buildings with high/modern seismic design (e.g., according to Eurocode 8) and low energy efficiency. The first energy codes emerged in Europe after 1980, with low energy demands. The modern codes (e.g., after 2000) require high insulation demands, resulting in buildings with high energy efficiency. A building designed according to Eurocode 8 (i.e., seismic design) and the first energy codes implemented in the 1980s can represent the HSD_LEE typology.

Two types of building structures are suggested to be analysed since they represent most of those existing in the EU building stock: masonry (M) structures and reinforced concrete (RC) structures. Tables 4 and 5 present the nomenclature adopted for each building typology that will be suggested for the different retrofitting scenarios. Other types of structures such as wood were excluded from this recommendation since they are not representative of the EU building stock.

Table 4. Definition of reinforced concrete building typologies proposed for retrofitting scenarios.

Nomenclature	Building Seismic Design	Building Energy Efficiency
RC_LSD_LEE	Low	Low
RC_HSD_LEE	High	Low
RC_LSD_HEE	Low	High

Table 5. Definition of masonry building typologies proposed for retrofitting scenarios.

Nomenclature	Building Seismic Design	Building Energy Efficiency
M_LSD_LEE	Low	Low
M_HSD_LEE	High	Low
M_LSD_HEE	Low	High

The present section aims to propose retrofitting scenarios, taking into account the type of building structure (i.e., M or RC), seismic design (i.e., LSD, MSD or HSD), energy efficiency (i.e., LEE, MEE or HEE) and seismic–climate zone (i.e., scenario SCZ_A, SCZ_B, SCZ_C, SCZ_D, SCZ_E, SCZ_F, SCZ_G, SCZ_H or SCZ_I).

For each retrofitting scenario (i.e., combination of building typology and seismic–climate zone), a datasheet containing information related to the seismic–climate zone (seismic and climate indicators), building performance type (seismic design, energy efficiency and particular building characteristics), possible retrofitting recommendations and examples of EU cities/regions is proposed. The framework of the datasheets produced for each retrofitting scenario is shown in Figure 10.

Moreover, apart from the suggestions included in each retrofitting scenario, special attention must be dedicated to isolated and aggregated buildings. It is recommended that different types of masonry are considered (regular or irregular arrangement of blocks, dry or mortared joints) in the case of masonry structures. It is also suggested that vertical and/or plan irregularities of structural and non-structural elements (such as infill walls) are considered since they are responsible for multiple failures on these types of structures.

The framework for defining the different retrofitting scenarios is based on a three-step procedure. Step 1 defines the seismic–climate zone. Step 2 defines the building performance type that will be located in the seismic–climate zone. The third step is dedicated to retrofitting recommendations based on the input data from Step 1 and Step 2. A maximum of 81 retrofitting scenarios can be extracted from this framework for each type of structure (i.e., masonry or reinforced concrete structure). Only six retrofitting scenarios are herein presented for each type of structure. They were selected based on the level of demand (i.e., seismic and/or climate) and the low performance of buildings. Six retrofitting scenarios are herein proposed (Figures 11–16).

As previously mentioned, the retrofitting scenarios herein proposed were developed assuming a seismic–climate matrix approach, i.e., by considering AALR and CI indicators. It should be stressed that this work needs to be performed by considering the economic losses estimated due to the energy consumption of buildings. The climate indicator (HDD plus CDD) gives a perspective of the heating and cooling needs but does not reflect the real energy efficiency of buildings in that zone.

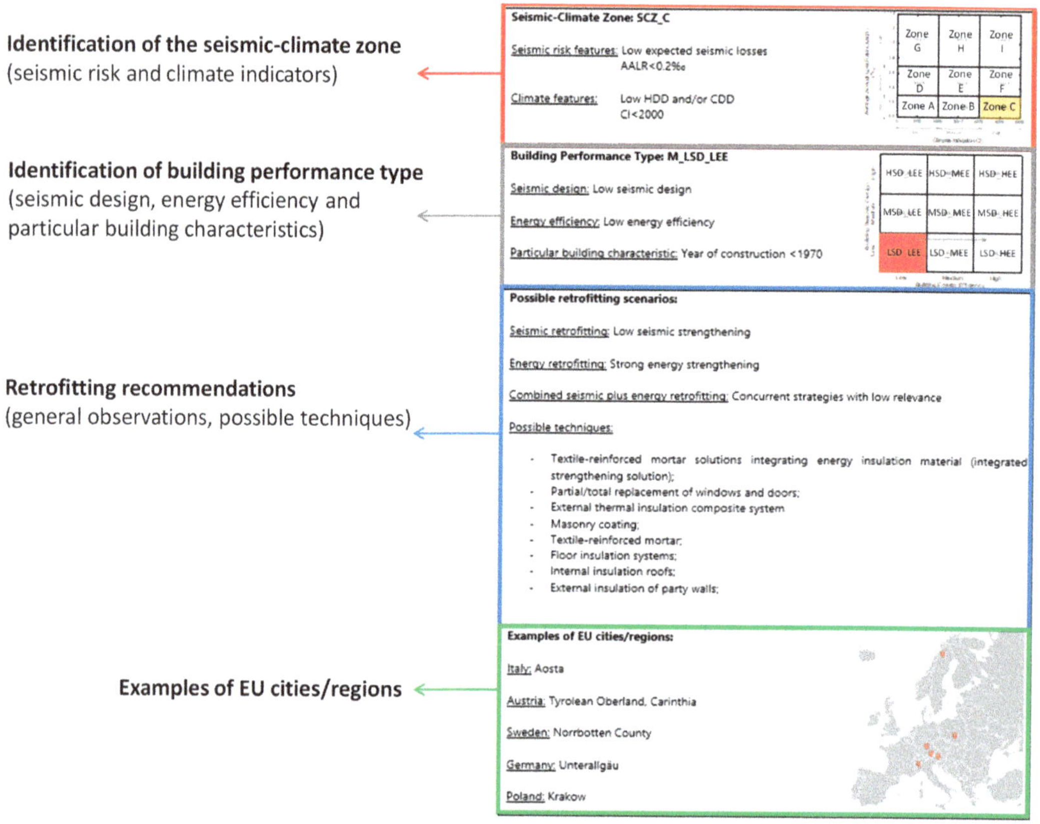

Figure 10. Framework of the datasheets produced for each combination.

On one hand, the AALR was estimated by performing an accurate assessment considering the actual seismic building vulnerability, exposure and hazard. On the other hand, energy efficiency is only associated with climate indicators. The energy codes are more demanding in countries subjected to high heating demands. Thus, the buildings designed according to these codes are more energy-efficient than others not designed according to modern energy codes but located in countries with lower heating (or cooling) demands.

Nonetheless, it is crucial to study the list of cities herein highlighted for future combined seismic plus energy retrofitting, namely, Vienna (Austria), Carinthia (Austria), Grad Sofiya (Bulgaria), Plovidv (Bulgaria), Dubrovacko-Neretvanska (Croatia), Grad Zagreb (Croatia), Splitsko-Dalmatinska (Croatia), Paphos (Cyprus), Larnaka (Cyprus), Lafkosia (Croatia), Ammochostos (Croatia), Lemesos (Croatia), Peloponnese (Greece), Arge (Greece), Crete (Greece), Thessaly and Central Greece (Greece), Epirus (Greece), Attica (Attica), Macedonia and Thrace (Greece), Emilia Romagna (Italy), Umbria (Italy), Molise (Italy), Abruzzo (Italy), Toscana (Italy), Friuli-Venezia Giulia (Italy), Marche (Italy), Veneto (Italy), Campania (Italy), Basilicata (Italy), Bucuresti (Romania) and Region of Murcia (Spain).

Seismic-Climate Zone: SCZ_C

Seismic risk features: Low seismic losses expected
AALR < 0.3‰

Climate features: High HDD and/or CDD
CI≥3000

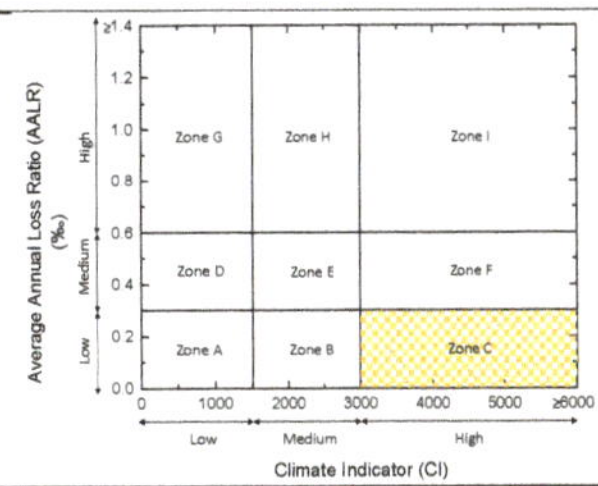

Building Performance Type: M_LSD_LEE

Seismic design: Low seismic design

Energy efficiency: Low energy efficiency

Particular building characteristic: Year of construction < 1970

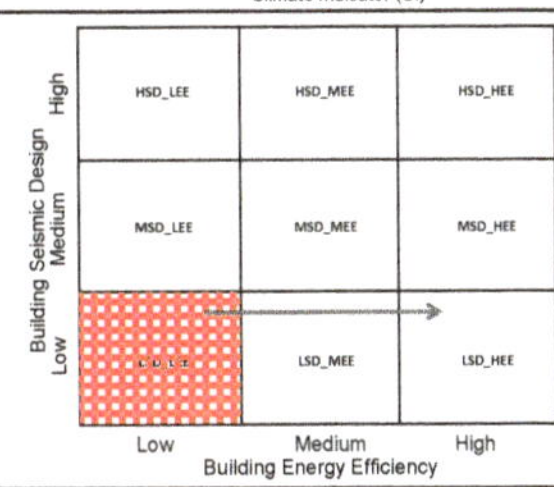

Possible retrofitting scenarios:

Seismic retrofitting: Low seismic strengthening

Energy retrofitting: Strong energy strengthening

Combined seismic plus energy retrofitting: Concurrent strategies with low relevance

Possible techniques:

- Textile-reinforced mortar solutions integrating energy insulation material (integrated strengthening solution);
- Partial/total replacement of windows and doors;
- External thermal insulation composite system
- Masonry coating;
- Textile-reinforced mortar;
- Floor insulation systems;
- Internal insulation roofs;
- External insulation of party walls;

Examples of EU cities/regions:

Italy: Aosta

Austria: Tyrolean Oberland, Carinthia

Sweden: Norrbotten County

Germany: Unterallgäu

Poland: Krakow

Figure 11. Retrofitting scenario 1: SCZ_C and M_LSD_LEE.

Seismic-Climate Zone: SCZ_G

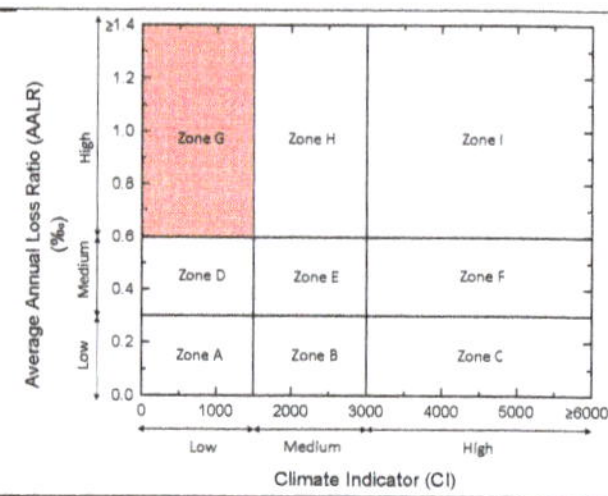

Seismic risk features: High seismic losses expected
AALR≥0.6‰

Climate features: Low HDD and/or CDD
CI < 1500

Building Performance Type: M_LSD_LEE

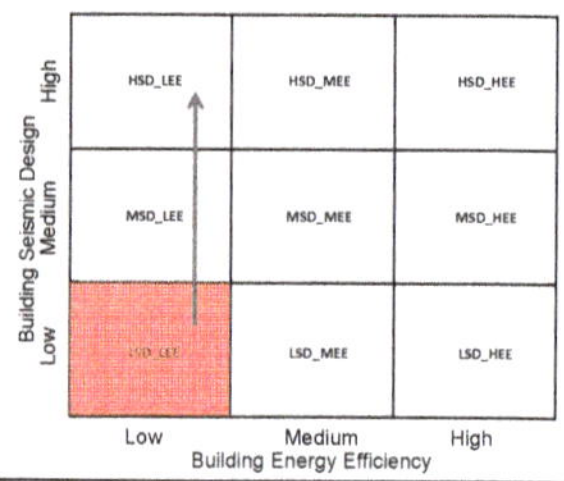

Seismic design: Low seismic design

Energy efficiency: Low energy efficiency

Particular building characteristic: Year of construction < 1970

Possible retrofitting scenarios:

Seismic retrofitting: Strong seismic strengthening

Energy retrofitting: Low energy strengthening

Combined seismic plus energy retrofitting: Concurrent strategies with low relevance

Possible techniques:

- Masonry coating
- Grout injection
- Steel dowels
- Horizontal and vertical steel ties
- Intramural tying
- Bed-joint reinforced repointing
- Stiffening of floors
- Strengthening of openings with steel frame
- Textile-reinforced mortar based solutions for external walls

Examples of EU cities/regions:

Italy: Calabria

Cyprus: Paphos, Larnaka, Lefkosia, Ammochostos, Lemesos

Greece: Crete

Figure 12. Retrofitting scenario 2: SCZ_G and M_LSD_LEE.

Seismic-Climate Zone: SCZ_I

Seismic risk features: High seismic losses expected
AALR≥0.6‰

Climate features: High HDD and/or CDD
CI≥3000

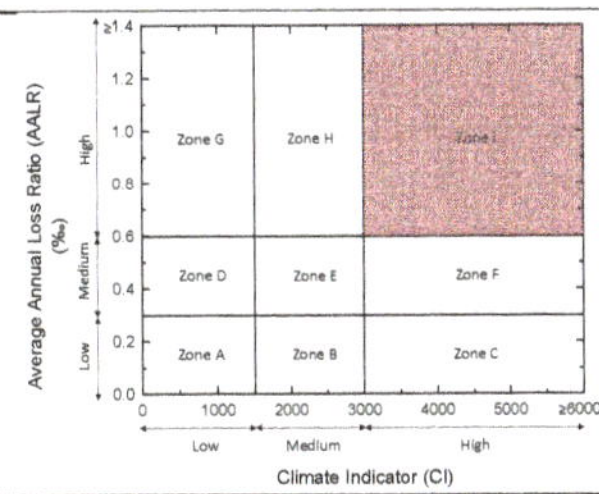

Building Performance Type: M_LSD_LEE

Seismic design: Low seismic design

Energy efficiency: Low energy efficiency

Particular building characteristic: Year of construction < 1970

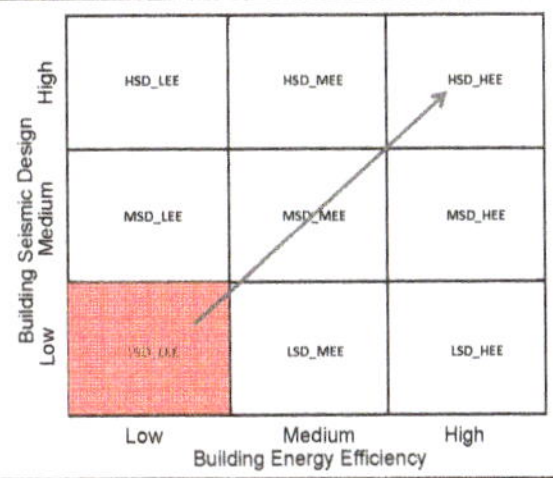

Possible retrofitting scenarios:

Seismic retrofitting: Strong seismic strengthening

Energy retrofitting: Strong energy strengthening

Combined seismic plus energy retrofitting: Concurrent strategies should be prioritized

Possible techniques:

- Masonry coating
- Grout injection
- Steel dowels
- Horizontal and vertical steel ties
- Intramural tying
- Bed-joint reinforced repointing
- Stiffening of floors
- Strengthening of openings with steel frame
- Textile-reinforced mortar based solutions for external walls
- Textile-reinforced mortar solutions integrating energy insulation material (integrated strengthening solution);
- Partial/total replacement of windows and doors;
- External thermal insulation composite system

Examples of EU cities/regions:

Italy: Veneto

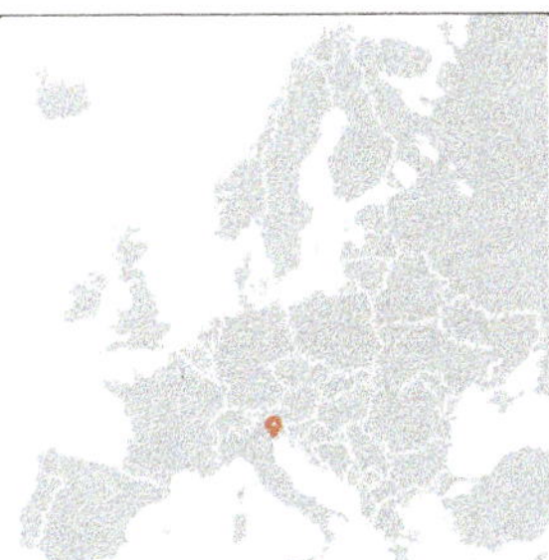

Figure 13. Retrofitting scenario 3: SCZ_I and M_LSD_LEE.

Seismic-Climate Zone: SCZ_C

<u>Seismic risk features:</u> Low seismic losses expected
AALR < 0.3‰

<u>Climate features:</u> High HDD and/or CDD
CI≥3000

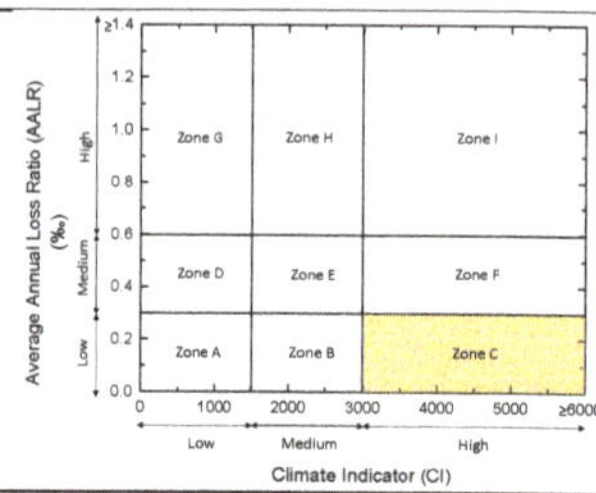

Building Performance Type: RC_LSD_LEE

<u>Seismic design:</u> Low seismic design

<u>Energy efficiency:</u> Low energy efficiency

<u>Particular building characteristic:</u> Year of construction < 1970

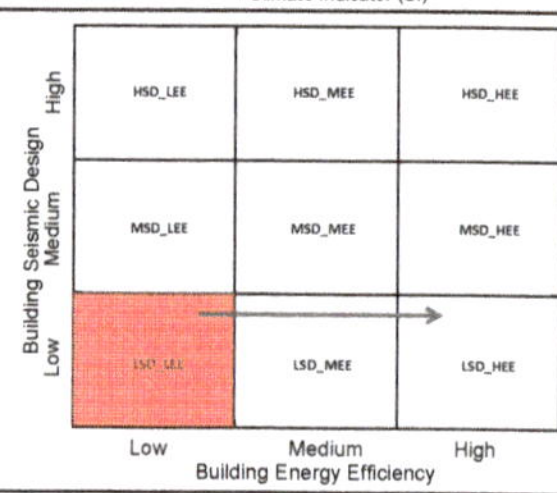

Possible retrofitting scenarios:

<u>Seismic retrofitting:</u> Low seismic strengthening

<u>Energy retrofitting:</u> Strong energy strengthening

<u>Combined seismic plus energy retrofitting:</u> Concurrent strategies with low relevance

<u>Possible techniques:</u>

- External thermal energy insulation
- Total replacement of windows and doors
- Floor insulation systems
- Internal insulation roofs
- External insulation of party walls
- Prefabricated units for external wall insulation
- System of facade refurbishment with cement panels sheathing
- External insulation of flat roofs
- Thermal insulation of external wall air chambers
- External insulation of sloped roofs

Examples of EU cities/regions:

<u>Italy:</u> Aosta

<u>Austria:</u> Tyrolean Oberland, Carinthia

<u>Sweden:</u> Norrbotten County

<u>Germany:</u> Unterallgäu

<u>Poland:</u> Krakow

Figure 14. Retrofitting scenario 4: SCZ_C and RC_LSD_LEE.

Seismic-Climate Zone: SCZ_G

Seismic risk features: High seismic losses expected
AALR≥0.6‰

Climate features: Low HDD and/or CDD
CI < 1500

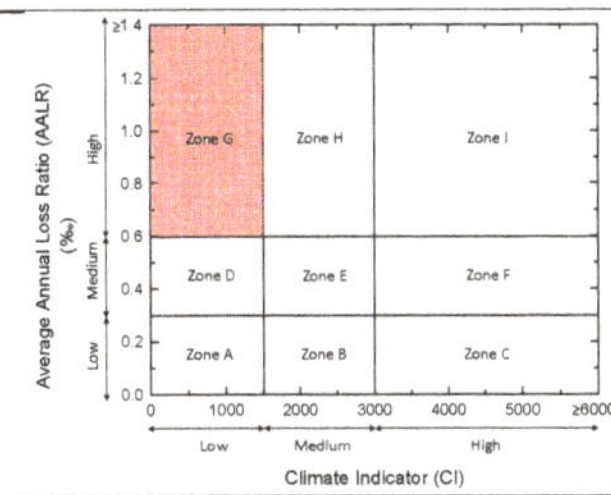

Building Performance Type: RC_LSD_LEE

Seismic design: Low seismic design

Energy efficiency: Low energy efficiency

Particular building characteristic: Year of construction < 1970

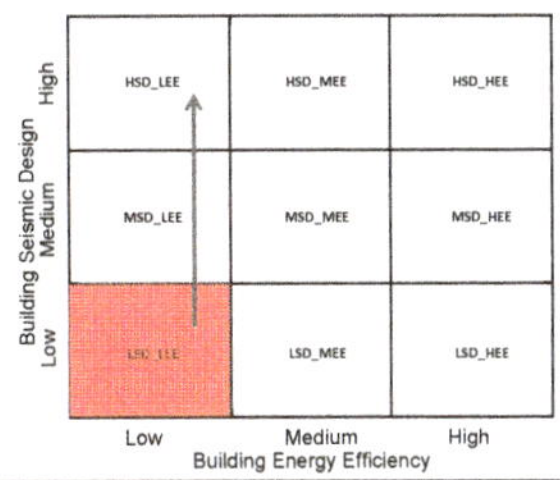

Possible retrofitting scenarios:

Seismic retrofitting: Strong seismic strengthening

Energy retrofitting: Low energy strengthening

Combined seismic plus energy retrofitting: Concurrent strategies with low relevance

Possible techniques:
- RC Jacketing of columns and beam-column joints
- Steel jacketing of columns and beam-column joints
- FRP jacketing of columns and beam-column joints
- RC shear walls
- Steel braces
- Retrofitting of masonry infill walls in the buildings envelope
- System of facade refurbishment with cement panels sheathing
- Textile-reinforced mortar solutions integrating energy insulation material
- Timber-based panels in the buildings envelope
- Engineered exoskeletons
- Replacement of existing masonry infill walls in the building envelope

Examples of EU cities/regions:

Italy: Calabria

Cyprus: Paphos, Larnaka, Lefkosia, Ammochostos, Lemesos

Greece: Crete

Figure 15. Retrofitting scenario 5: SCZ_G and RC_LSD_LEE.

Seismic-Climate Zone: SCZ_I

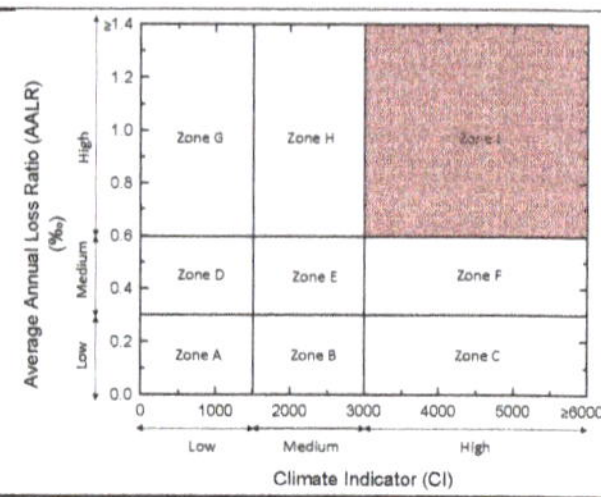

Seismic risk features: High seismic losses expected
AALR≥0.6‰

Climate features: High HDD and/or CDD
CI≥3000

Building Performance Type: RC_LSD_LEE

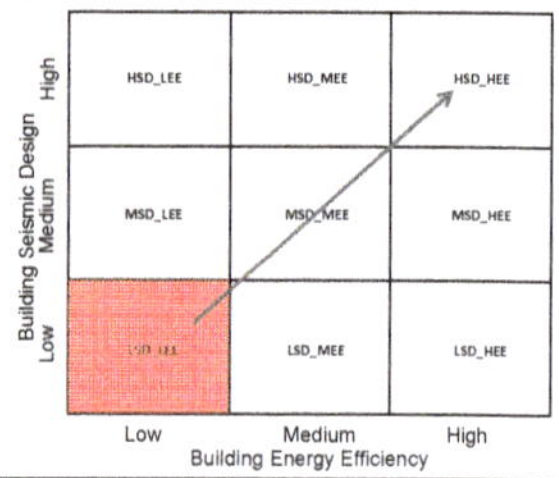

Seismic design: Low seismic design

Energy efficiency: Low energy efficiency

Particular building characteristic: Year of construction < 1970

Possible retrofitting scenarios:

Seismic retrofitting: Strong seismic strengthening

Energy retrofitting: Strong energy strengthening

Combined seismic plus energy retrofitting: Concurrent strategies should be prioritized

Possible techniques:

- System of facade refurbishment with cement panels sheathing
- Textile-reinforced mortar solutions integrating energy insulation material
- Timber-based panels in the buildings envelope
- Total replacement of windows and doors
- Floor insulation systems
- External insulation of flat roofs
- Internal insulation of roofs
- RC, steel and FRP Jacketing of columns and beam-column joints
- RC shear walls
- Steel braces
- Retrofitting of masonry infill walls in the buildings envelope
- Engineered exoskeletons
- External thermal insulation composite system

Examples of EU cities/regions:

Italy: Veneto

Figure 16. Retrofitting scenario 6: SCZ_I and RC_LSD_LEE.

Butenweg (2021) performed a simplified identification of areas with both kinds of exposure, i.e., seismicity (i.e., hazard) and climate conditions. The author proposed a study for the combined retrofitting of the following cities: Pleven (moderate seismic hazard), Sofia (moderate seismic hazard), Plovdiv (high seismic hazard) and Blagovgrad (high seismic hazard); Zadar (low seismic hazard), Osijek (low seismic hazard), Split (moderate seismic hazard), Primorje-Gorski-kotar (moderate seismic hazard), Dubrovnik (high seismic hazard) and Zagreb (high seismic hazard); Munich (low seismic hazard), Lindau (low seismic hazard) and Aachen (moderate seismic hazard); Andros (moderate seismic hazard), Kosani (low seismic hazard) and Dykiti Makedonia (low seismic hazard); Athens (medium/high seismic hazard) and Thessaloniki (medium/high seismic hazard); Sassari, Bari, Como, Verbano-Cusio-Ossola and Aosta (low seismic hazard), Salerno, Pisa, Vicenza and Bolzano (moderate seismic hazard) and Naples, Perugia, Bergamo and Trento (high seismic hazard); Cluj and Bistrita (low seismic hazard), Satu Mare and Hargita (moderate seismic hazard), and Buchurest, Vrancea and Covasna (high seismic hazard).

Butenweg (2021) pointed out that the area-related approach with the score is only a helpful indicator of prioritising countries and regions for combined actions. However, that does not mean that combined measures in countries with low to moderate seismicity are not required. Combined upgrading is also relevant for countries with lower scores, as a high percentage of buildings were built without sufficient seismic safety measures. Especially in countries with low to moderate seismicity, seismic design rules were often ignored and seismic codes were introduced during the 1990s.

5. Final Observations and Future Works

This research work provides a set of retrofitting scenarios for the EU based on the characteristics of the EU building stock and EU seismic and climate features. The proposed scenarios were defined based on a simplified approach, i.e., based on climate and seismic risk indicators. Apart from the climate and seismic risk inputs, the definition of priority regions for the combined seismic plus energy retrofitting should also be performed based on each country's energy consumption and energy poverty. The proposed scenarios help define priority regions for seismic, energy, or combined seismic plus energy retrofitting in the EU. A different set of building typologies are suggested for further study based on their seismic design and energy efficiency. Therefore, their characteristics must be adapted according to the typical characteristics of buildings in each country (i.e., architecture, structure, materials).

Other authors suggest that the impact of the existing building stock in Europe in terms of economic losses and emissions due to climate exposure needs to be addressed. The average annual costs due to heating/cooling buildings for the buildings in the exposure model could be estimated based on combining the heating/cooling degree days with models of heating/cooling costs for different building classes. The average annual CO_2 emissions due to heating/cooling of different building classes would also need to be estimated. Based on this information, it will be possible to obtain average annual economic losses due to energy consumption and/or CO_2 emissions. Therefore, this new evaluation will be merged with the average annual economic losses due to seismic exposure towards a new EU map. Efforts should be made to identify the most critical scenarios requiring intervention due to high seismic losses (economic and loss of life), energy costs, and CO_2 emissions. These new maps should be constructed across the NUTS3 regions in Europe.

As future works, a parametric study is suggested to evaluate the impact of the retrofitting measures suggested for each seismic–climate scenario. The effect of those retrofitting measures should be compared to the existing conditions of the building stock.

Author Contributions: Conceptualization, A.F.; Methodology, A.F.; Investigation, H.V.; Writing—original draft, A.F.; Writing—review & editing, H.R.; Visualization, H.R. All authors have read and agreed to the published version of the manuscript.

Funding: The first author is grateful for the Foundation for Science and Technology's support through funding UIDB/04625/2020 from the research unit CERIS. This work was also financially supported by: Project POCI-01-0145-FEDER-007457-CONSTRUCT-Institute of R&D In Structures and Construction funded by FEDER funds through COMPETE2020-Programa Operacional Competitividade e Internacionalização and by national funds through FCT-Fundação para a Ciência e a Tecnologia. This work was also supported by the Foundation for Science and Technology (FCT)-Aveiro Research Centre for Risks and Sustainability in Construction (RISCO), Universidade de Aveiro, Portugal [FCT/UIDB/ECI/04450/2020].

Data Availability Statement: Not applicable.

Acknowledgments: The authors acknowledge the very important contributions of the reviewers that improved the quality of the manuscript.

Conflicts of Interest: The authors declare no conflict of interest.

Abbreviations

AAELR	Average Annual Economic Losses Ratio
AAL	Average Annual Losses
AALL	Average Annual Life Loss
AALR	Average Annual Loss Ratio
BREEAM	Building Research Establishment Environmental Assessment Method
CDD	Cooling Degree Days
CI	Climate Indicator
CO_2	Carbon Dioxide
DGNB	German Sustainable Building Council
EU	European Union
HDD	Heating Degree Days
HEE	High Energy Efficiency
HSD	High Seismic Design
IRC	Infilled Reinforced Concrete
LC	Life Cycle
LEE	Low Energy Efficiency
LSD	Low Seismic Design
M	Masonry
MEE	Moderate Energy Efficiency
MIW	Masonry Infill Walls
MSD	Moderate Seismic Design
NS	Non-structural elements
OOP	Out-of-Plane
RC	Reinforced Concrete
SCZ	Seismic Climate Zone
SP	Primary members
SS	Secondary members
TRM	Textile-Reinforced Mortar
U-value	Thermal transmittance

References

1. Pohoryles, D.A.; Bournas, D.A.; Da Porto, F.; Caprino, A.; Santarsiero, G.; Triantafillou, T. Integrated seismic and energy retrofitting of existing buildings: A state-of-the-art review. *J. Build. Eng.* **2022**, *61*, 105274. [CrossRef]
2. Masi, A.; Chiauzzi, L.; Santarsiero, G.; Manfredi, V.; Biondi, S.; Spacone, E.; Del Gaudio, C.; Ricci, P.; Verderame, G.M. Seismic response of RC buildings during the Mw 6.0 August 24, 2016 Central Italy earthquake: The Amatrice case study. *Bull. Earthq. Eng.* **2017**, *17*, 5631–5654. [CrossRef]
3. Parliament, E. *Boosting Building Renovation: What Potential and Value for Europe*; Technical Report; European Parliament: Strasbourg, France, 2016.
4. Furtado, A.; Rodrigues, H.; Arêde, A.; Varum, H. Experimental tests on strengthening strategies for masonry infill walls: A literature review. *Constr. Build. Mater.* **2020**, *263*, 120520. [CrossRef]
5. Jelle, B. Traditional, state-of-the-art and future thermal building insultation materials and solutions-properties, requirements and possibilities. *Energy Build.* **2011**, *43*, 2549–2563. [CrossRef]

6. Posani, M.; Veiga, M.D.R.; de Freitas, V.P. Towards Resilience and Sustainability for Historic Buildings: A Review of Envelope Retrofit Possibilities and a Discussion on Hygric Compatibility of Thermal Insulations. *Int. J. Arch. Herit.* **2019**, *15*, 807–823. [CrossRef]
7. Vailati, M.; Monti, G.; Gangi, G.D. Earthquake-Safe and Energy-Efficient Infill Panels for Modern Buildings. In *Earthquake Engineering and Structural Dynamics in Memory of Ragnar Sigbjörnsson: Selected Topics*; Rupakhety, R., Ólafsson, S., Eds.; Springer International Publishing: Cham, Switzerland, 2018; pp. 233–261.
8. Papadopoulos, A. State of the art in thermal insulation materials and aims for future developments. *Energy Build.* **2005**, *37*, 77–86. [CrossRef]
9. Amaral, C.; Pinto, S.; Silva, T.; Mohseni, F.; Amaral, J.; Marques, P.; Barros-Timmons, A.; Vicente, R. Development of polyurethane foam incorporating phase change material for thermal energy storage. *J. Energy Storage* **2020**, *28*, 101177. [CrossRef]
10. Barreira, E.; de Freitas, V.P. External Thermal Insulation Composite Systems: Critical Parameters for Surface Hygrothermal Behaviour. *Adv. Mater. Sci. Eng.* **2014**, *2014*, 650752. [CrossRef]
11. Furtado, A.; Rodrigues, H.; Varum, H.; Costa, A. Evaluation of different strengthening techniques' efficiency for a soft storey building. *Eur. J. Environ. Civ. Eng.* **2015**, *21*, 371–388. [CrossRef]
12. Pohoryles, D.A.; Melo, J.; Rossetto, T.; Varum, H.; Bisby, L. Seismic Retrofit Schemes with FRP for Deficient RC Beam-Column Joints: State-of-the-Art Review. *J. Compos. Constr.* **2019**, *23*, 03119001. [CrossRef]
13. Sousa, L.; Monteiro, R. Seismic retrofit options for non-structural building partition walls: Impact on loss estimation and cost-benefit analysis. *Eng. Struct.* **2018**, *161*, 8–27. [CrossRef]
14. Erol, G.; Karadogan, H.F. Seismic strengthening of infilled reinforced concrete frames by CFRP. *Compos. Part B Eng.* **2016**, *91*, 473–491. [CrossRef]
15. Furtado, A.; Rodrigues, H.; Arêde, A.; Melo, J.; Varum, H. The use of textile-reinforced mortar as a strengthening technique for the infill walls out-of-plane behaviour. *Compos. Struct.* **2020**, *255*, 113029. [CrossRef]
16. De Risi, M.T.; Furtado, A.; Rodrigues, H.; Melo, J.; Verderame, G.M.; António, A.; Varum, H.; Manfredi, G. Experimental analysis of strengthening solutions for the out-of-plane collapse of masonry infills in RC structures through textile reinforced mortars. *Eng. Struct.* **2020**, *207*, 110203. [CrossRef]
17. Negro, P.; Mola, E. A performance based approach for the seismic assessment and rehabilitation of existing RC buildings. *Bull. Earthq. Eng.* **2015**, *15*, 3349–3364. [CrossRef]
18. Asteris, P.G.; Cotsovos, D.M.; Chrysostomou, C.Z.; Mohebkhah, A.; Al-Chaar, G.K. Mathematical micromodeling of infilled frames: State of the art. *Eng. Struct.* **2013**, *56*, 1905–1921. [CrossRef]
19. Furtado, A.; Rodrigues, H.; Arêde, A.; Varum, H. Simplified macro-model for infill masonry walls considering the out-of-plane behaviour. *Earthq. Eng. Struct. Dyn.* **2015**, *45*, 507–524. [CrossRef]
20. Gervásio, H.; Dimova, S.; Pinto, A.V. Benchmarking the Life-Cycle Environmental Performance of Buildings. *Sustainability* **2018**, *10*, 1454. [CrossRef]
21. Almeida, R.M.S.F.; Ramos, N.M.M.; Manuel, S. Towards a methodology to include building energy simulation uncertainty in the Life Cycle Cost analysis of rehabilitation alternatives. *J. Build. Eng.* **2015**, *2*, 44–51. [CrossRef]
22. Ramesh, T.; Prakash, R.; Shukla, K.K. Life cycle energy analysis of buildings: An overview. *Energy Build.* **2010**, *42*, 1592–1600. [CrossRef]
23. DGNB. *DGNB System: Buildings in Use, Criteria Set, Version 2020*; Deutsche Gesellschaft für Nachhaltiges Bauen (DGNB): Stuttgart, Germany, 2020.
24. BRE. *BREEAM USA in-Use Technical Manual, PD130 1.1:2016*; Building Research Establishment (BRE): Watford, UK, 2017.
25. European Commission. *Maintenance and Update of the EU Building Stock Observatory—Final Report (ENER/C3/2016-547/01)*; European Commission: Brussels, Belgium, 2020.
26. Valluzzi, M.R.; Sbrogiò, L.; Saretta, Y.; Wenliuhan, H. Seismic Response of Masonry Buildings in Historical Centres Struck by the 2016 Central Italy Earthquake. Impact of Building Features on Damage Evaluation. *Int. J. Arch. Herit.* **2021**, *16*, 1859–1884. [CrossRef]
27. Silva, V.; Amo-Oduro, D.; Calderon, A.; Costa, C.; Dabbeek, J.; Despotaki, V.; Martins, L.; Pagani, M.; Rao, A.; Simionato, M.; et al. Development of a global seismic risk model. *Earthq. Spectra* **2020**, *36*, 372–394. [CrossRef]
28. Vlachakis, G.; Vlachaki, E.; Lourenço, P.B. Learning from failure: Damage and failure of masonry structures, after the 2017 Lesvos earthquake (Greece). *Eng. Fail. Anal.* **2020**, *117*, 104803. [CrossRef]
29. Ferreira, T.; Costa, A.A.; Costa, A. Analysis of the Out-Of-Plane Seismic Behavior of Unreinforced Masonry: A Literature Review. *Int. J. Arch. Herit.* **2015**, *9*, 949–972. [CrossRef]
30. *Eurocode 8: Design of Structures for Earthquake Resistance—Part 1-1: General Rules, Seismic Actions and Rules for Buildings*; European Committee for Standardization: Brussels, Belgium, 2005.
31. Furtado, A.; Rodrigues, H.; Arêde, A.; Varum, H. A Review of the Performance of Infilled RC Structures in Recent Earthquakes. *Appl. Sci.* **2021**, *11*, 5889. [CrossRef]
32. Hermanns, L.; Fraile, A.; Alarcón, E.; Álvarez, R. Performance of buildings with masonry infill walls during the 2011 Lorca earthquake. *Bull. Earthq. Eng.* **2013**, *12*, 1977–1997. [CrossRef]
33. De Luca, F.; Verderame, G.M.; Gómez-Martínez, F.; Pérez-García, A. The structural role played by masonry infills on RC building performances after the 2011 Lorca, Spain, earthquake. *Bull. Earthq. Eng.* **2013**, *12*, 1999–2026. [CrossRef]

34. De Risi, M.T.; Del Gaudio, C.; Verderame, G.M. Evaluation of Repair Costs for Masonry Infills in RC Buildings from Observed Damage Data: The Case-Study of the 2009 L'Aquila Earthquake. *Buildings* **2019**, *9*, 122. [CrossRef]
35. Gkournelos, P.; Bournas, D.; Triantafillou, T. Combined seismic and energy upgrading of existing reinforced concrete buildings using TRM jacketing and thermal insulation. *Earthq. Struct.* **2019**, *16*, 625–639.
36. Economidou, M.; Todeschi, V.; Bertoldi, P.; D'Agostino, D.; Zangheri, P.; Castellazzi, L. Review of 50 years of EU energy efficiency policies for buildings. *Energy Build.* **2020**, *225*, 110322. [CrossRef]
37. Pohoryles, D.A.; Maduta, C.; Bournas, D.A.; Kouris, L.A. Energy performance of existing residential buildings in Europe: A novel approach combining energy with seismic retrofitting. *Energy Build.* **2020**, *223*, 110024. [CrossRef]
38. Gkatzogias, K.; Crowley, H.; Veljkovic, A.; Pohoryles, D.A.; Norlén, H.; Tsionis, G.; Bournas, D.A. Prioritising EU regions for building renovation: Seismic risk, energy efficiency, socioeconomic vulnerability. *Jt. Res. Cent.* **2022**. [CrossRef]
39. Crowley, H. Technical report on the impact of seismic retrofit in priority regions. In Pilot Project—Integrated Techniques for the Seismic Strengthening and Energy Efficiency of Existing Buildings. 2021.
40. Greek Interventions Regulation (KANEPE). 2015.

energies

MDPI

Article

Energy Design Synthesis: Algorithmic Generation of Building Shape Configurations

Tamás Storcz [1,*], Zsolt Ercsey [1], Kristóf Roland Horváth [2], Zoltán Kovács [3], Balázs Dávid [4,5] and István Kistelegdi [6]

1 Department of Systems and Software Technologies, Faculty of Engineering and Information Technology, University of Pécs, Boszorkány Street 2, 7624 Pécs, Hungary
2 Marcel Breuer Doctoral School, Faculty of Engineering and Information Technology, University of Pécs, Boszorkány Street 2, 7624 Pécs, Hungary
3 Optin Ltd., 6720 Szeged, Hungary
4 InnoRenew CoE, Livade 6a, 6310 Izola, Slovenia
5 Department of Information Sciences and Technologies, Faculty of Mathematics, Natural Sciences and Information Technologies, University of Primorska, Glagoljaška 8, 6000 Koper, Slovenia
6 Department of Simulation Driven Design, Ybl Miklós Faculty of Architecture and Civil Engineering, Institute of Architecture, Óbuda University, Thököly Street 74, 1146 Budapest, Hungary
* Correspondence: storcz.tamas@mik.pte.hu

Abstract: The building industry is responsible for a significant degree of energy consumption in the world, causing negative climate changes and energy supply uncertainties due to low energy efficiency as well as the high resource demand of construction. Consequently, energy design optimization has become an important research field. Passive design strategies are one of the most definitive factors concerning energy-related building development. The given architectural problem calls for a method that can create all potentially feasible building geometries, thus guaranteeing the optimal solution which is addressed in the current paper. To reach this requirement, the necessity of a modular space arrangement system and architectural selection rules were determined, focusing on the relationship between the rules and the generation of geometries with mathematical rigor. Next, the architecture-based congruency analysis performed, further reduced the number of simulation cases. With the simulations, it is illustrated how the building shape versions affect the heating energy demands: the performance of the configurations themselves. Results clearly illustrate the importance of the synthesis step of the architectural design.

Keywords: building energy design; heating energy assessment; geometry generation; modular space arrangement

Citation: Storcz, T.; Ercsey, Z.; Horváth, K.R.; Kovács, Z.; Dávid, B.; Kistelegdi, I. Energy Design Synthesis: Algorithmic Generation of Building Shape Configurations. *Energies* **2023**, *16*, 2254. https://doi.org/10.3390/en16052254

Academic Editors: Paulo Santos and Chi-Ming Lai

Received: 18 January 2023
Revised: 17 February 2023
Accepted: 24 February 2023
Published: 26 February 2023

1. Introduction

Buildings produce more than 40% of the world's total energy consumption and CO_2 emissions due to construction and operations [1,2]. One of the main reasons is represented by the planning methodology of buildings. The prevailing modern, conventional building design includes a typically tight project schedule, making only a very limited number of concepts possible to consider. Through a linear process, the architect and client agree on one concept to be worked out. The services systems and further design disciplines subsequently integrate their contents to cover the evolved needs of the plan and planner cooperation is minimized due to time limits. Solar gains are not considered sufficiently, in general, and high solar loads (overheating) cause high cooling costs due to large glass surfaces in the envelope together with local thermal discomfort phenomena. Passive and active design strategies rely solely on experience and theoretical, general knowledge. With the comprehensive use of passive design concepts (e.g., space organization, building body shaping, envelope design, etc.) the highest degree (up to 80%) of energy conservation is achievable [3]. However, architects who are experts in designing spaces (generating the highest amount of energy needed), at the same time, do not handle the building's physics aspects, calculations, etc. in detail. Under such conditions, optimization is obviously not conceivable.

The next development step is performed by inventing an integrated design methodology that, still today, belongs to the exceptional examples of professional design teams worldwide [4–6]. During this iterative planning process, the ideas and knowledge of each participant are taken into consideration from the beginning and the design concept is created through teamwork. The energy and building services system concept does not follow the architect's plan; rather it is carried out parallel to the architectural design. The Energia Design (ED) method, developed at the beginning of the 2010s, has become one of the iterative tools that is used in practice when considering both energy and comfort issues in building design [7]. This method includes systemized heuristic building simulations in each design stage at different levels, and as a result, unique buildings were constructed with significant improvement in energy and comfort. However, neither this methodology nor other integrated design methods consider all possible solutions, and thus the optimal concept may be overlooked.

To overcome this drawback—and additionally motivated by the growing negative climate and environmental statistics—new efforts are continuously made on the development of building design methodology in order to gain optimized solution(s) in the fields of energy, and comfort, as well as in environmental impact performance. Since the complex building design objectives are characteristically ill-defined and often contradictory [5], rigorous mathematical optimization is required.

In architecture so far, numerous papers have been dedicated to problems of optimization methods related to some specific targets. In a comprehensive review, more than 100 studies about diverse building energy, comfort, and environmental design optimization (BECEDO) issues were evaluated [8]. The analysis states that most BECEDO studies' focus is set on the HVAC system and envelope parameter, and occupancy behavior optimization [9–12]. Even though the building geometry's positive effect on comfort and energy performance is more and more proven [13–15], over 60% of energy demand [16], as well as 80% of environmental impact reduction [17] is achievable with the combined optimization of the shape and further variables. Nevertheless, building geometry is still an underestimated and more or less neglected design variable, since numerical variables of the active systems and envelope materials are considerably easier to handle in the algorithms than space organization and building form describing mathematical models. There exist two main directions in BECEDO research, which integrate building geometry variables: the modification of the shape [18–20] that represents approx. 90% of the available literature, as well as shape generation [21–23] to create case versions of buildings. Though the dimensional properties of the geometry describing variables (e.g., depth, height, roof as well as diverse ratios) are frequently used in BECEDO, however, they are not able to provide information about form assembly of the geometry i.e., the position of the spaces, walls, edges, etc. relative to each other.

Another problem of BECEDO lies in the stochastic nature of the most frequently applied evolutionary (genetic) algorithms (GA), whereas due to the random move through the search space, the optimum solution is not guaranteed [24–27].

A promising development is proposed by the extension of the mentioned ED method, i.e., the Energia Design Synthesis (EDS) method has been introduced by Kistelegdi [28]. This method is intended to ensure de facto optimal buildings performing the highest energy and comfort efficiency, together with the lowest LCA impact. The EDS starts with the generation of all potentially feasible building shape configurations and extends these with various architectural aspects, i.e., building structure, fenestration ratio, and orientation. In this way, in the first stage of the method, all buildings are considered with the fundamental, most important passive measures, since these can enable architects to ensure a high degree (up to 80%) of energy savings [29,30]. When the buildings with passive measures are available, these are equipped with artificial illumination, occupants, and equipment, and in further steps with HVAC systems to serve as input for complex comfort and energy performance simulations. The simulation results are then evaluated, and a ranked order of the building cases is determined including the optimal design. The application of the synthesis step within the field of architecture is a novelty. The idea to start the algorithmic solution of a specific problem is well-known in the chemical engineering field, for example, Ref. [31] presented the p-graph methodology to successfully represent chemical engineering problems to serve as the basis of mathematical programming models. Now, this algorithmic method has been successfully adapted in other fields from handling sustainability issues [32], manufacturing problems by [33], energy storage and distribution by [34], and bus transport optimization [35] to scheduling problems [36].

The present work focuses on the first step of the EDS methodology, namely on the synthesis step, when during a BECEDO process all possible feasible building geometry configurations are generated, while all other cases are left out of consideration. This phase of the EDS method is performed algorithmically and with mathematical rigor. To support the algorithmic building shape generation, some definitions and a theorem are also presented together with technical details. The applicability of the result for further steps inevitably includes optimality to be considered throughout the whole EDS method.

The following questions are therefore proposed to be addressed:

- Does energy efficiency depend on building geometry?
- Is it possible to algorithmically generate all building geometries?
- What kind of architectural rules are required to define geometry feasibility?
- Can the number of feasible geometries be reduced based on architectural congruency?
- What are the formal characteristics of the gained geometry configurations meeting the required architectural rules?
- What are the energy performance characteristics of the selected geometry configurations meeting the required architectural rules?
- How intensively does the building shape affect the most relevant annual energy demand, the heating requirement?

The paper is constructed in the following manner (see Figure 1). Section 2 describes modular space organization and based on its result, the search space of the considered problem is determined. Section 3 talks about the inevitable architectural design rules, i.e., principles, that result in habitable and sustainable family houses. Section 4 explains how the search space is explored, i.e., all building geometries are set. Sections 5–7 detail how those building geometries that satisfy the architectural rules can be selected and how architecturally congruent geometries are eliminated, furthermore, which building geometries are configurations. Section 8 provides results and an analysis of energy simulations which clearly demonstrates the impact of the geometry dependencies.

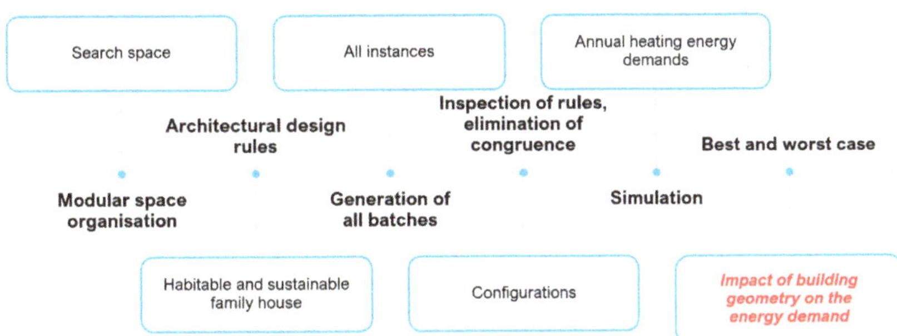

Figure 1. Process of configuration generation.

2. Modular Space Organization

The current problem from the architectural point of view is to develop all family house building configurations that must be considered when a house design is to be investigated, performing at as high energy and comfort performance as possible within the boundaries of the task. During this exemplary modeling, let us consider units or building blocks with the size of 5.5 m × 5.5 m × 3.0 m as general building elements; whereas the total of six blocks serves as the overall cubature of the housing. The size of the basic space unit provides generic room dimensions with a one-story height and the combination of the units results in larger spaces as living/dining/kitchen functionality or the division of the unit enables gaining smaller spaces such as bath, toilet, storage, etc. In other words, six blocks are to be placed next to each other according to architectural rules specified thereafter, thus, all family house building geometries of 181.5 m^2 floor area are sought, which meet the predefined architectural prerequisites (Figure 2).

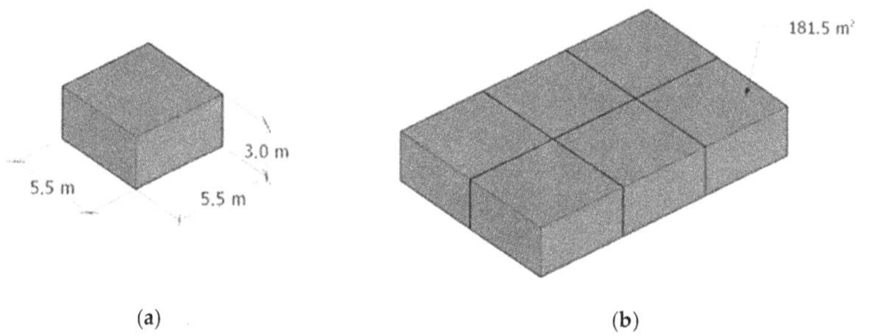

Figure 2. Basic space unit (**a**) and the combination of 6 units (**b**) to assemble a complete family house.

To easily describe the various building geometries and distinguish them from one another, let us consider an orthonormal coordinate system with X, Y, and Z axes and (5.5 m, 5.5 m, 3.0 m) unit size according to the architectural requirements. Moreover, let us consider a box of appropriate size within the positive orthant of the coordinate system, i.e., width × depth × height. In this specific family house generation case, where there are six building blocks under consideration, for the current problem let us consider 6 × 6 × 6 (width = depth = height = 6) as the size of this box. In other words, one vertex of this box is the origin of the coordinate (0, 0, 0) system and the opposite vertex is (6, 6, 6). Now, let us divide this box into 216 unit-sized cells. These will serve as cells to be used by the generation method. Any cell is to be identified by the triplet (x, y, z), which triplet now does not represent the coordinates of a point within the space but identifies the cells themselves. The triplet can be referred to as the coordinates of the vertex closest to the origin. In this particular case, there are 216 distinguished triplets, i.e., the set of triplets or the set of cells to be used during the algorithmic generation is the following: {(0, 0, 0), (1, 0, 0), … , (5, 5, 5)}. Now, let us place the blocks within these cells. It is worth mentioning that since the blocks are placed within cells of a coordinate system, it can be said that the blocks have floor, roof, and walls, which support the ease of understanding and the direct usability of the architectural rules. Should we place more than one block within the available cells of the box, then a batch is formed. For the specific architectural family house building geometry problem, let us consider batches of six blocks only since the total of six blocks serves as the overall cubature of the family house. As an example, the following description: {(0, 0, 0), (0, 1, 0), (1, 0, 0), (1, 1, 0), (1, 1, 1), (2, 0, 0)} corresponds to the building geometry depicted in Figure 3. Since the description can be given in a lexicographical order, the various structures can be easily ordered.

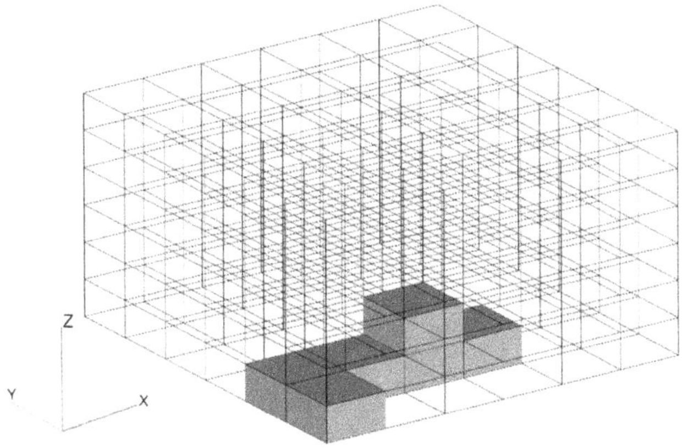

Figure 3. The batch.

3. Architectural Design Rules

In practice, no arbitrary batch can represent a building. A batch must satisfy certain characteristics to represent a family house appropriately. There are design principles that architects follow during the planning process of a family house; for instance functional dependencies between diverse spaces and use, shape design based on formal ideas, limitations due to legislation and regulation, visual view connections within the building and between inner and outer spaces, economical questions, site, and topography adaptation. However, these architectural design rules are diverse and often too complex to meet the abstraction level of the current investigation. Therefore, it is of utmost importance to adjust architectural rules to the requirements of the present scenario. Focus was set on avoiding sculpture-like shapes or futuristic, impractical batches. Shapes that are too high or too long are not supported either since these building shapes are impractical for family house use. In addition, building design regulations do not approve such solutions. Thus, the following set of rules was determined to express the necessary and sufficient architectural properties (elementary architectural requirements) to which a family house structure should conform:

(1) The blocks of the batch should be connected to each other with a minimum of one face, i.e., the batch should be compact.
(2) On the ground level, a minimum of two and a maximum of four blocks can be placed in a straight line. In this way, too short, too high as well as too long batches are avoided.
(3) There may be only one block within the batch, which is cantilevered. This means in architectural practice only one cantilevered space unit is allowed.
(4) The batch should not be more than three blocks high in order to exclude too high geometries, untypical for family housings.
(5) Above the ground floor, there should not be two blocks on top of each other with free walls, meaning at least one of the walls of the blocks should be connected to another block (no free-standing 'tower-like' building parts allowed).
(6) The length of the batches, which are taller than two blocks, should be a maximum of three blocks long in directions X and Y.

Please note, that for further architectural studies, only those batches are considered to be potentially feasible family house building configurations, which meet the above rules, while no other structures have to be considered. In other words, the search space is limited by the above-mentioned set of rules to fully control the optimization process by considering all possible and feasible solutions and thus ensuring that the optimal concept is also taken into account.

First, all batches within the search space should be assembled. When a batch is fully generated, it should be checked as to whether the batch satisfies the above-mentioned architectural rules. Should any of the rules be violated, the batch must be excluded from further consideration. Those building geometries that are not excluded, i.e., the remaining shapes satisfying the rules, are to be called building configurations hereinafter.

4. Generation Method

The given architectural problem specifies the generation of all potentially feasible building configurations, i.e., batches ofsix blocks satisfying the architectural rules. Please note that rule four limits the aforementioned original box size of $6 \times 6 \times 6$ cells to a maximum of three cells when considering the z-axis; while rules two and three limit the box size tofive cells when considering the axes x and y; resulting in the overall size of the box $5 \times 5 \times 3$, including 75 cells. This box will serve as the search space where batches of six blocks are sought that satisfy all of the above rules. It is worth mentioning that within this box there are $\binom{75}{6} = 201,359,550$ different ways to place the six blocks.

Now, as the first phase of the generation method, all possible placement combinations of the six blocks within this box are generated. Technically, a zero-based sequence number (SN) is assigned to each cell by formula or Equation (1):

$$SN = x + y \cdot width + z \cdot width \cdot depth \tag{1}$$

For example, the first cell (SN = 0) of the box within the positive orthant of the coordinate system is identified by the triplet (0, 0, 0), the 25th cell (SN = 24) is identified by the triplet (4, 4, 0) and the last one, i.e., the 75th cell (SN = 74) is identified by the triplet (4, 4, 2). This sequence number eases the generation process since the cells of the box can now be easily ordered, and it can be said that cells have predecessor and successor cells. The classical backtracking algorithm that incrementally builds partial candidates [37] finds all solutions to some computational problems. The candidates are represented in a tree structure, where the parent differs from the child by a single extension step. This backtracking serves as the guiding rule for the generation procedure in the present case in the following way (Figure 4): Let the initial state be, where the first block is in the first cell, the second block is in the second cell, and so on, and the sixth block is placed in the sixth cell. In other words, the six blocks are given as follows: {(0, 0, 0), (1, 0, 0), (2, 0, 0), (3, 0, 0), (4, 0, 0), (0, 1, 0)}. The final state of the procedure is given as follows: {(4, 3, 2), (0, 4, 2), (1, 4, 2), (2, 4, 2), (3, 4, 2), (4, 4, 2)}. Between the initial and the final states, the following is performed: the block in the cell with the highest sequence number is moved forward by one into the next cell. Should this not be possible, then the block with the predecessor's highest sequence number is moved forward by one, and all successors are moved behind. If the first block is in a cell from where it cannot be moved forward anymore, then the algorithm stops.

```
procedure Steps()
begin
    Size = width * depth * height
    for (i1=0; i1 < Size-5; i1++)
        for (i2 = i1+1; i2 < Size-4; i2++)
            for (i3=i2+1; i3 < Size-3; i3++)
                for (i4=i3+1; i4 < Size-2; i4++)
                    for (i5=i4+1; i5 < Size-1; i5++)
                        for (i6=i5+1; i6 < Size; i6++)
                            architectural compliance check
end
```

Figure 4. Generation procedure of the 6 blocks.

Each block is indexed with the zero-based sequence number (SN). Within the cycles, the blocks are placed into the cells of an index between 0 and 74 as mentioned earlier. The procedure given in Figure 3 represents a space-filling method based on SN. When all six blocks are placed, the batch is formed. The x, y, and z coordinates of the blocks must be calculated based on Equations (2)–(4).

In Equations (2)–(4) the // symbol denotes the integer division and % denotes the modulo division.

$$z = SN//(width \cdot depth) \tag{2}$$

$$y = (SN\%(width \cdot depth))//width \tag{3}$$

$$y = (SN\%(width \cdot depth))\%width \tag{4}$$

5. Meeting the Requirements

The above-described procedure generates all batches of six blocks within the search space. It is important that the procedure considers the whole search space and therefore it is sufficient to consider the generated batches only, and no other batches of six blocks should be taken into account. During the generation, when a new batch is available, it must be checked as to whether the architectural rules are satisfied or if any of the rules are violated. Should none of the rules be violated, then a potentially optimal configuration is found. To handle all rules the following process is performed.

Rule 1: corresponds to the elementary architectural requirement that the blocks, as general building elements of the house, should be connected, i.e., the configurations should be compact batches, see Figure 5. Obviously, a case where these blocks are separately scattered cannot be considered a family house. However, the connectivity requires further clarification, since the spatial connection of the blocks can be interpreted in various ways. For instance, blocks could be connected with their peaks or with their edges also, but these connections are also excluded from the current consideration since some space units become spatially separated from each other from an architectural point of view.

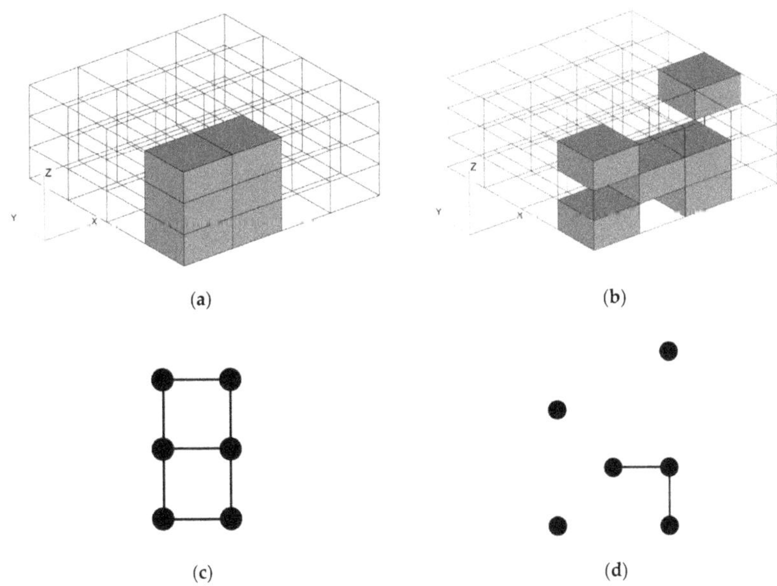

Figure 5. Compact batch (**a**) and a batch that is not compact (**b**). Underlying incidence graph (**c**) of the compact batch (**a**). Underlying incidence graph (**d**) of a batch that is not compact (**b**).

Hereinafter, a batch will be considered as connected, should the underlying incidence graph be connected. The underlying incidence graph can be formed as vertices of the graph are assigned to the blocks, namely, the node is assigned to the center of gravity of the block. Two vertices are linked together with an edge of the graph if, and only if, their corresponding blocks are connected to each other with their faces, i.e., the two blocks under consideration are connected through their floor, or roof, or one of their walls is common.

Rule 2: corresponds to the elementary architectural requirement that the blocks should not form a corridor-like long and narrow structure. Therefore, in the X or Y direction, in which the batch is longer, items on the ground and connected by walls, must form a 2–4 long line. This can be verified by counting connected elements in layers of the ground level of the incidence graph. Layers are created from the subgraph (ground level of the incidence graph) by slicing it parallel to X and Y axes. When counting the connected elements, the highest number must be in the range of 2–4.

Rule 3: corresponds to the elementary architectural prerequisite that accepts a cantilevered balcony under certain boundaries as a part of a family house. For example, a cantilevered shape with an open area below for cars, etc. may be accepted but this part of the building cannot become dominant, i.e., its size is limited to the size of only one block (Figure 6). This can be investigated by checking all blocks that are not on the ground level, i.e., whose z coordinate is not 0. Should there be no block below on the ground level, then the balcony index is increased by 1. All batches violate the rule where this index becomes greater than 1 at the end of the process. Moreover, should there be a block above a balcony-like block, the corresponding batch should be excluded from further considerations because due to the architectural request, unfeasible, impractical 'stacking' of the stories is also an unwanted solution.

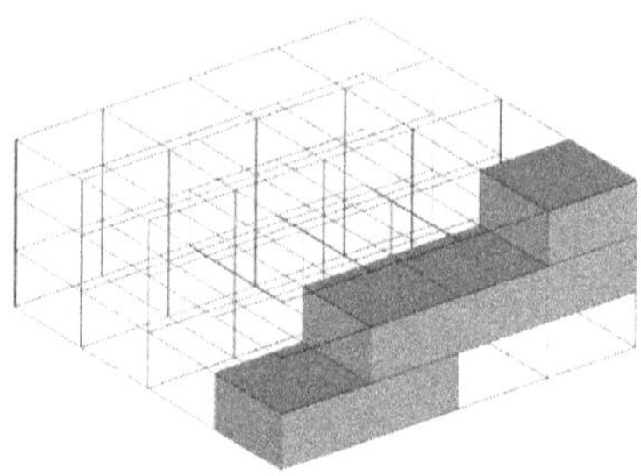

Figure 6. A batch of 6 blocks violating rule 3 as the cantilevered 'balcony' size exceeds the limit.

Rule 4: prescribes that only up to three blocks can be placed on top of each other, a higher building is not possible as a family house. In this way, unnecessary high-building shapes are avoided. This is obviously controlled by the search space, i.e., the height of the considered box is three.

Rule 5: does not allow the creation of a separate 2 story high (=stacked) stand-alone module, where each side of the module is free, and only the floor and roof sides of the blocks are connected to each other; in other words, no 'tower-like' shapes are allowed. That rule is satisfied when examining the underlying incidence graph without the ground level, there is no vertical subgraph with two nodes found. An example of violating rule 5, and an accepted configuration, are depicted in Figure 7.

Rule 6: buildings with a height of three blocks should not be longer than three blocks on the ground in any direction (X, Y). The total length of the batch along an axis could be longer. Figure 8 also violates rule 6.

It is important to emphasize that the rules include many other, elementary architectural requirements. For example, let us consider the requirement that a minimum of two blocks of the batch should stand on the ground floor. Should a batch satisfy the above rules, this requirement is also satisfied. It follows from the rules, namely, there can only be one cantilevered balcony and the batch cannot be higher than three blocks, therefore a minimum of two blocks should stand on the ground floor.

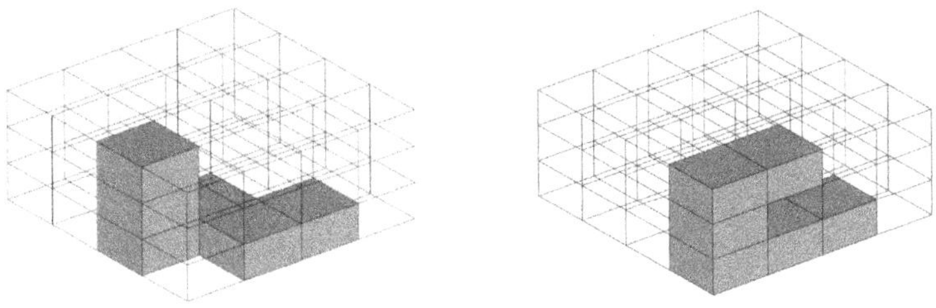

(**a**) (**b**)

Figure 7. Batch violating rule 5, 'towers' are not allowed (**a**); an accepted configuration (**b**).

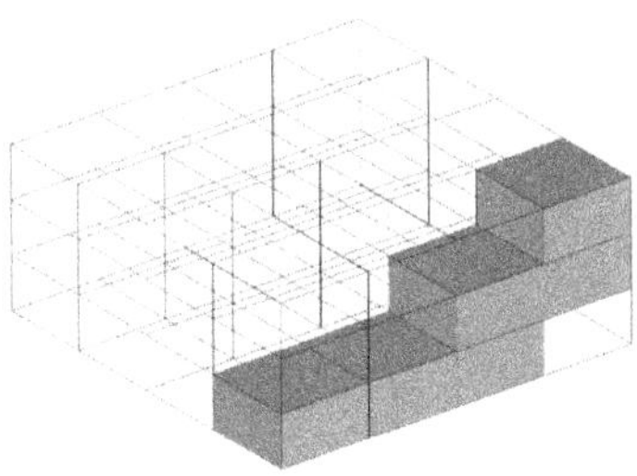

Figure 8. Batches of 6 blocks violating Rule 6. Total length is too long.

6. Identification of the Batches

It is easy to show that there is a large number of potentially feasible building configurations (batches ofsix blocks satisfying the rules), which have to be considered identical from the architectural point of view. Therefore, the problem of architectural identity or congruency must be solved. In the most common case, two batches can be considered to be identical, when one can become the other using diverse transformations such as shifts, rotations, and reflections. However, this is not appropriate for this specific architectural problem, since it is easy to transform a one-story batch into a two-story batch by only rotation around axis X, see Figure 9.

(**a**) (**b**)

Figure 9. Rotation around axis X is applied between a one-story batch (**a**) and a two-story batch (**b**).

Hereinafter, pursuant to an own developed transformation rule, two batches, or family house building configurations are considered to be identical or congruent, if they can become one another using rotation around a line parallel to axis *Z*, or reflection through a plane parallel to *XZ* and *YZ* planes, or shift parallel to the axes *X* or *Y*. All batches are accepted as identical, should they become one another by the application of any transformations or a series of the transformations below (Figures 10–12). The batch is considered to be in an initial position when any of its blocks touches the plane *XZ*, any of its blocks touches the plane YZ, and all of its blocks are within the positive orthant of the coordinate system.

Figure 10. The generated 1-level, ground floor batches (31 building shapes).

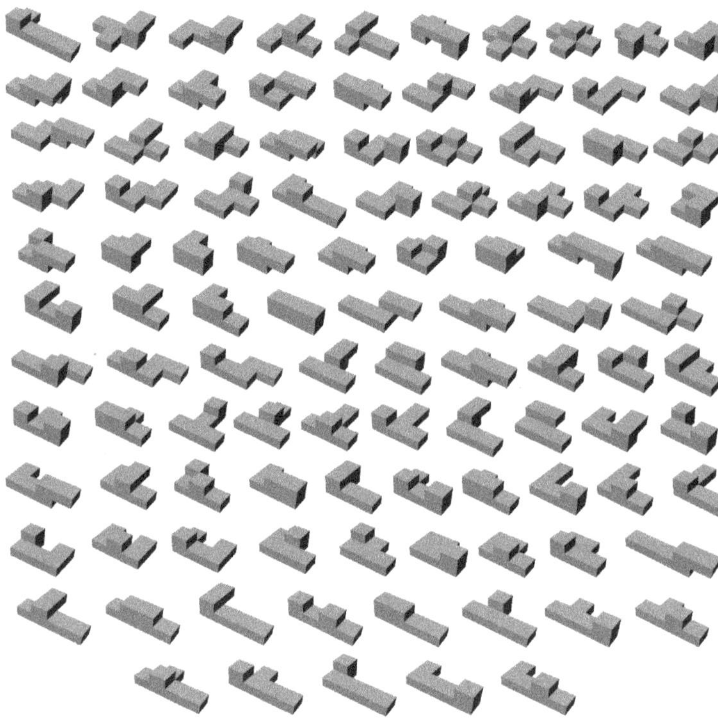

Figure 11. The generated 2-level batches (105 building shapes).

Figure 12. The generated 3-level batches (31 building shapes).

Transformation types:

Let there be a line e parallel to axis Z intersecting the point (2.5, 2.5, 0). Let us denote by R_1 the transformation that rotates the batch around the line e by -90°, R_2 the transformation that rotates the batch around the line *e* by $-180°$, and similarly, R_3 the transformation that rotates the batch around the line *e* by $-270°$. Please note that this transformation is very similar to the major orientation of the building, namely when the main façade of the building is turned in different directions. Nevertheless, it is worth mentioning that the orientation as well as the wall-to-window ratio (WWR) and other major architectural considerations are part of subsequent steps of the EDS methodology.

Let there be a plane *S* *(swap-plane)*, parallel to plane *XZ* intersecting the point (2.5, 0, 0). Let us denote by S_x the transformation reflection to the plane *S*. Similarly, let us denote by S_Y the transformation reflection to the plane that is parallel to the plane *YZ* intersecting the point (2.5, 0, 0). Therefore, the former transformation represents the swap between the front and the back of the building geometry, while the latter transformation represents the swap between the two sides of the batch.

Let us denote by T_x the transformation shift into the direction of the origin and parallel to axis *X*, and similarly T_Y the transformation shift into the direction of the origin and parallel to axis *Y*.

To support the algorithmic determination of whether two batches are identical or not, as well as to help further considerations, a new theorem is introduced.

Theorem 1. *Each and every identical batch can be constructed from an initial batch with a 1–2 steps long series of transformations of type R, and/or S, followed by at most one TX and at most one TY transformation. If the first part is 2 long, it has the form of* $R \circ S$.

Proof of Theorem 1. Let us perform all *R* and *S* transformations first and finally perform the required shift by T_x and T_y. First, the explicit form of the transformations is given.

$$R_{360} = R_0 : \; (x, y, z) \rightarrow (x, y, z) = I \tag{5}$$

$$R_{90} = R_1 : \; (x, y, z) \rightarrow (width - y, x, z) \tag{6}$$

$$R_{180} = R_2 : \; (x, y, z) \rightarrow (width - x, depth - y, z) \tag{7}$$

$$R_{270} = R_3 : \; (x, y, z) \rightarrow (y, depth - x, z) \tag{8}$$

$$S_X : \; (x, y, z) \rightarrow (x, depth - y, z) \tag{9}$$

$$S_Y : \; (x, y, z) \rightarrow (width - x, y, z) \tag{10}$$

The properties of used linear transformation are:

$$X \circ I = I \circ X = X \; where \; X \in \{R,\ S,\ T\} \tag{11}$$

$$(E \circ F) \circ G = E \circ (F \circ G) \; where \; E,\ F,\ G \in \{R,\ S,\ T,\ I\} \tag{12}$$

$$R_i = R_{i+k*4} \; where \; k \in N^+ \tag{13}$$

Based on Equations (5)–(10) and (11)–(13), the following identities can be defined:

$$R_i \circ R_j = R_k \; where \; k = (i + j) \; \% \; 4 \tag{14}$$

$$S_X \circ S_X = I \tag{15}$$

$$S_Y \circ S_Y = I \tag{16}$$

$$S_X \circ S_Y = S_Y \circ S_X = R_2 \tag{17}$$

$$S_Y = S_Y \circ I = S_Y \circ S_X \circ S_X = R_2 \circ S_X \tag{18}$$

$$S_X \circ R_i \circ S_X = R_{(4-i)\ \%\ 4} \quad (19)$$

$$S_X \circ R_1 \circ S_X = R_3 \quad (20)$$

$$S_X \circ R_2 \circ S_X = R_2 \quad (21)$$

$$S_X \circ R_3 \circ S_X = R_1 \quad (22)$$

$$S_X \circ R_i = R_{(4-i)\ \%\ 4} \circ S_X \quad (23)$$

$$S_X \circ R_1 = R_3 \circ S_X \quad (24)$$

$$S_X \circ R_2 = R_2 \circ S_X \quad (25)$$

$$S_X \circ R_3 = R_1 \circ S_X \quad (26)$$

Let us suppose that there exists an initial arbitrary series of transformations of type R and S. With the repeated application of the identities (14)–(17), a series of transformations can be constructed in which transformations of type R and type S are located alternately, i.e., there are neither two type R, nor two type S transformations beside each other. With the application of identity (18), S_Y transformations are eliminated from the sequence by being replaced by $R_2 \circ S_X$. After the application of identity (14), a series of transformations is made where transformations of type R and type S_X are located alternately. After the application of identities (19)–(22), the resulting series of transformations will be shortened to the length of one or two steps, or the resultant series will have the form of $R_i \circ S_X \circ R_j$. Should it have the latter type, one of identity (23)–(26) then Identity 1 has to be applied. Thus, the final series is an equivalent of the original series of transformations, has one step or two steps and if it is two steps long, it has the form $R \circ S$. After the application of the reduced sequence, the necessary shift must be applied by T_X and T_Y if required. Therefore, the theorem has been proved.

After the batch generation process, the identity of generated batches must be investigated. This investigation is performed pairwise. The new batch is compared to all previously generated unique batches collected in a list. First, the orientation and then the positioning of batches must be adjusted, then congruency can easily be checked by comparing batch elements. On orientation adjustment, instead of finding the transformation sequence which could transform one batch to match others' orientation, based on the above-presented theorem, we create one and two-step long transformation sequences from R and S transformations. That will result in 12 transformations, explained in (27) (including Identity to also check the generated batch), for which transformation matrices could be pre-generated to make transformation faster.

$$|Identity| + |R| + |S| + |R| \cdot |S| = 1 + 3 + 2 + 3 \cdot 2 = 12 \quad (27)$$

After the transformation of the new batch, shifting initial and all transformed cases to the X and Y axes handles positioning differences as described in Equations (28) and (29).

$$T_x = min(C_{i_x}) \text{ ; minimum of X coordinates of cells of batch} \quad (28)$$

$$T_Y = min(C_{i_Y}) \text{ ; minimum of Y coordinates of cells of batch} \quad (29)$$

If none of the 12 new cases are found in the list of previously generated batches, the shifted initial batch can be saved as a new unique batch by adding it to the list. □

7. Configurations

With the presented method, instead of generating the total number of 201,359,550 batches, only a total number of 167 batch configurations were necessary to be generated (Figures 10–12). The building geometries (=batches) satisfying the rules are presented in the following figures in form of an overview. 18.6% (31 pieces) of the geometries are one-level (ground floor) building bodies, 62.8% (105 pieces) consist of two stories and the remaining 18.6%

(31 pieces) include three levels. The ground-level configurations represent a marginal part of the complete geometry package, based on the multiple, transformation-based congruency instances. The three-level shapes contribute with a smaller number of examples, mainly due to the 'tower'-rule (Rule 6) that limits the possible, acceptable block-unit constellations above the ground level. A major part of the shapes is two-level configuration, where a higher number of batch arrangements of blocks were possible. This is mainly caused by the possibility to set blocks with unconnected walls (only floor-roof connection between the stories), next, the 'balcony'-rule (Rule 4) enables further variations as well. According to statistics [38], the greatest number of family housing are two-level buildings with a sloped roof (with or without attic rooms) or a flat roof. This shows an interesting correlation since in this particular study the largest number of generated geometry configurations possess two-levels as well. However, to see whether the optimal solution belongs to this configuration group or not is subject to further simulation-based investigation.

8. Simulations and Results

The total number of 167 batch configurations (Figures 10–12) gained were modeled in an indoor climate and energy simulation framework (IDA ICE 4.8 SP2) to assess the energy performance of each configuration. At this stage of the investigations, the shape versions do not possess glazed façade openings and orientation settings, therefore no solar radiation-based cooling and lighting demand permutations evolve among the versions. Although this simulation setup is not realistic, it was chosen to emphasize the effect of configuration (building shape) on energy demand. Consequently, only the annual heating energy demand is of particular interest, because under the chosen climate conditions, heating energy consumption plays the greatest role in the yearly energy balance of the residential building. This character was underlined in a preliminary study [39], whereas the heating energy demand was weighted at 80% of the total energy requirement (heating + cooling + lighting). The structures of the geometries were modeled from conventional construction materials, and they all meet the minimum requirements of the Hungarian building energy regulations (7/2006 V.24. TNM) as follows

- External wall structure (from inside to outside), $U_{wall} = 0.24$ W/m^{2}K:
  - render 1.5 cm
  - brick wall 30.0 cm
  - external thermal insulation 8.75 cm
  - render 1.5 cm
- Slab on ground structure (from inside to outside), $U_{floor} = 0.283$ W/m^{2}K:
  - ceramic tiles 0.8 cm
  - concrete screed 8.0 cm
  - thermal insulation 10 cm
  - reinforced concrete 8.0 cm
  - consolidated layer of gravel (hardcore) 15 cm
- Flat roof structure (from top to bottom), $U_{roof} = 0.17$ W/m^{2}K:
  - consolidated layer of gravel (hardcore) 15 cm
  - thermal insulation 20 cm
  - reinforced concrete 20.0 cm
- Cantilever slab (floor) structure, $U_{roof} = 0.17$ W/m^{2}K:
  - ceramic tiles 0.8 cm
  - concrete screed 8.0 cm
  - thermal insulation 5.66 cm
  - reinforced concrete 15.0 cm
  - external thermal insulation 15 cm

Figure 13 displays the ordered list of simulated annual heating energy demands (grey and black colored graphs) of the configurations; in other words, it is visible how the building shape versions affect the results. The maximum value is 18.461 kWh/year while the minimum is 13.163 kWh/year. Thus, the deviation range is 5297 kWh/year. The plus-minus deviation of the heating demand values related to the average heating value of all configurations amounts to 16%. In addition, a 40% difference between the minimum (optimal) energy demand and the worst-case heating value can be observed, demonstrating how high the increment of heating demand develops by the suboptimal solution(s). Both results indicate a significant measure of how building shape permutation affects annual building heating energy performance.

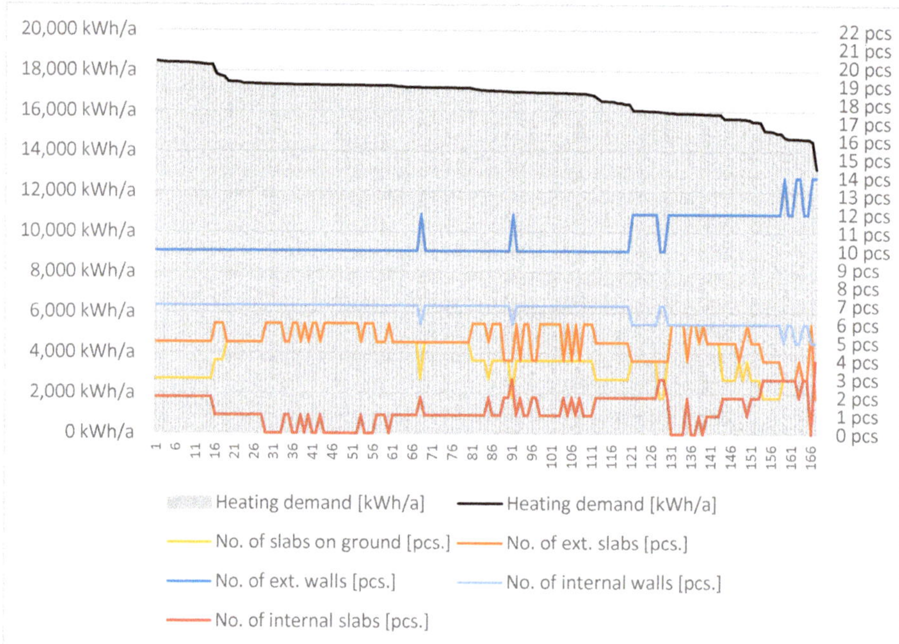

Figure 13. Heating energy demand [kWh/m^{2}a] of the 167 batches with indicated amount of external heat loss surface and internal connection surface in each configuration.

Figure 13 includes the heat loss surfaces of the building envelope structures, separated into vertical (external walls) and horizontal (external slabs and slabs on the ground) components. The number of external wall components (blue colored graphs) possesses no significant effect on the energy results, moreover, it has a counterproductive effect: the higher the number of external walls, the lower the heating demand develops. However, the reason for that is clearly seen in the graph of external slabs (yellow- and orange-colored graphs). Compared to an external wall, one slab measures a 40% larger transmission heat loss surface, therefore with decreasing number of external slab surfaces (more stories and hence a higher number of external walls in the shape configurations), the heating demand improves (becomes lower) as well. In other words, by increasing the number of internal slabs, the number of internal walls and transmission losses decrease as well. This interdependency offers the potential to elaborate an energy performance-related geometry generation rule, whereby for the number of horizontal internal connections of the space units (number of internal walls), as well as for the vertical connections of the space units (number of internal slabs) defined threshold values can be linked. Under a certain minimum and/or maximum threshold value, the given shape configuration is considered an energy-inefficient case during the geometry generation procedure. However, further investigation is required to gain precise statements about the modular connectivity rules, especially when in further simulation-based investigations the geometries are combined with further passive design elements (wall-to-window ratio, structures, materials, shading, orientation), as well as active planning factors such as HVAC and energy supply systems and operation control mechanisms. Interesting aspects may arise, for example, when the Weather Research and Forecast meteorological model (WRF) is used to predict future climate scenarios [40], or when state-of-the-art visualization techniques are applied to the simulation results [41].

9. Conclusions

In the framework of the given architectural problem, by using an appropriate algorithmic method, it is possible to specify the process of the guaranteed generation of all potentially feasible building configurations. The current study proposed a mathematically verified method that includes a modular space unit system (blocks) to arrange them in building shapes (batches) according to predefined architectural rules. The space organization codes served the goal to create only such shape configurations, which are feasible from the architectural point of view (e.g., appropriate compactness, etc.). All possible connection solutions of the space units can be ensured by a classical back-tracking algorithm to meet the prerequisites. To avoid congruency of the configurations and thus to further reduce the number of simulation cases, an algorithmic congruency detection is applied to check the shapes via a transformation analysis. Through changing the models by rotation, reflection, and shifting, the transformation of one batch into the other can be successfully proven. To further reduce computation time, the series of transformations can be shortened to the length of one or two steps by using an own developed theorem.

By using the above method, a significant reduction of the possible geometry variants can be carried out, in this particular case it is more than 99%. It can be stated that the guaranteed number (167) of geometry configurations was successfully generated, satisfying the given architectural rule system. The generation of all configurations, which meet the predefined rules, would logically result in the case of a changed predefined rule system another number of batch configurations.

Since scientific studies focus mainly on the numerically easy-to-define design variables such as heating, ventilation, and air conditioning (HVAC) system parameters, energy management system values as well as system operation data, and only a few efforts tried to integrate geometry-related architectural design variables, the present paper focuses on the effect of building shape on the annual heating energy demand. Thus, simulations were performed in a limited manner, i.e., without taking into account, for example, solar radiation by not using fenestration and/or main façade orientation. By simulating the most dominant (heating) energy performance of all gained batches, the energy-related significance of the building geometry modifications was decisively underlined.

In a future step, the result of this research enables the integration of the proposed generative step in a comprehensive sustainable building performance optimization system: by simulating the geometries equipped with glazed façade partitions (windows), diverse structures, materials, orientation, as well as varieties of HVAC and electrical systems or seasonal operation strategies (natural vs. mechanical ventilation, passive and active lighting, passive and active heating-cooling, etc.) all relevant passive, shape-related building design strategies, as well as active, system related solutions can be evaluated. This gives a helping hand to the architects to fully control their pursuit towards designing energy, and comfort-optimal residential buildings.

Author Contributions: Conceptualization, I.K. and Z.E.; Mathematical modeling, Z.E., T.S. and Z.K.; Software, Z.K.; Validation, B.D.; Architecture and simulation, K.R.H. and I.K.; Visualization, T.S.; Supervision, Z.E. and I.K. Founding, Z.E., I.K. and B.D. All authors have read and agreed to the published version of the manuscript.

Funding: The publication was granted by the Faculty of Engineering and Information Technology, University of Pécs, Hungary, within the framework of the "Call of Grant for Publication".

Data Availability Statement: The building configuration data can be generated using the procedure described in this article. The stated thesis can be verified by the presented proofs and does not require a concrete implementation. The simulated energy data mentioned in the introduction are only used data, not relevant to the actual content, and therefore do not need to be shared.

Acknowledgments: This work was partially supported by the [2019-2.1.11-TÉT Bilateral Scientific and Technological Cooperation]. B.D. gratefully acknowledges the European Commission for funding the InnoRenew CoE project (Grant Agreement #739574) under the Horizon2020 Widespread-Teaming program, and the Republic of Slovenia (Investment funding of the Republic of Slovenia and the European Union of the European Regional Development Fund). He is also grateful for the support of the Slovenian National Research Agency (ARRS) through grant BI-HU/21-22-004.

Conflicts of Interest: The authors declare no conflict of interest.

Abbreviations

BECEDO	Building Energy: Comfort and Environmental Design Optimization
ED	Energy Design architectural design method
EDS	Energia Design Synthesis architectural design method
GA	Genetic Algorithm
HVAC	Heating, Ventilation and Air Conditioning
LCA	Building Life-Cycle Assessment
SN	Sequence Number
WRF	Weather Research and Forecast
WWR	Wall-Window Ratio

References

1. Cao, X.; Dai, X.; Liu, J. Building energy-consumption status worldwide and the state-of-the-art technologies for zero-energy buildings during the past decade. *Energy Build.* **2016**, *128*, 198–213. [CrossRef]
2. IEA; UNEP. *2019 Global Status Report for Buildings and Construction*; UNEP: Gigiri Nairobi, Kenya, 2019; Volume 224.
3. Hausladen, G.; Saldanha, M.; Liedl, P.; Sager, C. *Climate Design*; Birkhäuser: Basel, Switzerland, 2005.
4. Lemke, C.R. *Architectur Form & Solarenergie*; Cuvillier: Gottingen, Germany, 2009.
5. Nguyen, A.-T.; Reiter, S.; Rigo, P. A review on simulation-based optimization methods applied to building performance analysis. *Appl. Energy* **2014**, *113*, 1043–1058. [CrossRef]
6. Szalay, Z. Modelling building stock geometry for energy, emission and mass calculations. *Build. Res. Inf.* **2008**, *36*, 557–567. [CrossRef]
7. Baranyai, B.; Póth, B.; Kistelegdi, I. Planning and research of smart buildings and constructions with the 'Energydesign Roadmap' method. *Pollack Period.* **2013**, *8*, 15–26. [CrossRef]
8. Kistelegdi, I.; Horváth, K.R.; Storcz, T.; Ercsey, Z. Building Geometry as a Variable in Energy, Comfort, and Environmental Design Optimization—A Review from the Perspective of Architects. *Buildings* **2022**, *12*, 69. [CrossRef]
9. Mert, Y.; Saygın, N. Energy efficient building block design: An exergy perspective. *Energy* **2016**, *102*, 465–472. [CrossRef]
10. Granadeiro, V.; Duarte, J.P.; Correia, J.R.; Leal, V.M. Building envelope shape design in early stages of the design process: Integrating architectural design systems and energy simulation. *Autom. Constr.* **2013**, *32*, 196–209. [CrossRef]
11. Rodrigues, E.; Fernandes, M.; Soares, N.; Gomes, A.; Gaspar, A.R.; Costa, J.J. The potential impact of low thermal transmittance construction on the European design guidelines of residential buildings. *Energy Build.* **2018**, *178*, 379–390. [CrossRef]
12. Lopes, M.A.R.; Antunes, C.H.; Reis, A.; Martins, N. Estimating energy savings from behaviours using building performance simulations. *Build. Res. Inf.* **2016**, *45*, 303–319. [CrossRef]
13. Depecker, P.; Menezo, C.; Virgone, J.; Lepers, S. Design of buildings shape and energetic consumption. *Build. Environ.* **2001**, *36*, 627–635. [CrossRef]
14. AlAnzi, A.; Seo, D.; Krarti, M. Impact of building shape on thermal performance of office buildings in Kuwait. *Energy Convers. Manag.* **2009**, *50*, 822–828. [CrossRef]
15. Yu, Z.; Gou, Z.; Qian, F.; Fu, J.; Tao, Y. Towards an optimized zero energy solar house: A critical analysis of passive and active design strategies used in Solar Decathlon Europe in Madrid. *J. Clean. Prod.* **2019**, *236*, 117646. [CrossRef]
16. Ciardiello, A.; Rosso, F.; Dell'Olmo, J.; Ciancio, V.; Ferrero, M.; Salata, F. Multi-objective approach to the optimization of shape and envelope in building energy design. *Appl. Energy* **2020**, *280*, 115984. [CrossRef]
17. Kiss, B.; Szalay, Z. Modular approach to multi-objective environmental optimization of buildings. *Autom. Constr.* **2020**, *111*, 103044. [CrossRef]
18. Waibel, C.; Evins, R.; Carmeliet, J. Co-simulation and optimization of building geometry and multi-energy systems: Interdependencies in energy supply, energy demand and solar potentials. *Appl. Energy* **2019**, *242*, 1661–1682. [CrossRef]

19. Caldas, L.G.; Norford, L.K. Genetic Algorithms for Optimization of Building Envelopes and the Design and Control of HVAC Systems. *J. Sol. Energy Eng.* **2003**, *125*, 343–351. [CrossRef]
20. Saratsis, E.; Dogan, T.; Reinhart, C.F. Simulation-based daylighting analysis procedure for developing urban zoning rules. *Build. Res. Inf.* **2017**, *45*, 478–491. [CrossRef]
21. Michalek, J.; Choudhary, R.; Papalambros, P. Architectural layout design optimization. *Eng. Optim.* **2002**, *34*, 461–484. [CrossRef]
22. Javanroodi, K.; Nik, V.M.; Mahdavinejad, M. A novel design-based optimization framework for enhancing the energy efficiency of high-rise office buildings in urban areas. *Sustain. Cities Soc.* **2019**, *49*, 101597. [CrossRef]
23. Gan, V.J.; Wong, H.; Tse, K.; Cheng, J.C.; Lo, I.M.; Chan, C. Simulation-based evolutionary optimization for energy-efficient layout plan design of high-rise residential buildings. *J. Clean. Prod.* **2019**, *231*, 1375–1388. [CrossRef]
24. Yi, Y.K.; Malkawi, A.M. Optimizing building form for energy performance based on hierarchical geometry relation. *Autom. Constr.* **2009**, *18*, 825–833. [CrossRef]
25. Wang, W.; Rivard, H.; Zmeureanu, R. Floor shape optimization for green building design. *Adv. Eng. Informatics* **2006**, *20*, 363–378. [CrossRef]
26. Tuhus-Dubrow, D.; Krarti, M. Genetic-algorithm based approach to optimize building envelope design for residential buildings. *Build. Environ.* **2010**, *45*, 1574–1581. [CrossRef]
27. Bichiou, Y.; Krarti, M. Optimization of envelope and HVAC systems selection for residential buildings. *Energy Build.* **2011**, *43*, 3373–3382. [CrossRef]
28. Kistelegdi, K.H.; Ercsey, Z. Energia design synthesis. In Proceedings of the VOCAL 2018. 8th VOCAL Optimization Conference: Advanced Algorithms, Esztergom, Hungary, 10–12 December 2018.
29. Sadoughi, A.; Kibert, C.; Sadeghi, F.M.; Jafari, S. Thermal performance analysis of a traditional passive cooling system in Dezful, Iran. *Tunn. Undergr. Space Technol.* **2019**, *83*, 291–302. [CrossRef]
30. Ochoa, C.; Capeluto, I.G. Strategic decision-making for intelligent buildings: Comparative impact of passive design strategies and active features in a hot climate. *Build. Environ.* **2008**, *43*, 1829–1839. [CrossRef]
31. Friedler, F.; Tarján, K.; Huang, Y.; Fan, L. Graph-theoretic approach to process synthesis: Axioms and theorems. *Chem. Eng. Sci.* **1992**, *47*, 1973–1988. [CrossRef]
32. Cabezas, H.; Heckl, I.; Bertok, B.; Friedler, F. Use the P-graph framework to design supply chains for sustainability. *Chem. Eng. Prog.* **2015**, *111*, 41–47.
33. Ercsey, Z. Process network solution of a clothing manufacturer's problem. *Pollack Period.* **2017**, *12*, 59–67. [CrossRef]
34. Bertok, B.; Bartos, A. Renewable energy storage and distribution scheduling for microgrids by exploiting recent developments in process network synthesis. *J. Clean. Prod.* **2020**, *244*, 118520. [CrossRef]
35. Nagy, Z.; Ercsey, J.T.; Kovács, Z. Bus transport process network synthesis. *Acta Polytech. Hung.* **2019**, *16*, 25–43. [CrossRef]
36. Holczinger, T.; Ősz, O.; Hegyháti, M. Scheduling approach for on-site jobs of service providers. *Flex. Serv. Manuf. J.* **2019**, *32*, 913–948. [CrossRef]
37. Russel, S.; Norvig, P. *Artificial Intelligence A Modern Approach*, ISBN-13: 978-0-13-604259-4, 3rd ed.; Pearson Education, Inc.: London, UK, 1995.
38. American Housing Survey. Available online: https://www.census.gov/programs-surveys/ahs/data.html (accessed on 12 April 2022).
39. Kistelegdi, K.R.; Kovács, Z.; Ercsey, Z. Energy Design Synthesis For Buildings: An Example. In Proceedings of the 22nd Conference on Process Integration, Modelling and Optimisation for Energy Saving and Pollution Reduction, Crete, Greece, 20–23 October 2019.
40. Giama, E.; Chantzis, G.; Kontos, S.; Keppas, S.; Poupkou, A.; Liora, N.; Melas, D. Building Energy Simulations Based on Weather Forecast Meteorological Model: The Case of an Institutional Building in Greece. *Energies* **2022**, *16*, 191. [CrossRef]
41. Vera-Piazzini, O.; Scarpa, M.; Peron, F. Building Energy Simulation and Monitoring: A Review of Graphical Data Representation. *Energies* **2022**, *16*, 390. [CrossRef]

energies

Review

Energy Efficiency of Tall Buildings: A Global Snapshot of Innovative Design

Mir M. Ali [1], Kheir Al-Kodmany [2,*] and Paul J. Armstrong [3]

1 Structures Program, School of Architecture, University of Illinois at Urbana-Champaign, Champaign, IL 61820, USA
2 Department of Urban Planning and Policy, University of Illinois at Chicago, Chicago, IL 60607, USA
3 School of Architecture, University of Illinois at Urbana-Champaign, Champaign, IL 61820, USA
* Correspondence: kheir@uic.edu

Abstract: Design priorities for tall and supertall buildings have for some time shifted to achieving more energy efficiency to address the energy needs of the increasing global population. Engineers and architects aim to achieve energy conservation through active and passive approaches, pursuing technological innovations and adopting climate-responsive design. Because of the green movement currently dominating the building industry, tall buildings that need a massive amount of energy to build and operate, and the practical desire to switch from non-renewable to clean renewable energy resources, intense attention has been given to the energy efficiency of tall buildings in the recent past. Due to the vast array of energy-efficient design features, equipment, and applications available now, it is timely to examine the pros and cons of these issues. This review paper is an attempt to comprehensively present and deliberate these issues. It illustrates and discusses the concepts and applications through a few case studies from several continents worldwide. The review shows that the design of tall buildings focusing on energy conservation is an evolutionary process and there is a need for further research about how to face the associated challenges to improve energy efficiency by developing creative solutions and strategies, as well as applying additional innovative technologies.

Keywords: passive design; technology; non-renewable energy; renewable energy; sustainable tall buildings; architectural design

Citation: Ali, M.M.; Al-Kodmany, K.; Armstrong, P.J. Energy Efficiency of Tall Buildings: A Global Snapshot of Innovative Design. *Energies* **2023**, *16*, 2063. https://doi.org/10.3390/en16042063

Academic Editor: Maria Founti

Received: 19 January 2023
Revised: 16 February 2023
Accepted: 17 February 2023
Published: 20 February 2023

1. Introduction

In ancient times, buildings were massive in the absence of sophisticated structural analysis and design methods, as well as mechanical and electrical systems. They were designed using materials with a large amount of thermal mass and appropriate building orientation that considered the direction and movement of the Sun and wind, provision of natural ventilation, etc. In other words, the design was primitive yet climate-responsive that builders developed from their ingenuity, empirical observations, trial and error, and experience. Following the industrial revolution, as modern technologies emerged, engineers devised Heating, Ventilation, and Air-Conditioning (HVAC) systems, and architects designed buildings drawing upon primitive notions but more erudite. The energy crisis of the 1970s was a revelation for energy consumers and building designers to appreciate the importance of fuel efficiency, not only for tall buildings but also across the board for the entire building enterprise. It functioned as a strong impetus to do something about energy conservation.

The notion of sustainable development, or sustainability, did not come so much from the academic discussion as from the international political process [1,2]. In 1983, the United Nations established the World Commission on Environment and Development (WCED), aiming to resolve conflicts cropping up in the developed and developing worlds. The organization published the 1987 "Our Common Future" Report, also known as the Brundtland

Report [2,3], which launched the expression "sustainable development", later strengthened by the Earth Summit in Rio de Janeiro, Brazil, in 1992. The leading organization in the US promoting and educating building professionals is the Green Building Council (USGBC), which sponsored a series of rating systems called Leadership in Energy and Environmental Design (LEED) for evaluating a building's sustainability [4,5].

The construction of tall buildings has proliferated in many cities in the past few decades. Several reasons have resulted in the emergence of this building type [3,6]. While "tall building" is a generic term for buildings exhibiting tallness and verticality, there is a gradation of these buildings based on height. Buildings over 50 m (164 ft) but less than 300 m (984 ft) are considered tall, over 300 m (984 ft) are supertall, and over 600 m (1968 ft) are megatall. The taller the building, the more energy demand becomes pronounced because the building is more exposed to the environment at greater heights, particularly to the effect of greater wind intensity.

Energy is consumed in tall buildings in three phases, i.e., during construction, operation, and demolition. As a massive amount of energy is consumed by these buildings during the long-term operational phase spanning their entire life, the scope of this paper is limited to this phase. As the energy consciousness of architects and engineers began in the 1970s and was reinforced in the 1990s following the Brundtland Report of 1989, many architectural and engineering/technological innovations and developments occurred since then [3,7]. The UN predicts that there will be an average global temperature increase of 10 °C (50 °F) from now (2023) until the end of this century [4,8]. The effect of climate change because of increasing temperatures and heat waves results in the rise in air-conditioning costs. This paper aims to develop a narrative based on a review of these developments. The literature offers scattered efforts on energy efficiency [9–14]. For example, González-Torres et al. [9] investigated causes of energy efficiencies in the residential sector, including irresponsible tenants' behavior, leading to increased CO_2 emissions, and exacerbating climate change. Yang et al. [10] explained how electric vehicle (EV) charging in prominent tall buildings had increased power grid load and the role of renewable energy in offsetting the demand. Li and Colombier's article [11] examined China's Building Energy Efficiency (BEE) involvement in climate change mitigation, assessed BEE technology in Chinese cities, and discussed economic and institutional barriers to the large-scale implementation of sustainable, low-carbon, and carbon-free construction technologies. Al-Kodmany's book [12] offers a comprehensive account of the environmental design of tall buildings. Ali's paper [13] examined low-energy buildings and architectural technologies that use climate and environment as allies and turns buildings into batteries. Closely related, this article focuses on the architectural design, engineering, and technologies that improve the thermal performance of office buildings. It lays out the theories and principles of sustainable design, surveys prominent buildings, and examines a half dozen buildings in depth with illustrations (Figure 1).

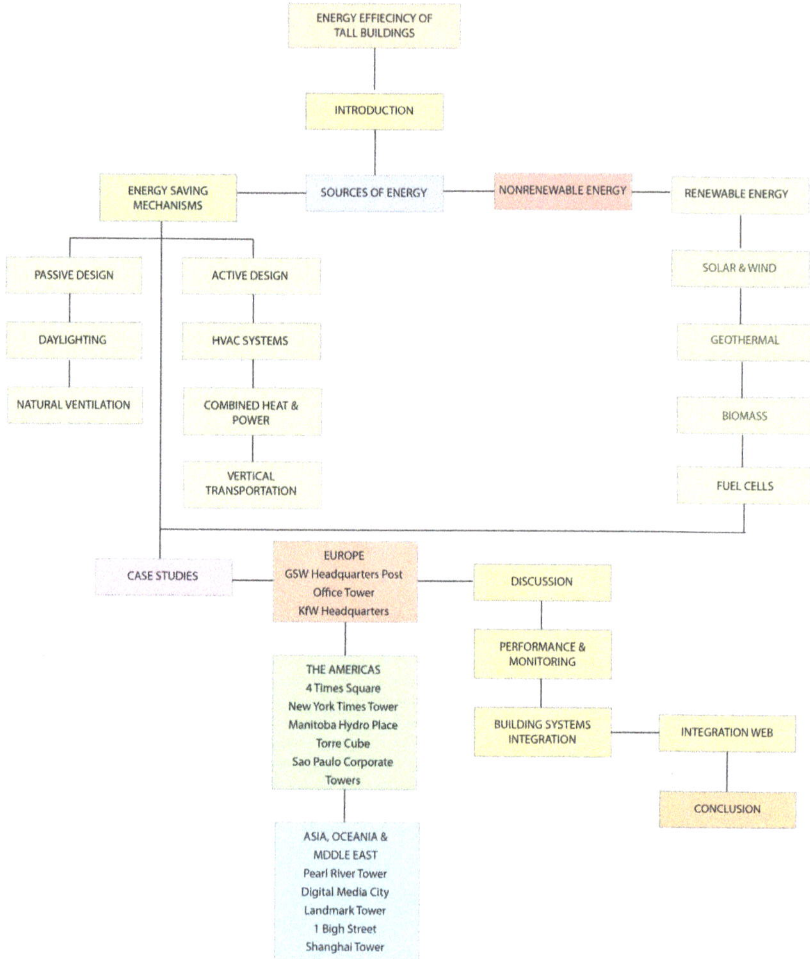

Figure 1. Illustration of the structure and layout of the paper. (Diagram by P. Armstrong).

2. Sources of Energy

The primary sources of energy generation can be broadly categorized as non-renewable and renewable. The non-renewable sources are natural fossil fuels such as coal, crude oil, and natural gas, formed underground in the remote geological past from the remains of living organisms. These are depleting due to their non-stop consumption, and the depletion is exacerbated by the growing world population, raising the concern for energy deficiency. Unlike fossil fuels, renewable energy is free from depletion. This type of energy is also called green or clean energy. While these last two terms have the same meaning, they have some minor distinctions.

2.1. Non-Renewable Energy

Natural fossil fuel contains hydrocarbon formed from the remains of dead plants and animals and is extracted and burned as fuels. Pollution is a significant problem of fossil fuels as they give off carbon dioxide when burned, which causes a greenhouse effect, leading to global warming. Coal is the worst of the three fossil fuels (i.e., coal, crude oil, and natural gas) as it produces more carbon dioxide and gives off sulfur oxide, creating acid rain. The mining of coal destroys vast areas of land. Capturing the carbon and diverting it to the green forest, the greenery, plants, vegetation, and crops that could absorb it will go a long way toward reducing global warming. The byproduct of this might be to strengthen the trees, plants, and crops. The crops will produce better foods and positively contribute to the food chain [5,11,14,15].

Oil causes pollution and poses environmental hazards such as oil spills in oceans and seas from oil tankers and releases toxic chemicals when combusted, causing air pollution. Using natural gas can cause unpleasant odors during transportation and accidents due to explosions. Methane is a potent greenhouse gas emitted from oil and gas infrastructure. When non-renewable energy is spent less in buildings, it naturally leads to less carbon emission into the atmosphere and, hence, less global warming [11,16–18].

2.2. *Renewable Energy*

Unlimited renewable energy sources are derived from natural sources such as solar and wind that do not run out. Furthermore, hydro, biomass, i.e., plant and waste material, and geothermal energy are other sources. These do not result in carbon emissions but may impact the environment. Nuclear energy is clean, but if a meltdown occurs due to human error, it will be disastrous. This discourages the use of atomic energy for power generation. Biomass causes deforestation, and hydropower creates land use problems and affects marine life. Green energy is a renewable energy causing no carbon emissions and has a minimal environmental impact. It includes solar, wind, low-impact hydro, and limited types of biomasses. Likewise, clean energy has zero carbon emissions, but many biogases from organic matter, household wastes, and manure are clean but not completely renewable. The capture of biofuel and landfill gas can produce clean energy [19–21].

The most innovative new building designs are those that double as energy generators. Many adjustments are being made to the way electricity is distributed. Coal-fired power facilities and large hydroelectric projects are used to account for most of the electricity generation in the US, for example, and distribution to customers throughout the country has resulted in significant losses. "Distributed resources" such as rooftop solar PVs are increasingly being used to generate electricity, rather than traditional centralized power facilities. In the past, buildings have been passive users of the electric grid, even though they account for 70% of all electricity use in the US. With distributed energy systems at the helm, buildings are increasingly taking on a more proactive role in the energy network, acting as generators as well as consumers. In some areas, renewable energy is already mandated for brand new buildings. "Solar-ready" building codes are becoming increasingly mandatory in the new construction industry. As climate change becomes a more pressing concern, it stands to reason that stringent regulations such as these will always be necessary [8,9].

2.2.1. Harnessing Solar and Wind Energy

There are two types of solar energy: active and passive. Active solar energy is implemented through technological installations such as solar collectors and photovoltaic (PV) panels. Researchers are working hard and advancing PV to make it a practical solution for the sustainable energy supply in buildings. PV cells convert light into electrical energy. Commercial PV cell performance has been steadily improving depending upon the type of cell and density to permit the transmission of sunlight. The application of PV technology for high-rise buildings can be significant as these structures offer an opportunity for direct sunlight if neighboring buildings do not over-tower them. The disadvantage of PV technology is the massive power required to produce them, the source of which is fossil-based fuels. Moreover, strict management and recycling assessments are needed to produce toxic and flammable gases containing phosphate and cadmium. Extensive research is continuing to overcome these difficulties. Whereas active solar energy is applied via technological installations, the passive solar energy concept is applied in practice as a design strategy to realize space heating, daylighting, etc. [22–25].

To exploit wind energy, wind turbines can be installed on tall buildings to generate electricity. At higher heights of tall buildings where wind speed is particularly strong, wind can be used as a source of energy. Tall buildings can be shaped to funnel wind into a zone containing wind turbines without negatively affecting the structure, the surroundings, and the occupants. In fact, by such form-giving of the building, wind speed can be magnified to enhance energy production. A disadvantage of wind turbines is that, together with other mechanical components, they can cause vibration in slender buildings. It is, for this reason, that the initial employment of a windmill at the top of the 542 m (1778 ft) tall One World Center in New York was subsequently dropped. Wind turbines, however, do not impact tall concrete buildings much, unlike steel buildings, as these buildings have considerable mass and damping characteristics [12,26–28].

2.2.2. Geothermal Energy

Tall buildings are often built deep into the ground; hence, this form of plentiful thermal energy from the ground source can be exploited to support the HVAC system of the buildings. The common surface manifestation of this type of energy is the hot water from springs. Geothermal heat pumps can exploit the high temperature of the upper layers of the planet's crust [29–31].

The rate of temperature increase in the ground, or the "geothermal gradient", averages 2.5 to 3 °C (36.5 to 37.5 °F) every 100 m (330 ft) of depth [13,32,33]. Modern drilling methods can reach depths of up to 9.5 km (6 mi). Hot water from springs is the most frequent surface manifestation of geothermal energy. Since the 19th century, natural hot water has been utilized in industrial settings. Built in 1913, the first geothermal power plant generated 250 J/s (250 kW) of power [14,34–36]. The heated, dry rock with a high temperature is another source. It is necessary to bring geothermal heat to the surface. The method of heating a place by pumping water via boreholes and then returning it to the surface is referred to as borehole heat exchange. Geothermal energy has the advantage of being unaffected by seasonal variations and climatic changes [37].

A significant area of innovation is the pairing of geothermal energy with heat pump technology. This technology has incrementally been upgraded, especially in the US. During the last few years, the number of geothermal ground-source heat pumps has grown significantly, with most of the development in the US and Europe. As the foundation of tall buildings necessitates deep excavation, its application to these buildings could prove more relevant than any other building type. The technology needed to tap into this energy warrants considerable advancement, refinement, and expertise. More research on this technology is necessary to make geothermal energy economically competitive with conventional energy sources [38].

2.2.3. Biomass Energy

Another renewable energy source is biomass energy derived from biomass fuel, i.e., living and once-living things or organic materials, such as wastepaper, which is available in abundance in office buildings. Biomass is the sum of all the Earth's living matter within the biosphere. More specifically, it refers to the concept of growing plants as a source of energy. When biomass is converted to fuel as a source of chemical energy, the process is carbon-neutral. This energy can be used for generating electricity. Humans have historically made fires from wood for cooking and staying warm in cold climates. Now, biomass is utilized to fuel electric generators and machinery. Common biomass materials are from plant sources that can be burned to generate heat and transformed into electricity. Biomass is a clean, renewable energy source. Its initial energy comes from the Sun that plants need to grow. Trees, crops such as corn and soybean, and municipal solid wastes are generally available and can be managed sustainably [39].

Substantial amounts of biomass are abundant in tall office buildings in the form of paper, most of which is used only briefly and then trashed. Biomass fuel, such as wastepaper in office buildings, can be used for generating electricity and steam for tall buildings. The Illinois Institute of Technology investigated a 73-story Chicago multi-use high-rise project in this research area [15]. Based on the investigative data, the study assumed a wastepaper production of 0.110 kg/m^2 (0.022 lb/ft^2) per day for offices and commercial space. The study concluded that a biomass-integrated gasifier/steam-injected gas (BIG/STIG) turbine would be the most efficient system for using biomass fuel. The Princeton Center for Energy and Environmental Studies researched using gas turbines with biomass fuels [16,40,41]. Bioenergy is renewable due to the continuously growing botanical sources, and its generation does not contribute to global warming, as it is essentially carbon-neutral.

Biomass, however, has some disadvantages. For example, if biomass feedstocks are not restocked as rapidly as they are used, they can be converted to non-renewable energy. Biomass generally requires arable land to grow. This means that the land used for biofuel crops are unavailable for other uses. In addition, burning biomass releases carbon, nitrogen oxides, and other pollutants and particulates. If these pollutants are not captured and reprocessed, they can produce smog. Moreover, the amount of wastepaper generated by office buildings has decreased due to the digitization of documents and their internet transmission [42].

2.2.4. Fuel Cells

Another source of renewable energy is the fuel cell. A fuel cell is an electromagnetic device that generates electricity like batteries and that can be considered an electrochemical internal combustion engine. It is a reactor that combines hydrogen and oxygen to produce electricity, heat, and water. It is used for spacecraft, airplanes, and other mechanical transportation systems. An example is 4 Times Square in New York City, which employs two 200 J/s (200 kW) fuel cells utilizing natural gas to generate power. The cells provide 100% of the nighttime electric demand without combustion, and hot water and carbon dioxide are the only byproducts. Similarly, the One World Trade Center Tower is partly powered by 12 hydrogen fuel cells, which produce 4.8 MW of power. Notably, the waste heat output from the fuel cell system is used for hot water and heating, amounting to 73,899 kJ (70,000 BTU of high-grade heat and 527,550 kJ (500,000 BTU) of low-grade heat. At present, its cost is high, but with future mass production, it is bound to go down [12]. Soon, fuel cells will provide heat and electricity for many offices and residences. More research is needed to make fuel cells economically competitive and to improve their performance to broaden their application on a larger scale [43].

3. Energy-Saving Mechanisms

3.1. Passive Design

3.1.1. Façades, Daylighting, and Electric Lighting

Daylighting is a crucial aspect of façade design for sustainable tall buildings. A façade acts as a building's "skin". The energy loss or gain of a tall building depends much upon the materiality and technology employed in the façade treatment. Glazed façades were considered a weak link in tall buildings for energy performance because of their insignificant insulating capability other than preventing the inflow of outside air and altering the inside temperature. These have now become sophisticated with the application of innovative technologies. The total-building concept, including the systems of HVAC, electricity, structure, and the façade, promotes the notion of integrated design in which all the systems are interdependent [26]. A high-performance façade engineered in the early stages of design development has become critical as the energy efficiency to achieve sustainability has become an indispensable performance criterion for buildings [27].

In the latter half of the 20th century, the usage of double-skin façades—two glass layers separated by an air space—and occasionally triple-skin façades with a natural ventilation system gained popularity [28]. A double-skin façade lessens the heat input in the summer and heat loss in the winter; this type of glazing serves as a barrier between internal and external conditions. Likewise, double-glazing with argon-filled cavities, triple-glazing, and glass coatings can increase U-values and screen ultraviolet rays of light [29]. The internal temperature is maintained via passive thermal processes, which include heating and cooling without needing electricity, in conjunction with ventilation of the space between the skins. The extra cost incurred for materials with a higher embodied energy to improve the thermal insulating capabilities of façades is typically recovered through lower energy usage over the course of the building's life.

Using energy-efficient lighting is vital in tall buildings, in which optimal solutions can result in substantial savings in energy consumption. Lights in a tall building have a relatively low embodied energy cost and, thus, these savings are real. Energy-efficient lamps and lighting control systems can be integrated with daylight to provide reductions in total consumption. Lighting in tall buildings can make up 10 to 25% of the total electrical load, depending upon the power requirements and the amount of heat load that is transferred back into the cooling load [24]. When daylighting is combined with electric illumination produced from renewable and non-renewable sources, it can significantly reduce the amount of energy needed for indoor lighting. According to estimates, daylighting combined with continuous dimming of electric lighting can reduce the energy required for interior illumination by 25 to 40% [25]. Automatic management of electric lighting in response to ambient daylight levels offers the highest benefit.

3.1.2. Natural Ventilation

An analysis of the site conditions is vital to natural building ventilation, especially concerning sufficient air circulation in sheltered areas (e.g., atria, sky gardens, and sky lobbies) [17]. A basic assessment of wind directions and intensities can be made using wind roses, which display the predominant winds throughout the seasons. Through vector analyses of virtual models, a more thorough study can be accomplished. Engineers and architects can now comprehend how wind affects structures and optimize their designs by utilizing a range of computerized programs. Wind tunnel studies using physical models fitted with sensors that provide accurate measurements of the effects of the wind on the structure and urban setting can result in a more thorough and complex experimental investigation.

Tall buildings must be constructed in a way that makes use of both positive and negative pressures acting on their exteriors to enhance natural ventilation and prevent internal building issues such as wind pressure on doors and windows. When outdoor temperatures and air velocity are above 29 km/h (18 mi/h) during the transitional seasons (autumn and spring) in temperate climates, high-rises can benefit from natural ventilation. When the outside temperature is above 22 °C (72 °F), additional mechanical ventilation is needed, especially in the summer. When the outdoor temperature drops below 5 °C (40 °F), mechanical ventilation and heat recovery devices are also advised [18].

3.2. *Active Design*

3.2.1. HVAC Systems

Manufacturers of air-conditioning appliances have developed high-performance, energy-efficient appliances for saving energy. These have EPA-certified ENERGY STAR labels on them. For any HVAC, the airflow distribution system is of vital importance. When the air distribution is well designed, it will result in energy efficiency and occupant comfort. The use of an underfloor air distribution system (UFAD) is one of the most efficient HVAC systems. It has several advantages over the traditional overhead system, such as raised-access floors for environmental control and improved access to building services. It also improves the ventilation system, causing high-quality airflow. Its layout reduces frictional losses in the system and causes a consistent temperature throughout the building. Furthermore, the UFAD permits complete flexibility for office areas and improved individual temperature control. Using highly efficient system components and an ice storage system, the building's central cooling and heating plant lowers the power demand. A LEED-certified design is something that architects are increasingly looking for.

3.2.2. Combined Heat and Power

Combined Heat and Power (CHP) (or a cogeneration system) is a very efficient energy-saving technology. In addition, its application is suitable for tri-generation systems. CHP allows concurrent production of heat, power, and, infrequently, chilled water for air-conditioning. An advantage of it is that, unlike conventional systems, it averts transmission losses as electricity is produced near the point of use. The simultaneous generation of power and heat enables overall thermal efficiencies, reducing a significant amount of fuel. Efficiency and operational time are enhanced further by adding a boiler system or another heat storage medium. The generated electricity can be used by a facility or transferred into a public power grid. The heat energy can be utilized for heating water and producing steam. Consequently, this system reduces fuel consumption, costs, and CO_2 emission. The CHP system is increasingly popular in several European cities for servicing commercial and institutional buildings. Stockholm, Helsinki, Copenhagen, etc., provide much of their electricity and heating from CHP systems [14]. Thus, CHP employment is an attractive option as most energy generated is valuable. It is a flexible system that can be adapted to low-to-zero-carbon applications.

3.2.3. Vertical Transportation

Elevators and escalators represent vertical mobility, which affects how efficiently energy is consumed and how effectively people move around. Vertical transportation makes up 5 to 10% of the building's energy use. Elevator technology is a persistent area of research and development as a result of the recent construction of numerous supertalls and some megatall buildings. This mode of transportation is currently more efficient due to technological advancements. In a "regenerative elevator", for instance, energy from descending vehicles is absorbed and used to power ascending ones. Additionally, design choices can increase the effectiveness of the net-to-gross-floor-area ratio. For instance, to increase energy efficiency, the double-decker elevators at the Petronas Towers in Kuala Lumpur have doubled the ridership of single-decker elevators. Another way to cut energy use by up to 50% is to use variable-speed gearless elevators with sophisticated programming and energy recovery [23].

Destination-oriented elevators can optimize the trip schedule and further save energy. In traditional systems, users wait after pressing an up-or-down call button. They then rush to choose their location and board the first available car, stopping at each floor they have chosen. The building's residents and visitors enter their destination floor at the central lobby booth using the destination-oriented system, which groups people traveling to the same floor together and cuts down on travel time. By reducing the number of elevators and conserving energy, this configuration creates more floor space that may be rented out or used for other purposes. Energy-efficient machine-room-free elevators that use permanent-magnet synchronous motor technology are also a significant breakthrough.

4. Case Studies

This section reviews case studies of tall buildings that feature energy-efficient designs. These projects are among the "greenest" towers in the world's continents, including Europe, North America, South America, Asia, Australia, and the Middle East. The review highlights the salient architectural and engineering aspects incorporated to improve energy performance. 13 critical projects are listed here as examples of energy-efficient tall buildings (Table 1 and Figure 2). A critical review discussing the case studies is presented in Section 5.

Table 1. Tall Buildings with Energy-Design Focus. (Compiled by authors).

	#	Name/Location	Year/Number of Floors	Function/Architect	Thumbnail Image
Europe	1	GSW Headquarters Berlin, Germany	1999 22	Office Sauerbruch Hutton	
	2	Post Office Tower Bonn, Germany	2002 41	Office Murphy/Jahn	
	3	KfW Headquarters Frankfurt, Germany	2010 14	Office Sauerbruch Hutton	
North America	4	4 Times Square NYC, USA	2000 43	Office Fox & Fowle Architects	
	5	New York Times Tower NYC, USA	2007 52	Office Renzo Piano	
	6	Manitoba Hydro Place Manitoba, Canada	2009 21	Office Kuwabara Payne McKenna Blumberg Architects	
	7	Torre Cube Guadalajara, Mexico	2005 20	Office Carme Pinós	

Table 1. *Cont.*

	#	Name/Location	Year/Number of Floors	Function/Architect	Thumbnail Image
South America	8	São Paulo Corporate Towers São Paulo, Brazil	2016 30	Office Pelli Clarke Pelli	
Asia	9	Pearl River Tower Guangzhou, China	2013 71	Office SOM/AS + GG	
	10	Digital Media City Landmark Tower Seoul, South Korea	Proposed 133	Mixed-use SOM	
	11	Shanghai Tower Shanghai, China	2015 128	Mixed-use Gensler	
Oceania	12	1 Bligh Street Sydney, Australia	2011 27	Office Ingenhoven Architects of Germany and Architectus of Australia	
Middle East	13	Al Bahar Towers Abu Dhabi, UAE	2012 26	Office AHR/Arup	

Figure 2 displays the energy-design features of the case study buildings. The red dots in the figure indicate the features that tall buildings have been provided with. For example, GSW Headquarters shown in the top row has been provided with solar shading, a double-skin façade, automated environmental controls, and adequate daylighting features. While each case-study tall building incorporates multiple energy-saving systems, the systems vary according to climate. For example, the Al Bahar Towers use passive solar shades attached to the façade that actively trace the Sun's path during the day. This system is applicable to the hot, arid climate of the Middle East. In Europe and North America, where climate conditions vary seasonally, acquiring daylight and natural ventilation is important and tall office buildings are likely to use active double-curtain walls for thermal comfort as well.

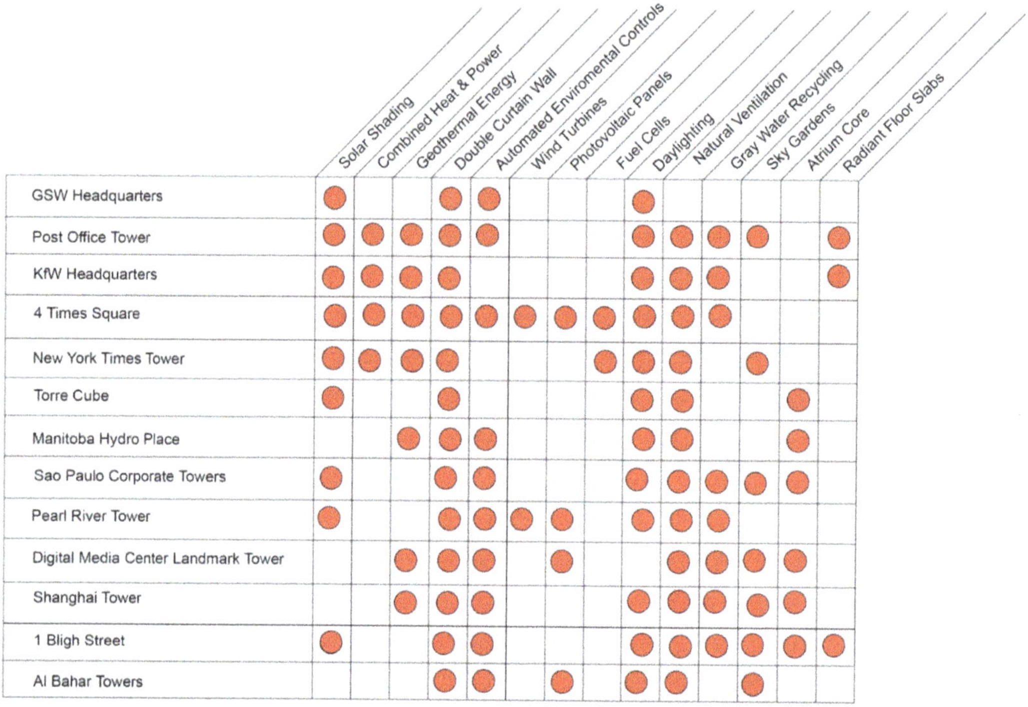

Figure 2. Matrix of energy-design features of case study buildings. (Chart by P. Armstrong).

Due to the lack of empirical data, the analysis does not quantify energy efficiency. Instead, the examination focuses on identifying innovative ideas that facilitate energy efficiency through the interaction between form, structure, function, and technology. The investigated case studies identified key energy design elements. For the purpose of this paper, the authors selected six significant cases out of those in Table 1 for illustration in the following section, ensuring geographic representations, height variation, and design diversity. Therefore, a single case study was selected representing each continent/region. For Europe, KfW Headquarters was chosen for it integrates the latest technology compared to the other two candidates i.e., GSW Headquarters and Post Office Tower. For North America, 4 Times Square was selected because it is widely considered a model of green skyscrapers. For Asia, Shanghai Tower was selected because it incorporates the most up-to-date technologies and innovative ideas, symbolizing China's advanced scientific achievements. Similarly, for Oceania and the Middle East 1 Bligh Street and Al Bahar Towers were selected, respectively.

4.1. Europe

KfW Headquarters, Frankfurt, Germany

KfW Headquarters of 2010 is a 14-story, 39,000 m^2 (419,793 f^2) extension to the KfW Banking complex in Frankfurt, Germany. The building has sustainable design features, including a double-layered wind-pressurized façade, radiant and thermally activated floor slabs, and a geothermal heating and cooling system. In addition, what distinguishes this building is that, unlike a conventional double-skin system, the cavity is not ventilated using a stack effect but instead by exploiting wind pressure and suction at the leeward side of the tower. The building achieves an outstanding low energy consumption of 98 kWh/m^2 per year—it is the first tall office building that consumes an energy below 100 kWh/m^2 worldwide.

The double-skin façade buffers against high winds and significantly cuts the incoming heat gain while providing ample natural light. Therefore, the computer-controlled ventilation panels accurately regulate the air entering the building during different seasons. In the winter, the façade functions as a passive thermal solar collector, as the fresh air flow is pre-tempered by solar radiation within the façade. This way, heat loss is minimized, and energy is conserved (Figure 3). Further, the enclosed sawtooth-shaped cavity, 0.7 m wide at its deepest point, contains automated blinds that help block solar gain and control glare [44,45].

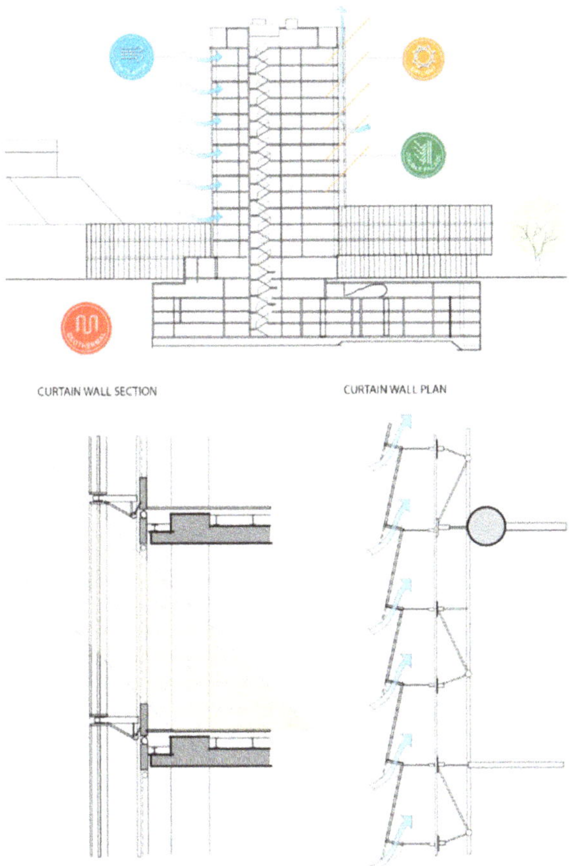

Figure 3. KfW Headquarters in Frankfurt, Germany. The building features sustainable design elements, including a double-layered wind-pressurized façade. (Illustration by P. Armstrong).

Additional energy-saving strategies include a thermal activation of the slabs, a heat recovery system, and a supplemental raised-floor ventilation system. Thermally activated slabs comprise a system of pipes built into the solid floors, which convey water that serves as a heating and cooling medium. This creates energy-efficient, comfortable, and constant room temperatures. A supplemental raised-floor ventilation system is used only when outside temperatures are below 10 °C (50 °F) or above 25 °C (77 °F). Sauerbruch Hutton—Architects designed the tower. It received the CTBUH Award of Best Tall Buildings in Europe in 2011.

4.2. North America

4.2.1. Times Square, NYC, USA

4 Times Square (also known as The Condé Nast Building) of 2000 is a 48-story, 247 m (810 ft) tall skyscraper in Manhattan's most renowned commercial and financial district. The building was designed by FXFOWLE (Fox & Fowle) Architects. Nicknamed the "Green Giant", it is the first building of its size in the USA to adopt high standards for energy efficiency. A critical step in reducing electric usage was the selection of environmentally friendly natural-gas-powered absorption chillers/heaters, which supply chilled and hot water to cool and heat the building. Variable-speed drives on pumps, fans, and motors optimize equipment efficiency and minimize energy use. Further, the building reduces energy consumption by primarily applying low-E glazing, which optimizes daylight transfer while reducing the passage of heat, minimizing the need for heating and cooling (Figure 4). The building incorporates an energy-efficient artificial lighting system, exit signs that use light-emitting diodes, and light sensors in unoccupied areas, including stairwells [46,47].

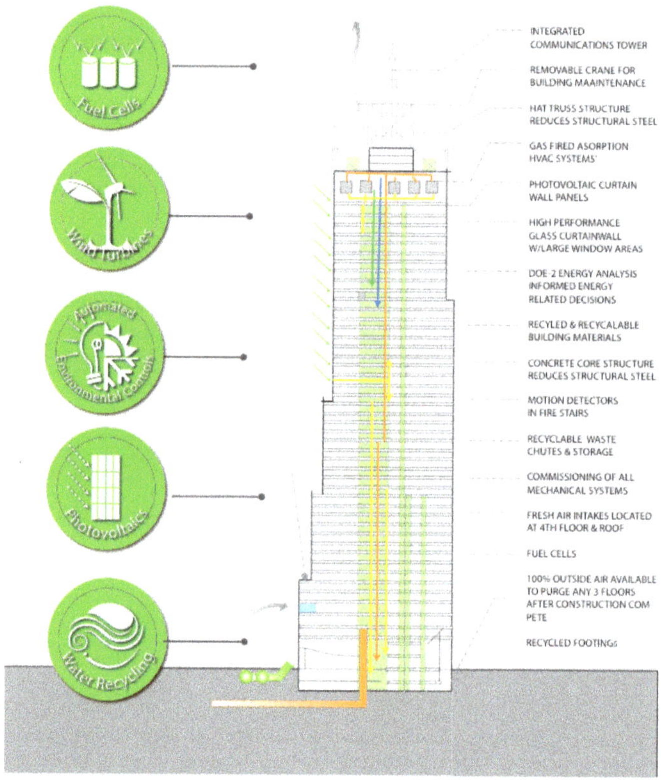

Figure 4. 4 Times Square in NYC. Nicknamed the "Green Giant", it is the first building of its size in the US to adopt high standards for energy efficiency. (Illustration by P. Armstrong).

In terms of energy production, the building generates much of the electricity it uses on-site through two 200 kW PureCell Systems on the fourth floor. In the USA, 42% of all energy produced is lost as waste heat in combustion and transmission; therefore, generating energy on-site and without combustion averts substantial energy loss. Thin-film photovoltaic (PV) panels replace glass spandrels from the tower's 37th to 43rd floors on the tower's south and east façades. Energy from the PV panels and the chiller is stored in two 200 kW fuel cells that provide continuous electrical power to the building. A web-based metering system allows for tracking energy consumption for tenants [48]. Completed in 2000, FXFOWLE Architects designed 4 Times Square with sustainable features at a time when the LEED accreditation was in its infancy. Shortly afterward, it earned the Gold Certification. Additionally, FXFOWLE Architects received a National Honor Award for the structure from the American Institute of Architects (AIA) in 2001.

4.3. South America

São Paulo Corporate Towers, São Paulo, Brazil

Located in the heart of the Southern Hemisphere's largest city, the São Paulo Corporate Towers are two mixed-use buildings, each thirty stories tall. They sit atop an amenity podium and underground parking, surrounded by spacious 19,000 m^2 (204,514 ft^2) lush botanical gardens and hundreds of trees, creating a "green oasis in the center of a dense metropolis." The established contrast between a natural landscape and the rising towers aligns with the city's close relationship with parks and gardens. The park uses more permeable soils to reduce run-off in the water-stressed city and is irrigated with rainwater collected by the towers' system. This water is also utilized to chill the structures.

The curtain walls were designed to make dynamic, sculptural shapes gradually narrow as they ascend. The façade's design, as well as its choice of glass and shading, was informed by extensive daylight studies. To maximize energy efficiency, the project included high-thermal-performance glass and external shading, efficient air conditioning and electric light, a stormwater capture and reuse system, and a sustainable landscape design. The project also incorporates a power plant with four 2 MW generators powered by diesel or natural gas. Overall, the project achieves a 48% energy consumption reduction yearly (Figure 5). It earned the highest environmental standard of LEED Platinum. Pelli Clarke Pelli Architects designed the building, which was completed in 2016 [49,50].

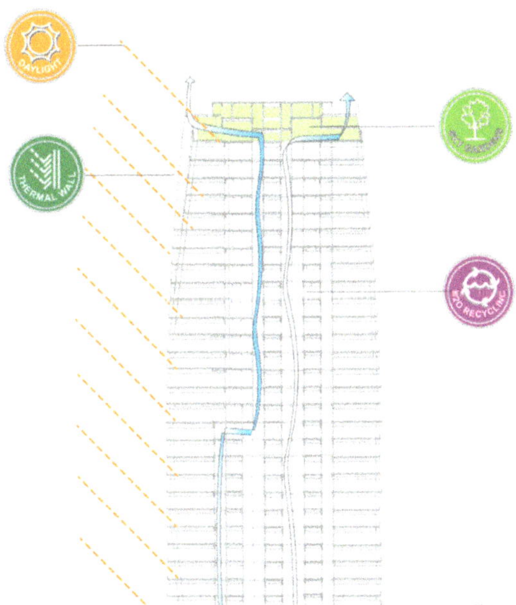

Figure 5. São Paulo Corporate Towers. The project comprised a high-thermal-performance façade and external shading, efficient air conditioning and electric light, stormwater capture and reuse, and a sustainable landscape design to maximize energy efficiency. (Illustration by P. Armstrong).

4.4. Asia

Shanghai Tower, Shanghai, China

The Shanghai Tower of 2015 in Shanghai rises to 632-m (2073 ft). It is China's tallest building and the world's third, after the Burj Khalifa and Merdeka 118 Tower. The building is divided into nine tiers, stacked vertically; each comprises 12 to 15 floors and contains a sky garden, providing energy-saving and ventilation advantages. By absorbing heat from the interior of the building in the summer and warming the cool air in the winter, they serve as buffer zones between inside and outside. Used indoor air is spilled into each garden before being exhausted from the building. The double-skin façade, advanced lighting controls, and an efficient central plant, among other features, help Shanghai Tower to save energy (Figure 6).

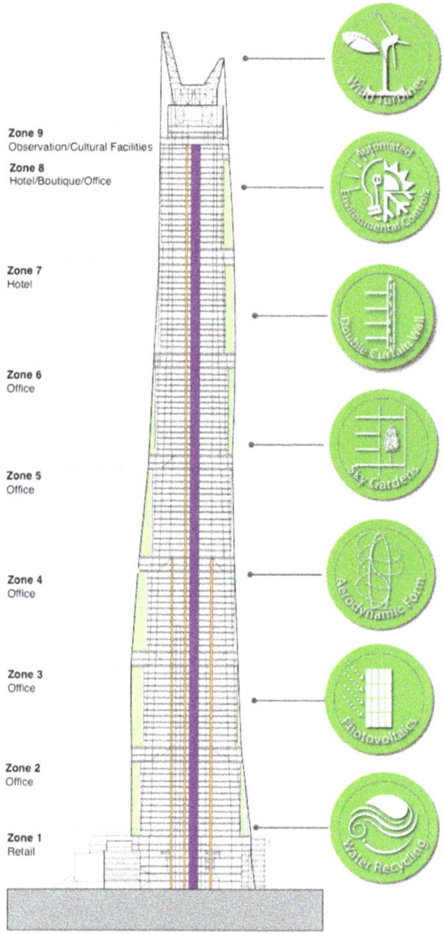

Figure 6. Shanghai Tower, Shanghai, China. Shanghai Tower represents a new paradigm of rethinking the sustainable vertical city. (Illustration by P. Armstrong).

With a mix of updraft, controlled top exhausts, and spill air on the last zone, the atrium is naturally ventilated in large portions, resulting in a 21% energy efficiency, 12.5% more than China's Three-Star Rating. The outside glass wall is staggered to reflect sunlight upward and away from the street below, lowering the light reflectance level to 12% [50–52], preventing severe glare [50–52].

On-site energy is provided using renewable energy features. The upper floors are powered by wind turbines that are situated right beneath the parapet, and the lower floors are powered by a 2130 kW natural-gas-fired cogeneration system. The spiraling parapet of the building collects rainwater, which is then used for the tower's heating and cooling systems. By recycling non-potable greywater, the tower consumes 40% less water. This equates to a yearly water savings of 178 million gallons (674 million liters), which is the same as 450 Olympic-sized swimming pools [53]. To save on pumping energy, water treatment facilities are built within the podium, basement, and shaft of the tower. Shanghai Tower has received a LEED Platinum Certification from the U.S. Green Building Council as well as a China Green Building Three Star rating. The tower was designed by Gensler.

4.5. Oceania

4.5.1. Bligh Street, Sydney, Australia

The 28-story tall 1 Bligh Street of 2011 is a premium-grade office tower. The building's sustainable features include a complete double-skin façade with movable blinds in the cavity to minimize solar heat gain and glare. The 1774 motorized Venetian blinds are automatically controlled using an intelligent motor controller that optimizes their operations to exterior weather conditions and the geometric shape of the building. Sydney's prevailing breezes enhance the louvers' performance and make the temperature inside the cavity close to the temperature outside the building, reducing the need for using the HVAC system. The façade allows most offices to receive adequate daylight, reducing the need for artificial lighting (Figure 7).

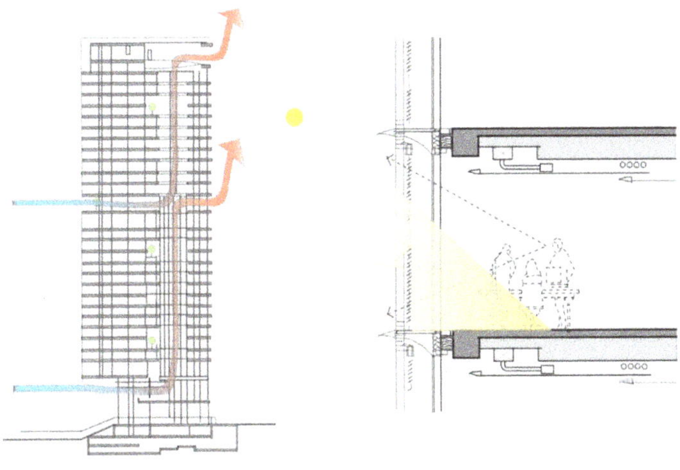

Figure 7. 1 Bligh Street, Sydney, Australia. The tower establishes new standards for site consciousness and innovation and sets a new benchmark for sustainable office design. (Illustration by P. Armstrong).

To further capitalize on Sydney's mild climate, a dramatic 120-m (394-ft) tall atrium, dubbed "the heart of glass", is located at the epicenter of the building and extends from the ground floor to the glazed rooftop. As such, the atrium provides natural ventilation and light to the surrounding corridors, meeting spaces, and balconies.

A hybrid tri-generation plant, which generates cooling, heating, and electricity, is placed on the site. It utilizes gas-fired power, absorption chillers, and solar-powered cooling to increase efficiency and minimize peak loads. An array of 500 m^2 (5382 ft^2) of roof-mounted solar panels provides the energy to drive the cooling systems. The tri-generation plant and solar panels reduce the load on the CBD grid power by more than 27% [54]. The building received a Platinum certification. It has also achieved a 6 Green Star performance rating—Office Design V2 Certified Rating, the highest rating by the Green Building Council of Australia (GBCA). Architectus designed the tower.

4.6. The Middle East

Al Bahar Towers, Abu Dhabi, UAE

Al Bahar Towers of 2012 comprise two 25-story-tall towers with elliptical forms. The towers' design modernized the traditional mashrabiya, a perforated wooden-lattice screen of geometric patterns found in vernacular Islamic architecture. The mashrabiya skin partially wraps around the towers to protect them from the harsh desert sun southward, exposing the northern portion of the façades. It comprises transparent umbrella-like units that open and close in response to external solar conditions. Sensors on the façades communicate solar conditions to the BMS (Building Management System), i.e., a centralized automated control system to manage the operation of different building systems, including HVAC and vertical transportation. The south-facing roof of each building incorporates PV cells to generate adequate power to operate the mashrabiya system.

Geometrically, the mashrabiya system follows a hexagonal pattern that simulates traditional Arabic-Islamic design. The system avoids using dark-tinted glass, which inevitably restricts incoming light. Instead, it lets daylight in for part of the day, reducing the need for artificial lighting. The system provides a 50% solar gain reduction, reducing energy consumption and CO_2 emissions [55] (Figure 8). The towers are among the first buildings in the Gulf region to receive the USGBC's LEED Silver rating. Aedas Architects designed the towers.

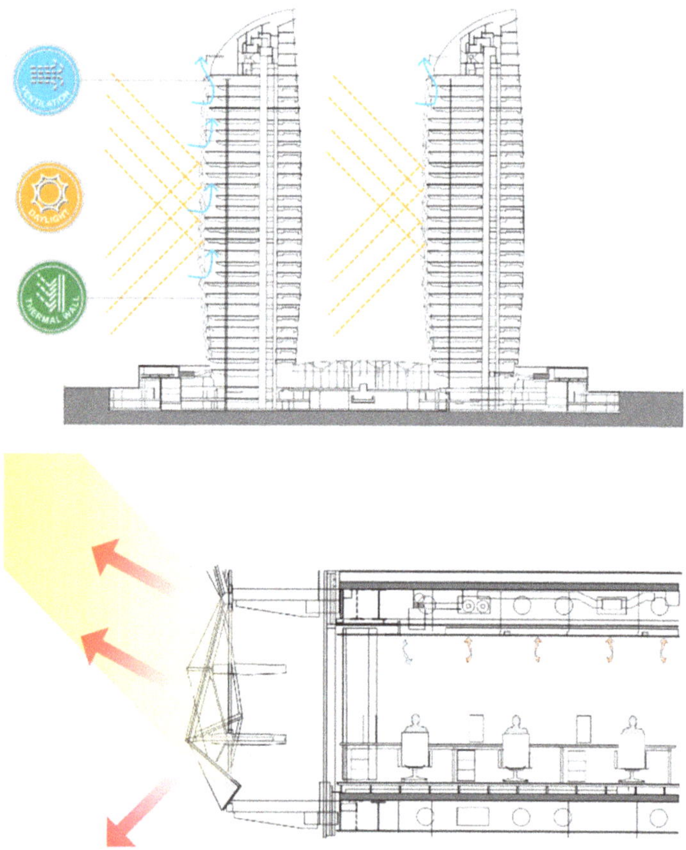

Figure 8. Al Bahar Towers, Abu Dhabi, UAE. Al Bahar's modified mashrabiya skin shields the towers from the south but leaves the northern portion of the façade exposed. Its automated translucent umbrella-shaped units open and close responding to the Sun. (Illustration by P. Armstrong).

5. Discussion

The examined case studies represent engineering and architectural feats. They exhibit the courage to break out of the status quo and explore ways to improve building design. As explained in this paper's earlier "theory" section, DSFs are excellent for insulating the building from exterior climatic conditions. KfW Headquarters in Frankfurt, Germany, and Shanghai Tower in Shanghai employ DSFs, and 1 Bligh Street in Sydney uses DSF. Interestingly, Shanghai Tower "inserts" sky gardens between the two building's skins, providing extra insulation. 1 Bligh Street places motorized blinds between the two skins to offer shade from the Sun. When the local temperature does not vary a lot, a DSF may not be needed. For example, in Sao Paulo, the temperature fluctuates from 33 °C (55 °F) to 28 °C (83 °F) and is rarely below 9.5 °C (49 °F) or above 32 °C (89 °F). As such, São Paulo Corporate Towers did not employ DSFs.

In contrast, some places experience extreme climates. For example, in the summer, Abu Dhabi experiences scorching weather with average daytime temperatures of around 45 °C (113 °F). Consequently, Al Bahar Towers employs an innovative, dynamic second skin that opens when the Sun is away and closes when the Sun is on the building. In all cases, high-performance glass, such as low-E glass, is recommended for it improves the thermal performance of the building. This is true, particularly in office buildings that demand maximum natural daylight to reduce the reliance on artificial lighting. In addition, incorporating artificial lighting control systems enhances energy saving.

Taking advantage of renewable energy is exhibited differently in the examined cases. For example, KfW Headquarters employs a geothermal heating and cooling system, 4 Times Square uses thin-film photovoltaic (PV) panels, Al Bahar Towers use PV cells, and Shanghai Tower uses wind turbines and harvests rainwater. Generating energy on-site reduces power loss as it travels from the source of production to the place of consumption. In the case of 4 Times Square, two 200 kW PureCell Systems generate most of the building's electricity. São Paulo Corporate Towers incorporate a power plant with four 2 MW generators powered by diesel or natural gas. Shanghai Tower uses a 2130 kW natural-gas-fired cogeneration system that provides electricity and heat energy, and 1 Bligh Street uses a hybrid tri-generation plant.

The prime task of architects and engineers becomes figuring out the set of design strategies most appropriate for a specific building. For example, KfW Headquarters in Frankfurt, Germany, uses activated radiant slabs that improve thermal comfort in both summer and winter. All its workplace areas receive optimal daylight to reduce the use of artificial lighting and employ techniques for sun protection, glare reduction, and daylight redirection to reduce heat gain. The primary energy consumption of the facility was further decreased by connecting KfW's with an effective heating and cooling network and utilizing trigeneration, biomass heating, excess data center heat, and free cooling energy from evaporative cooling towers. The Condé Nast Building (also known as Four Times Square or 4 Times Square) used numerous energy-saving architectural elements. With the help of solar panels and hydrogen fuel cells, the structure produces most of the electricity it needs. An improved air delivery system, fuel-efficient absorption chillers, and additional insulation in the walls and roof allow the building to maintain good interior air quality while using little energy.

Similarly, Shanghai Tower is a benchmark for energy-efficiency tall buildings. It has 270 wind turbines embedded into its outside that use Shanghai's strong winds to power the building's exterior lighting, and it also collects rainwater for internal use. It employs state-of-the-art control technologies that monitor power usage to reduce expenditures by $556,000 annually. Additionally, it incorporates ground source heat pumps (GSHPs) and DSFto further reduce energy consumption. In the same vain, 1 Bligh Street in Sydney uses innovative technologies. A cutting-edge tri-generation system, powered by gas and solar energy, provides cooling, heating, and electricity to the building. An innovative hybrid of variable air volume (VAV) and chilled-beam air conditioning technology runs on curved solar thermal collectors.

5.1. Energy Performance and Monitoring

The actual performance of tall buildings regarding energy consumption is difficult to assess and verify [5,13]. Although it appears to be a straightforward task that entails collecting data on the total annual energy usage of a tall building, the challenge lies in determining the individual energy expenditure and, hence, the performance of different systems, such as HVAC, lighting, pumping, computing, and elevators, as well as the contribution of passive systems. Utility companies collect data on total energy consumed and not necessarily the power supplied to individual systems and subsystems. Knowing the breakdown of the energy demand of a tall building for its systems can enable designers to design the building more efficiently. Computer software packages offering designers tools for determining the energy performance and lifecycle are developed and constantly upgraded. However, extensive research and monitoring devices are necessary to measure the relative energy performance of individual systems and subsystems. This will allow architects and engineers to determine the extent of the performance of these individual components for improving them, thereby optimizing the overall performance to minimize the energy consumption of tall buildings.

5.2. The Role of Building Systems Integration

The precise coordination and integration of all critical building systems are necessary to achieve sustainable construction. All design team members, including architects, engineers, planners, contractors, and others, must work together to complete this process, which starts in the early stages of planning. Physical integration, visual integration, and performance integration are the three categories of integration that Bachman lists [34].

Physical Integration: The sharing of space between systems and components is known as physical integration. In many structures, the floor–ceiling section is separated into several zones for structural, ductwork, and lighting. This region, called the "power membrane", aims to be energy- and space-efficient.

Visual integration: The development of visual harmony between numerous structural components, energy performance, and their compliance with the intended visual impacts of design is known as visual integration. This could comprise openly displayed and formally expressive building elements that work together to shape the building's appearance. An illustration is the exterior of Abu Dhabi's Al Bahar Towers, which features dynamic solar "parasol" sunscreens that save energy for cooling the building while giving a modernized vernacular visual appearance [56].

Performance integration: This concerns "shared functions" while saving energy. For example, a load-bearing wall serves as a structure and an envelope, combining two functions into a single unit while improving thermal performance. Another example is the ventilated double-skin curtain wall of Deutsche Post Tower in Bonn, Germany, which combines the duties of two components without joining the parts together.

Integration, which leads to simplicity, produces discipline and order and long-term economy because it achieves more significant energy efficiency and, hence, cost reduction throughout the building's lifecycle. The A/E firm Hellmuth, Obata & Kassabaum (HOK) conducted an internal study and discovered that projects with sustainable energy-conscious design were, on average, 25% more profitable for the company than those with conventional design [43].

Advanced façade applications that are appropriately linked with cutting-edge HVAC systems lead to significant energy savings and enhance indoor air quality as a byproduct. Not only will performance be improved if the HVAC and façade systems are designed as two parts of a single solution, but initial and ongoing expenses may also be considerably decreased. This shows that a new strategy is required, bringing mechanical and electrical experts and façade specialists together in the early stages of the design process.

5.3. Integration Web

Ali and Armstrong [57] created the Tall Building Systems Integration Web as a tool to aid architects and engineers in making decisions about integrating the physical systems for environmentally friendly tall buildings (Figure 9). Because they need to coordinate various intricate, interconnected systems to operate well and use less energy during operation, sustainable tall buildings demand greater integration from the start of the design process. It is more beneficial for a team to concentrate on long-term objectives and prospects at initial sessions [43]. The Integration Web improves decision-making at critical points by describing the connections between each important physical system and subsystems of a tall building. The creation of a technique for comparing the performance of an integrated sustainable building to a conventional structure that was planned with no consideration for sustainability can also result from it. The Integration Web demonstrates the network of interrelated building system items amenable to integration.

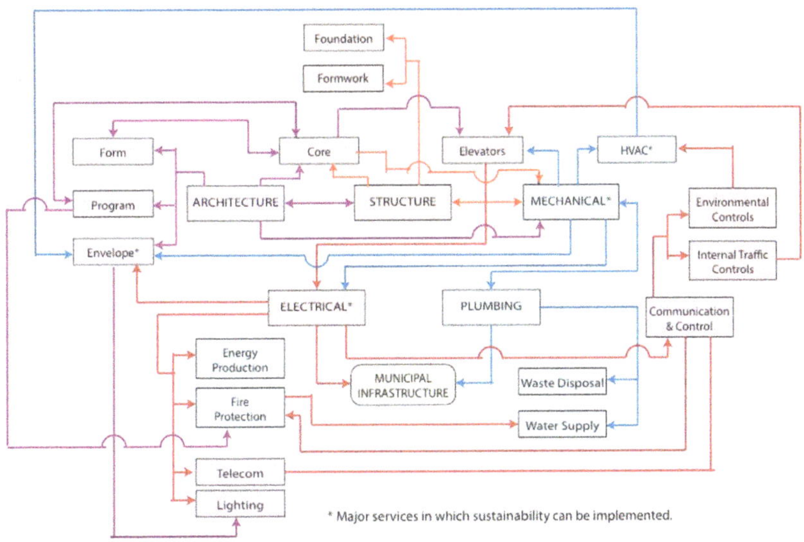

Figure 9. Integration Web displaying a network of interconnected tall building systems (Illustration by P. Armstrong).

The methodologies and strategies utilized by engineers in partnership with architects to construct sustainable, intelligent buildings will determine the future of the built environment. This is something that engineers and architects owe to society, and they need to lead by example while creating new structures. Energy-efficient structures may initially cost more than traditional buildings, but the longer they last, the less they will cost to operate. Assume that one learns that it is possible to use cutting-edge concepts and technology in building design to make the environment and climate work for the building rather than the building working against them by releasing harmful gases. In that circumstance, creating zero-energy buildings is very possible. To completely avoid utilizing any energy from the city's power system, the goal is to build structures with zero net energy use. Increasingly plausible, any structure might be planned to be a source of energy production. Not only can the energy use be reduced, but by taking things a step further, the building might even become a self-sufficient power source. By integrating the physical systems, or creating a symbiotic structure, high-performance design can result in a zero-energy tall building. Although integrated building systems initially cost more, long term cost reductions are possible due to tall structures' buildings' ability to maximize their overall entire energy performance through each different system's' synergistic interaction with each other and synergy [57].

6. Conclusions

Numerous facets of tall structures' energy efficiency have been reviewed in this paper. Several project examples have been presented to demonstrate the ideas and uses of energy efficiency. Intense research is continuing in this critical field as the consumption of energy is of crucial importance for the survival of the increasing world population. Unless the idea is firmly embraced and sustainable development inclusive of energy consciousness is put into practice, energy shortage, global warming, urban sprawl, air pollution, overflowing landfills, water shortage, disease, and international conflicts will continue to be the ills of the 21st century as one can see them today, and leave a legacy for the future. The survival of humans depends upon the survival of the cities—their built environment experiencing the proliferation of tall building construction for human habitation and work. The energy supply is the lifeline of human survival. Assurance of this warrants vision and action through the partnership and commitment of engineers, architects, allied professionals, policymakers, and other stakeholders. Comparing renewable energy sources to fossil-fuel-fired power plants, the former emits far fewer emissions. Every effort must be made to reduce emissions of greenhouse gases, which are a significant contributor to climate change. Clearly then, from the perspective of carbon emission, exclusive use of renewable and "cleansed" nonrenewable energy should be the goal of having as the singular source of energy for not only tall buildings but also other energy-consuming applications. In addition, new untapped sources of energy must be explored. This paper was intended to be an overview of existing modes and applications rather than to propose any new methods or novelties in the field of energy efficiency. This is left for future research.

Author Contributions: Conceptualization, M.M.A. and K.A.-K.; methodology, K.A.-K.; software, P.J.A.; validation, K.A.-K.; formal analysis, M.M.A., K.A.-K. and P.J.A.; investigation, M.M.A. and K.A.-K.; resources, M.M.A., K.A.-K. and P.J.A.; data curation, K.A.-K. and P.J.A.; writing—original draft preparation, M.M.A. and K.A.-K.; writing—review and editing, M.M.A., K.A.-K. and P.J.A.; visualization, P.J.A. All authors have read and agreed to the published version of the manuscript.

Funding: This research received no external funding.

Data Availability Statement: Not applicable.

Acknowledgments: We thank anonymous reviewers for insightful comments.

Conflicts of Interest: The authors declare no conflict of interest.

References

1. Newman, P. *Sustainability and Cities: The Role of Tall Buildings in the New Global Agenda*; CRC Press: Boca Raton, FL, USA, 2001; pp. 76–109.
2. Western Cape Education Department. *World Commission on Environment and Development. Our Common Future*; WCED: Berlin, Germany, 1987; Volume 17, pp. 1–91.
3. Al-Kodmany, K.; Ali, M.M. *The Future of the City: Tall Buildings and Urban Design*; WIT Press: Southampton, UK, 2013.
4. Tavernor, R. Visual and cultural sustainability: The impact of tall buildings on London. *Landsc. Urban Plan.* **2007**, *83*, 2–12. [CrossRef]
5. Beedle, L.S.; Ali, M.M.; Armstrong, P.J. *The Skyscraper and the City: Design, Technology and Innovation*; The Edward Mellen Press: Lewiston, UK, 2007.
6. Elotefy, H.; Abdelmagid, K.S.; Morghany, E.; Ahmed, T.M. Energy-efficient tall buildings design strategies: A holistic approach. *Energy Procedia* **2015**, *74*, 1358–1369. [CrossRef]
7. El-Darwish, I.; Gomaa, M. Retrofitting strategy for building envelopes to achieve energy efficiency. *Alex. Eng. J.* **2017**, *56*, 579–589. [CrossRef]
8. US Department of Energy. Available online: https://www.energy.gov/ (accessed on 15 January 2023).
9. González-Torres, M.; Pérez-Lombard, L.; Coronel, J.F.; Maestre, I.R.; Paolo, B. Activity and efficiency trends for the residential sector across countries. *Energy Build.* **2022**, *273*, 112428. [CrossRef]
10. Yang, Y.; Jia, Q.S.; Deconinck, G.; Guan, X.; Qiu, Z.; Hu, Z. Distributed coordination of EV charging with renewable energy in a microgrid of buildings. *IEEE Trans. Smart Grid* **2017**, *9*, 6253–6264. [CrossRef]
11. Li, J.; Colombier, M. Managing carbon emissions in China through building energy efficiency. *J. Environ. Manag.* **2009**, *90*, 2436–2447. [CrossRef] [PubMed]
12. Al-Kodmany, K. *The Vertical City: A Sustainable Development Model*; WIT Press: Southampton, UK, 2018.

13. Ali, M.M. Energy efficient architecture and building systems to address global warming. *Leadersh. Manag. Eng.* **2008**, *8*, 113–123. [CrossRef]
14. Smith, P.F. *Architecture in a Climate of Change*; Routledge: London, UK, 2006.
15. Moon, K.S. Sustainable structural engineering strategies for tall buildings. *Struct. Des. Tall Spec. Build.* **2008**, *17*, 895–914. [CrossRef]
16. Larson, E.D.; Williams, R.H. Biomass-gasifier steam-injected gas turbine cogeneration. *J. Eng. Gas Turbines Power* **1990**, *112*, 157–163. [CrossRef]
17. Ali, M.M.; Dimick, P.G. Structural Sustainability of High Performance Buildings. In *Challenges, Opportunities and Solutions in Structural Engineering and Construction*; CRC Press: Boca Raton, FL, USA, 2009; pp. 901–906.
18. Elnimeiri, M.; Gupta, P. Sustainable structure of tall buildings. *Struct. Des. Tall Spec. Build.* **2008**, *17*, 881–894. [CrossRef]
19. Ghoreishi, A.H.; Ali, M.M. Contribution of thermal mass to energy performance of buildings: A comparative analysis. *Int. J. Sustain. Build. Technol. Urban Dev.* **2011**, *2*, 245–252. [CrossRef]
20. Khan, F.R. New Structural Systems for Tall Buildings and their Scale Effects on Cities. In Proceedings of the Tall Building, Planning, Design and Construction Symposium, Nashville, Tennessee, 14–15 November 1974; pp. 14–15.
21. Gonçalves, J.C.S. *The Environmental Performance of Tall Buildings*; Routledge: London, UK, 2010.
22. Li, W.; Koo, C.; Hong, T.; Oh, J.; Cha, S.H.; Wang, S. A novel operation approach for the energy efficiency improvement of the HVAC system in office spaces through real-time big data analytics. *Renew. Sustain. Energy Rev.* **2020**, *127*, 109885. [CrossRef]
23. Foraboschi, P.; Mercanzin, M.; Trabucco, D. Sustainable structural design of tall buildings based on embodied energy. *Energy Build.* **2014**, *68*, 254–269. [CrossRef]
24. Raman, M. Aspects of energy consumption in tall office building. *CTBUH Review. Online J. Counc. Tall Build. Urban Habitat* **2001**, *1*, 17–25.
25. Deshmukh, N.P. Energy Conservation of Moderately Tall Office Buildings. Master's Thesis, School of Architecture, The University of Illinois at Urbana-Champaign, Champaign, IL, USA, 1992.
26. Hart, S. Facade engineering emerges as a highly specialized science and striking art form. *Archit. Rec.* **2002**, *190*, 163–170.
27. Herzog, T.; Lang, W. Using multiple glass skins to clad buildings. *Archit. Rec.* **2000**, S171–S182. Available online: https://www.semanticscholar.org/paper/Using-multiple-Glass-Skins-to-clad-Buildings-Herzog-Lang/bd77a5667dc00df47b7caf71586ac093aca7773c (accessed on 16 February 2023).
28. Behr, U. *Double Facades for High-Rise Buildings, Tall Buildings Beyond 2000, Vol. a, Urban Habitat, and High-Rise, TBMa, 1994–1999*; Awang, A., Ed.; Institute Sultan Iskandar: Malaysia, Asia, 2001.
29. Pank, W.; Girardet, H.; Cox, G. *Tall Buildings and Sustainability*; Corporation of London; London, UK, 20 March 2002.
30. Agathokleous, R.A.; Kalogirou, S.A. Double skin facades (DSF) and building integrated photovoltaics (BIPV): A review of configurations and heat transfer characteristics. *Renew. Energy* **2016**, *89*, 743–756. [CrossRef]
31. Trabucco, D. An analysis of the relationship between service cores and the embodied/running energy of tall buildings. *Struct. Des. Tall Spec. Build.* **2008**, *17*, 941–952. [CrossRef]
32. Ilgin, E. Core Design and Space Efficiency in Contemporary Supertall Office Buildings. In *Sustainable High Rise Buildings: Design, Technology, and Innovation*; The Institution of Engineering and Technology: London, UK, 2022; pp. 243–264.
33. Wood, A.; Du, P.; Safarik, D. Skybridges: State of the Art. *CTBUH J.* **2020**, 36–44. Available online: https://www.ctbuh.org/resources/papers/download/4340-skybridges-state-of-the-art.pdf (accessed on 16 February 2023).
34. Bachman, L.R. *Integrated Building Systems: The Systems Basis of Architecture*; John and Wiley and Sons: New York, NY, USA, 2003.
35. Meresi, A. Evaluating daylight performance of light shelves combined with external blinds in south-facing classrooms in Athens, Greece. *Energy Build.* **2016**, *116*, 190–205. [CrossRef]
36. Hansen, V.G. Innovative daylighting systems for deep-plan commercial buildings. Ph.D Dissertation, Queensland University of Technology, Brisbane, Australia, 2006.
37. Al-Kodmany, K. *Eco-Towers: Sustainable Cities in the Sky*; WIT Press: Billerica, MA, USA, 2015.
38. Shaviv, E. The influence of the thermal mass on the thermal performance of buildings in summer and winter. In *Science and Technology at the Service of Architecture*; Kluwer Academic Publishers: Dordrecht, The Netherlands, 1989; pp. 470–472.
39. Kamal, M.A. The Study of Thermal Mass as a Passive Design Technique for Building Comfort and Energy Efficiency. *J. Civ. Eng. Archit.* **2011**, *5*, 84–88.
40. Ghoreishi, A.H.; Ali, M.M. Parametric study of thermal mass property of concrete buildings in US climate zones. *Archit. Sci. Rev.* **2013**, *56*, 103–117. [CrossRef]
41. Yeang, K.; Richards, I. *Eco Skyscrapers I*; Images Publishing: Melbourne, Australia, 2007.
42. Allouhi, A.; El Fouih, Y.; Kousksou, T.; Jamil, A.; Zeraouli, Y.; Mourad, Y. Energy consumption and efficiency in buildings: Current status and future trends. *J. Clean. Prod.* **2015**, *109*, 118–130. [CrossRef]
43. Malin, N. A Group Effort. In *Green Source*; 2006; pp. 46–51.
44. Headquarters, G.S.W.; Hutton Architect, S. *Intelligent Skins*; GSW Headquarters: Berlin, Germany, 2002; p. 49.
45. Winterstetter, T.; Sobek, W. Innovative und energiesparende Fassadentechnik am Beispiel der KfW Westarkade, Frankfurt/Main. *Glasbau* **2013**, *82*, 35–46.
46. Sagalyn, L.B. *Times Square Roulette: Remaking the City Icon*; MIT Press: Cambridge, MA, USA, 2003.

47. Hedrick, M.; Horst, K.; Leman, C.; Perez, A. The New York Times Building. Master's Thesis, Massachusetts Institute of Technology, Cambridge, MA, USA, 2009.
48. Kuwabara, B.; Auer, T.; Gouldsborough, T.; Akerstream, T.; Klym, G. Manitoba Hydro Place: Integrated design process exemplar. In Proceedings of the 26 International Conference on Passive and Low Energy Architecture: PLEA 2009: Architecture Energy and the Occupant's Perspective, Quebec City, QC, Canada, 22–24 June 2009.
49. Murray, S.; Murray, S.C. *Contemporary Curtain Wall Architecture*; Princeton Architectural Press: Princeton, NJ, USA, 2009.
50. Flain, E.P.; Righi, R. Stone Claddings in the Corporate Real Estate Enterprises Citicorp Center and Rochaverá Corporate Towers in Sao Paolo, Brazil. *J. Civ. Eng. Archit.* **2018**, *12*, 716–726.
51. Fortmeyer, R. SOM's Pearl River Tower. *Archit. Rec.* **2006**, *194*, 172–173.
52. Song, J. The Making of a Creative City: Urban Cultural Policy and Politics in the Digital Media City (DMC) Seoul. Ph.D. Thesis, University of Warwick, Coventry, UK, 2015.
53. Xia, J.; Poon, D.; Mass, D. Case Study: Shanghai Tower. *CTBUH J.* **2010**, *2010*, 12–18.
54. Oldfield, P.; Vivian, P. Toward a generous skyscraper. *Archit. Aust.* **2019**, *108*, 74–76.
55. Alotaibi, F. The role of kinetic envelopes to improve energy performance in buildings. *J. Archit. Eng. Technol.* **2015**, *4*, 149–153. [CrossRef]
56. Ray, S. *The Increase in Carbon Dioxide in Earth's Mid-Troposphere from 2002 to 2013, Jet Propulsion Laboratory News*; California Institute of Technology: Pasadena, CA, USA, 2013.
57. Ali, M.M.; Armstrong, P.J. The Role of Systems Integration in the Design of Sustainable Skyscrapers. *Int. J. Sustain. Build. Technol. Urban Dev.* **2012**, *1*, 95–106. [CrossRef]

energies

Article

Application of Machine Learning to Assist a Moisture Durability Tool

Mikael Salonvaara [1,*], Andre Desjarlais [1], Antonio J. Aldykiewicz, Jr. [1], Emishaw Iffa [1], Philip Boudreaux [1], Jin Dong [1], Boming Liu [1], Gina Accawi [1], Diana Hun [1], Eric Werling [2] and Sven Mumme [2]

[1] Oak Ridge National Laboratory, Oak Ridge, TN 37830, USA
[2] Building Technologies Office, U.S. Department of Energy, Washington, DC 20585, USA
* Correspondence: salonvaaramh@ornl.gov; Tel.: +1-865-341-0022

Abstract: The design of moisture-durable building enclosures is complicated by the number of materials, exposure conditions, and performance requirements. Hygrothermal simulations are used to assess moisture durability, but these require in-depth knowledge to be properly implemented. Machine learning (ML) offers the opportunity to simplify the design process by eliminating the need to carry out hygrothermal simulations. ML was used to assess the moisture durability of a building enclosure design and simplify the design process. This work used ML to predict the mold index and maximum moisture content of layers in typical residential wall constructions. Results show that ML, within the constraints of the construction, including exposure conditions, does an excellent job in predicting performance compared to hygrothermal simulations with a coefficient of determination, R^2, over 0.90. Furthermore, the results indicate that the material properties of the vapor barrier and continuous insulation layer are strongly correlated to performance.

Keywords: building envelope; moisture; durability; design; machine learning; optimization; artificial intelligence

check for updates

Citation: Salonvaara, M.; Desjarlais, A.; Aldykiewicz, A.J., Jr.; Iffa, E.; Boudreaux, P.; Dong, J.; Liu, B.; Accawi, G.; Hun, D.; Werling, E.; et al. Application of Machine Learning to Assist a Moisture Durability Tool. *Energies* **2023**, *16*, 2033. https://doi.org/10.3390/en16042033

Academic Editor: Paulo Santos

Received: 26 January 2023
Revised: 15 February 2023
Accepted: 16 February 2023
Published: 18 February 2023

1. Introduction

The built environment accounts for almost 40% of global carbon dioxide emissions [1,2]. More than 25% of those emissions, or 11% of total global emissions, is attributed to manufacturing building materials such as steel, cement, plastic, and glass. Between 2013 and 2016, emissions from the built environment were flat. However, after 2016, emissions started to increase again. The emissions growth rate for the past two years was 2% annually in response to increased construction to meet the growing population's demand. These trends are not sustainable and are certainly not aligned with the goals established in the Paris Agreement, which include a reduction in carbon dioxide emissions of 50% by 2030 [3].

An effort is underway to help architects and engineers reduce energy consumption and carbon emissions associated with the built environment. In the past, the emphasis was on carbon dioxide emissions related to heating, cooling, lighting, and plug loads. However, the amount of carbon dioxide emitted that is associated with the manufacture of building materials is significant. For example, the production of Portland cement alone accounts for almost 8% of global greenhouse gas emissions [4]. As a result, tools and databases are being developed to help architects and engineers carry out life cycle analysis of building materials to minimize carbon dioxide emissions across all segments of the buildings' value chain. Tools such as EC3 [4], One Click LCA [5], GaBi [6], and the ICE Database [7] are just a few examples of the resources available to provide information regarding the embodied carbon of building materials.

Another example is the use of plastic foam insulation on the exterior of buildings to reduce heating and cooling loads. However, to what extent will the negative impact of the embodied carbon of plastic foam outweigh the energy savings related to the increased insulation value? Addressing this question requires well-characterized material and manufacturing processes to, in part, quantify the embodied carbon. Manufacturers are using international standards to codify this information into what is defined as an Environmental Product Declaration (EPD) [8]. According to the International EPD System, the EPD is an independently verified and registered document that quantifies product life cycle environmental impact in a standardized form that enables a direct comparison [9]. Most importantly, the EPD contains information regarding a material's embodied carbon that can then be used as part of a life cycle assessment for the building.

This realization has also driven the growth and development of sustainable building materials. In response to environmental regulations such as California's Buy Clean California Act, manufacturers have begun modifying existing products and developing products with lower global warming potentials [10]. The challenge accompanying these changes is the long-term performance and durability of these materials incorporated in the building envelope. The time to develop and characterize the long-term performance of building materials is certainly not aligned with adopting these new regulations. The iterative process of material development, characterization, and systems testing is complex. The number of elements in residential and commercial wall systems coupled with different environments or climate zones results in tens of thousands of combinations that must be analyzed to evaluate suitability and performance. This process is not only time-consuming but also expensive.

One approach to facilitate material development and deployment in building enclosure systems from the roof to the foundation is machine learning (ML). ML is a subset of artificial intelligence (AI) that can learn by using statistical models and algorithms to recognize patterns in data. This learning is accomplished without an explicit set of instructions or rules.

An extensive literature review has revealed more than 9000 publications that have a connection between advanced data analytics and building performance. Hong et al. focused on 150 publications that applied machine learning to the design, operation, and control of buildings [11]. Machine learning is finding utility in the built environment because it offers several benefits or advantages compared to conventional simulation tools. Speed seems to receive the most attention, provided the data sets are large enough to train the models. Monitoring building performance lends itself to the collection of large data sets. The challenge is what to do with it beyond, for example, temperature control. One case in point is the work by Tzuc et al. They were able to use weather data to train a neural network to model the hygrothermal performance of a vegetative façade [12]. What makes ML attractive when it comes to modeling the performance of systems with large data sets is that there are options. For example, Tijskens and coworkers evaluated or compared three neural networks' performance to predict a masonry wall's hygrothermal performance [13]. In this case, they found that a convolutional neural network required less training time and was the best at predicting performance [14]. To facilitate the selection of moisture-durable constructions, Salonvaara and coworkers implemented ML [15]. They demonstrated that an artificial neural network and gradient-boosted decision trees could be used to simulate hygrothermal performance with reasonable accuracy compared to hygrothermal simulations.

Machine learning has been applied to other problems beyond hygrothermal performance. For example, Kim and coworkers applied ML to optimize a double skin façade for performance and aesthetics [16]. In addition, they compared the results to conventional simulations and found that the two were comparable. Another example of an application is predicting the performance of enclosures integrated with phase change materials (PCMs) [17]. In this example, several models were compared to understand the difference in speed and accuracy when selecting suitable PCMs for this application.

ML was also used to optimize the energy performance of buildings [18]. Using design variables similar to those used in EnergyPlus [19], the model determined the optimal design in less than one minute instead of hours using traditional simulation approaches. Furthermore, the deviation between the optimum value obtained using ML and the simulation results using EnergyPlus was less than 3%, indicating good agreement between the two methods.

Bansal et al. [20] used ML to create a metamodel to forecast long-term hygrothermal responses and the moisture performance of light wood frames and massive timber walls. However, the authors note that unless the input data include the full range of variability of the climate variables, the prediction could be inconsistent. That demonstrates the obvious limitation of ML methods: extrapolating outside the ranges of the input variables can but might not necessarily produce erroneous results.

The benefits of ML applied to the built environment are apparent across a wide range of applications. For example, when designing energy-efficient constructions, it is important to account for the hygrothermal performance or the relationship between the movement of heat, air, and moisture through the building envelope and its effect on durability. The two are not mutually exclusive. Depending on the temperature and relative humidity, condensation within the building envelope can occur, resulting in durability problems such as mold and rot. For example, because insulation materials made from natural fibers are inherently a food source, they are more susceptible to mold growth than synthetic materials such as plastic foam insulation. To avoid the issue, hygrothermal simulations are used to evaluate the effect of material properties and wall designs on durability. The approach is complex and requires expert knowledge to implement. However, the benefits of ML can be used not only to facilitate the implementation of new materials in wall construction but also to provide guidance regarding material properties required to achieve a certain level of performance in a wall design. More importantly, such work is possible in the absence of expert knowledge.

The conventional approach to material development and deployment consists of the following steps: (1) new material is developed, (2) hygrothermal properties are characterized, (3) the building envelope material layout is adjusted so the new material can be incorporated, and (4) hygrothermal simulations are carried out to assess performance and durability.

In a wall system, the number and combination of variables that describe its performance are vast, more than tens of thousands when breaking down the levels with respect to material properties and the material layers that make up the wall assemblies. Coupled with climate, the number increases by more than an order of magnitude. Optimizing the materials and systems using an iterative process, as described earlier, requires significant resources, time, and cost.

Another limitation of using hygrothermal simulations is that they can only be carried out in one direction. That is, first, the user determines the properties of the material and then constructs the wall assembly and runs the simulation. Then, the materials and design are refined based on expert knowledge to improve performance. Unfortunately, the simulations cannot be run in the opposite direction where the architect or designer specifies a level of performance and the simulation generates the design or material properties required to meet that level of performance.

Because ML looks at patterns in data sets, it is irrelevant whether the starting point is the material properties or a level of performance at the assembly level. Herein lies the advantage of using ML over conventional simulation methods, and the present paper demonstrates this.

2. Building Science Advisor—New Construction

To help architects and engineers select durable enclosure designs with minimal effort, Oak Ridge National Laboratory developed the Building Science Advisor (BSA). The new construction tool guides the user to input the following data for a proposed wall: climate zone, exterior cladding, water-resistive barrier, continuous insulation, sheathing, wall structure type, cavity insulation, and interior vapor barrier. After the user inputs these data, the tool searches a database for similar walls. Each entry into the database includes a set of material properties for some or all the wall layers and an associated moisture durability rating. The moisture durability ratings for the condition described in the entry are pass, inconclusive, or fail. These ratings are based on hygrothermal simulations or expert opinions. The user's wall can match more than one entry in the database. If this happens, the most conservative durability rating and relevant durability guidance are displayed. The durability guidance will tell the user why the wall did not pass the moisture durability assessment and what can be done to improve the performance of the wall.

Figure 1 shows a presentation of this information for a user-selected wall. The user can see and change the wall construction. The wall's moisture durability and thermal performance, as well as the wall's construction visualization and durability guidance, are displayed.

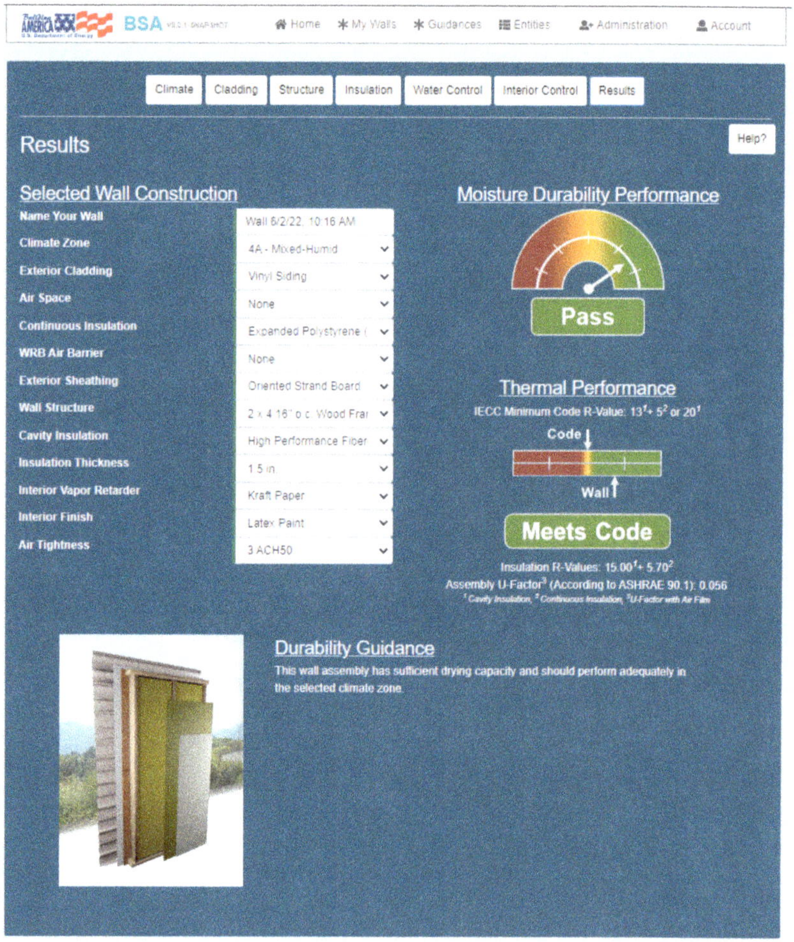

Figure 1. Results screen of the BSA new construction tool showing the user-selected wall with predicted performance and guidance for improving the moisture durability of the selected wall.

The moisture durability performance is based either on the mold index calculated from the hygrothermal modeling outputs as outlined in ASHRAE Standard 160 or on the consolidated expert opinion [21]. Typically, the entries in the database from hygrothermal simulations specify the material properties of every layer in the wall, and the entries from expert opinion specify only the material properties of select layers. The expert opinion-based entries focus on common issues found in walls that are known to cause long-term moisture durability problems. A more thorough description of how this tool works can be found in the literature [22].

3. Building Science Advisor—Retrofit

The retrofit web tool has similar inputs for the wall construction as the new construction tool. In the retrofit web tool, the user describes the existing wall to be retrofitted and then selects the desired retrofit: exterior, gut, or interior retrofit. Figure 2 shows the materials in the wall that can be changed depending on the chosen retrofit approach. After the user selects the preferred retrofit approach, the tool searches a database of hygrothermal simulation results for matches on the unchanged materials. For example, suppose the user selected an exterior retrofit. In that case, entries in the database are found with the same wall structure, interior vapor retarder, and interior finish as the existing wall.

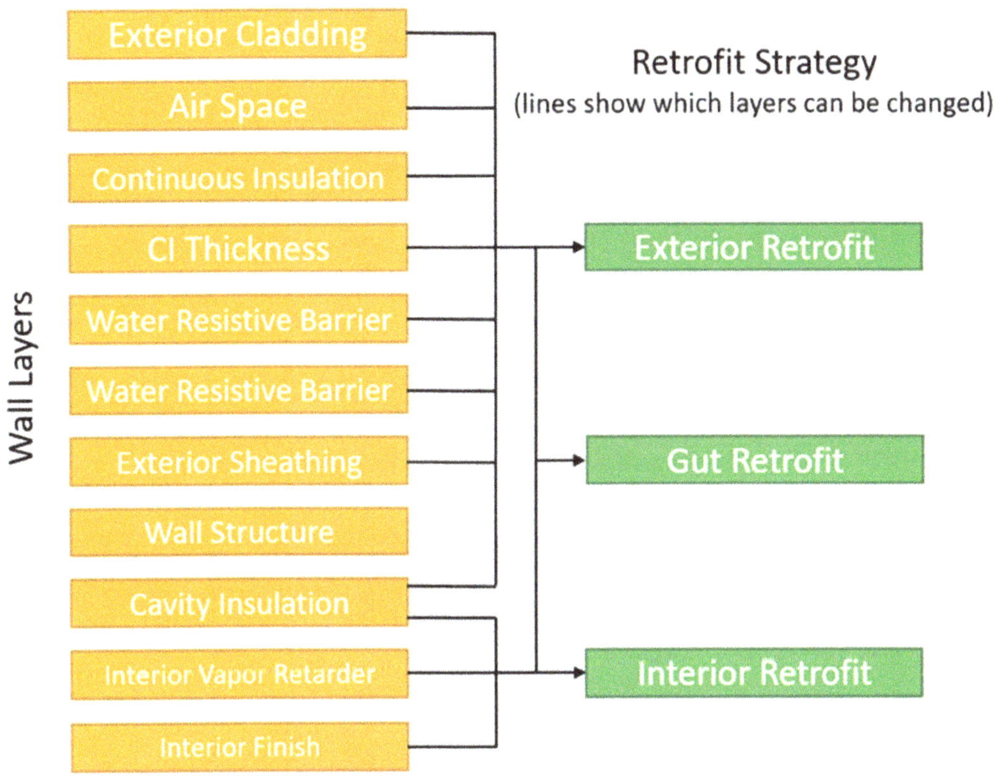

Figure 2. Wall materials that can be changed depending on the type of wall retrofit. CI = Continuous Insulation.

All matches with a mold index of less than two are presented to the user. The matches can be filtered for desired materials, and then one wall can be selected to view its moisture and thermal performance, which can be compared with the existing wall. Figure 3 shows the result screen for the tool, where the existing and retrofit wall materials, moisture durability, and thermal performance are compared.

The core of these webtools is in the database used to assess the moisture durability of a user's wall. Most entries for the retrofit and new construction databases are based on hygrothermal simulation results. However, to cover the wide range of layer and material combinations for walls in each climate zone requires millions of simulations, which is expensive in both simulation computing time and server speed when hosted for the webtool. Furthermore, using a static database does not allow flexibility for the user to input new or custom materials to be evaluated in a wall.

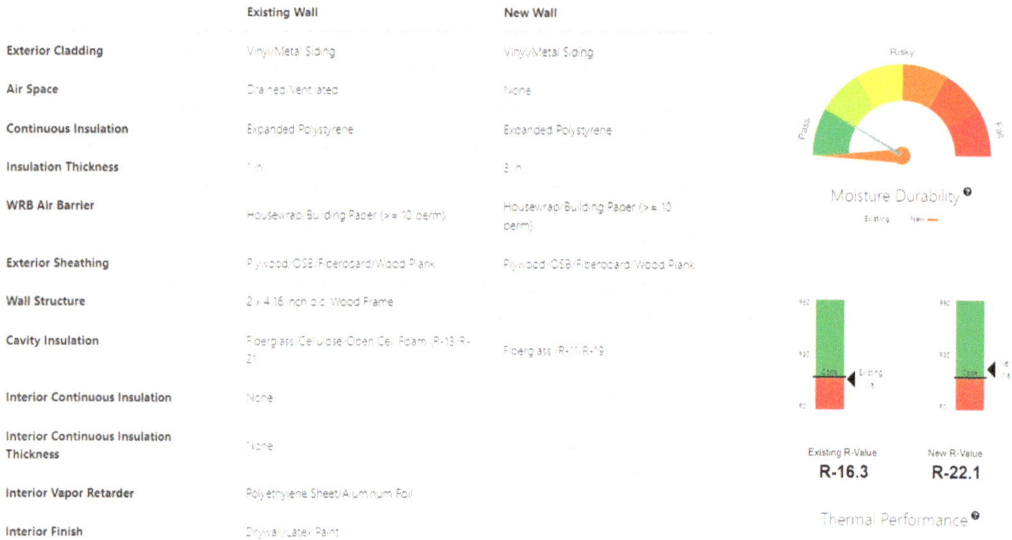

Figure 3. The final screen of the BSA retrofit tool comparing the materials and performance of the existing and proposed retrofit wall.

4. Hygrothermal Simulations to Create Inputs for Machine Learning

The BSA tool helps users answer design questions regarding thermal performance and durability. Hygrothermal simulations are time-consuming and require expertise not only in heat and mass transfer but also in setting up numerical models to evaluate the performance and risks of the designs. Furthermore, once the simulations have been carried out for a specific design, the remaining question is whether the solution is optimal and what should be completed to make it perform better. Therefore, a quicker and simpler tool is needed to evaluate and optimize the design.

The number of simulations to create a database to cover all possible design options the user might select can rise to millions of cases. That is because each layer in the wall can have multiple options. For example, the input options for the cladding could be different types of brick, fiber cement siding, vinyl siding, wood siding, and others. The water-resistive barrier can be almost water vapor impermeable, very vapor open, or somewhere in between. Wall assemblies with just three options for a 7-layer wall assembly in 15 climates result in more than 30,000 simulation cases in just 1 orientation. Adding options to select would increase the number of cases exponentially.

A standard or typical database includes climate zone, wall orientation, the identity of the layers (e.g., "brick" or "mineral fiber insulation"), description or property of the layers (e.g., "thickness" or "vapor permeability"), and performance parameters (e.g., "mold index" or "maximum moisture content"). Named inputs, or categories, are sufficient in a database that is used only to select the performance for the cases the database includes. In a previous paper, the authors used machine learning to predict performance using categorical inputs [15]. However, when using categorical inputs, the user would not be able to evaluate the performance of a wall with a material layer that is not included in the original data set. Furthermore, interpolating between the simulated cases in the database would not be possible with categorical or named inputs. Therefore, we elected to use the material properties of the layers in the database instead of names or categories of materials, which is what is performed in BSA. These inputs would then be analyzed with ML techniques to interpolate new results between the simulated cases.

Even though the simulations for the BSA tool include mass walls, this paper focuses on lightweight wood frame construction.

4.1. Design Process

Hygrothermal simulations are post-processed per ASHRAE Standard 160 [21] to consider the maximum acceptable moisture content in a critical layer and the mold growth index in the wall. The exterior sheathing was selected as the critical layer. The mold growth index (MI) was calculated on both the exterior and interior sides of the insulated cavity to account for performance in cold and hot-humid climates. Mold growth in the building assembly was predicted by running a five-year simulation and taking the maximum MI. The MI can range from between 0 and 6 [23].

An additional criterion for some materials, such as wood-based sheathing boards, is typically the limit for moisture content. Wood begins to suffer damage if its moisture content remains at 20% for days or longer. High moisture content for a prolonged period can allow the wood to begin to rot. In addition, dimensional changes due to high moisture can further create damage and impact the integrity of the building envelope assembly.

4.2. Wall Assemblies and Climates Included in the Study

One-dimensional simulations were carried out for lightweight and masonry walls using a hygrothermal simulation tool [24]. The simulation parameters used in the training set included 19 climate locations covering all 8 US climate zones and the wall structures, as depicted in Figure 4. The simulations included the locations and climate zones listed in Table 1. Mobile, AL, Grand Island, NE, and Burlington, VT, were used to test the model's ability to predict performance; hence, they were not part of the training set.

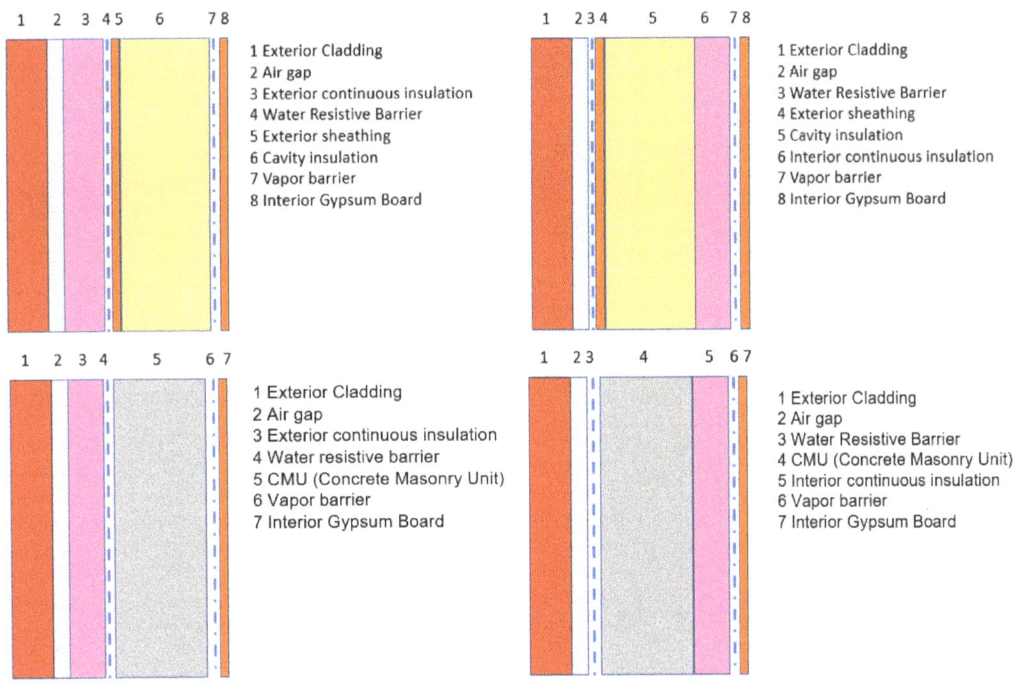

Figure 4. Wall assemblies evaluated in simulations with and without continuous insulation.

Table 1. Locations and climate zones used in simulations for hygrothermal performance. Italicized locations were used as testing climates and were not part of the training set for ML.

Location	Climate Zone	Climate Characteristic
Miami, FL	1A	Hot humid
Houston, TX	2A	Hot humid
Mobile, AL	*2A*	*Hot humid*
Phoenix, AZ	2B	Hot dry
Tucson, AZ	2B	Hot dry
Atlanta, GA	3A	Mixed humid
Los Angeles, CA	3B	Hot dry
San Francisco, CA	3C	Marine
Baltimore, MD	4A	Mixed humid
Knoxville, TN	4A	Mixed humid
Albuquerque, NM	4B	Mixed dry
Seattle, WA	4C	Marine
Chicago, IL	5A	Cold wet
Madison, WI	5A	Cold wet
Syracuse, NY	5A	Cold wet
Grand Island, NE	*5A*	*Cold wet*
Flagstaff, AZ	5B	Cold dry
Minneapolis, MN	6A	Cold wet
Burlington, VT	*6A*	*Cold wet*
Boise, ID	6B	Cold dry
Anchorage, AK	7	Very cold
Fairbanks, AK	8	Subarctic

The layer details are listed in Table 2.

Table 2. Layer options for the lightweight and the masonry walls used in parametric simulations, adapted from ref. [15].

Layer	Lightweight Wall	Masonry Wall
Brick, buff matt brick, vinyl, fiber cement, stucco, wood	X	
Vinyl, stucco		X
Ventilated, nonventilated air gap	X	X
Continuous insulation, extruded polystyrene (XPS), expanded polystyrene, mineral fiber, cork (0/16/25/38/51/76/102 mm)		
Water-resistive barrier, permeance: 0.5, 5, 10, or 50 perms	X	X
OSB (oriented strand board), plywood, chipboard, wood fiberboard, wood, exterior grade gypsum board	X	
89 mm or 140 mm wood frame	X	
203 mm CMU (concrete masonry unit), grouted or ungrouted		X
Cavity insulation: Fiberglass, closed-cell spray foam, no insulation	X	
Interior continuous insulation: None, 25 mm XPS, 25 mm mineral fiber	X	
Vapor retarder: None, 0.5/1/5 perm, variable permeance	X	X
Gypsum wallboard and latex paint	X	X

4.3. Example of Simulation Results for Machine Learning Analyses

The performance data for ML analyses were created by running hygrothermal simulations for wall assemblies in different climates. An example of a wall system, its layers, and locations of interest are shown in Figure 5. The simulations were run for five years using the same single-year weather file. The moisture design reference year (second most severe year) per the ASHRAE Research Project 1325 [25] was used in each climate location. Three key values are taken from each simulation: (1) temperature and relative humidity on the exterior side of the insulated cavity (Point 1), (2) temperature and relative humidity on the interior side of the cavity (Point 2), and (3) average moisture content of the exterior sheathing (Point 3). The hourly temperature and relative humidity were used in postprocessing to calculate the predicted mold index for Points 1 and 2.

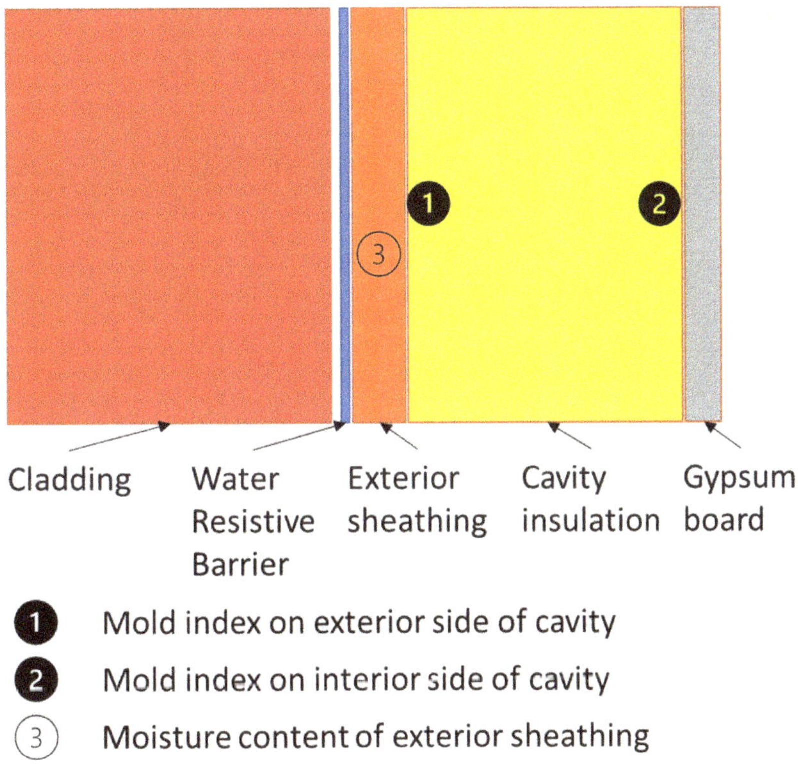

Figure 5. Wall assembly used in simulations with points of interest to develop post-processed results for ML.

The maximum mold index and moisture content values over the five years were then recorded for each case as the performance value and inputs for ML analyses. Figure 6 shows the hourly temperature and relative humidity on the exterior side of the insulated cavity. Figure 7 shows the post-processed mold index and the average moisture content of the exterior sheathing. The maximum values of the mold index and the moisture content over the simulated period are selected as the final values. The mold index is calculated both for the exterior and the interior side of the insulated wall cavity. The highest mold index of the two is selected as the final performance value.

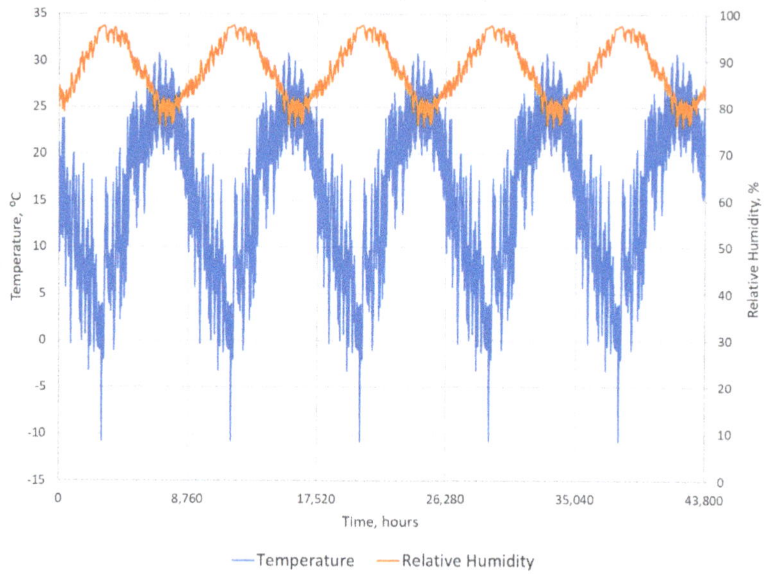

Figure 6. Temperature and relative humidity on the exterior side of the wall cavity in an example wall case.

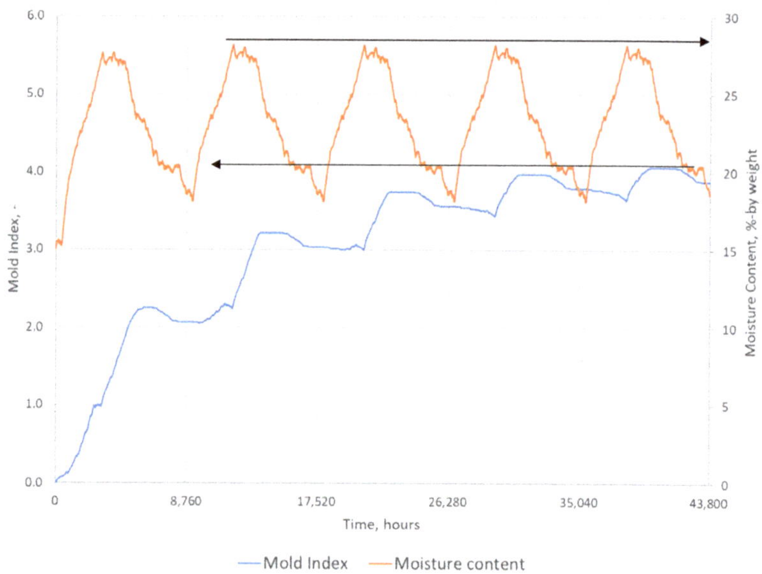

Figure 7. Mold index on the exterior side of the insulated cavity in an example wall based on hourly temperature, relative humidity, and the average moisture content of the exterior sheathing. Arrows are pointing to the maximum value over the period.

The data for performance analysis based on the results in Figure 7 would be a mold index slightly above 4 and a maximum moisture content of about 28% by weight.

4.4. Creating Inputs from Simulation Cases for Machine Learning Analysis

This paper focuses on the analysis of wood-framed walls. These cases form 87% of the cases in the database.

Named materials and thicknesses describe a wall in each functional layer of the wall, such as cladding, continuous insulation, and water-resistive barrier. In this study, we converted the named layers in the simulated wall assemblies into numeric values describing each layer with relevant material properties for heat and moisture transport. Using regression analysis with ML, these numeric values allow for a correlation between the performance values and the inputs (material properties and weather parameters).

Cladding materials are exposed to rain and ambient air temperature and humidity. Therefore, the cladding materials were characterized by the liquid uptake value (A-value, $kg/m^2 \cdot s^{05}$), water vapor permeance (1 U.S. Perm = 57 $ng/s \cdot Pa \cdot m^2$) for so-called dry-cup and wet-cup conditions (0%RH-50%RH and 50%RH-100%RH, respectively), moisture storage (sorption at 80%RH, W80, kg/m^3), and thermal resistance R ($m^2 \cdot K/W$). The exterior sheathing materials were characterized with the same parameters but without the liquid uptake value. The water-resistive barriers (WRBs) and vapor retarders had only the water vapor resistance values for dry- and wet-cup conditions. Instead of giving the permeance values of layers as inputs for the model, we used the water vapor resistance, Z = 1/permeance. Finally, the insulation layers were characterized by the water vapor permeance (given in input as water vapor resistance, Z) and thermal resistance values. The properties of materials used in the simulations are shown in Table 3.

Table 3. Materials and properties.

Materials	k-Dry (W/m·K)	A-Value, $kg/m^2 \cdot s^{0.5}$	Permeance, Dry Cup, Perm	Permeance, Wet Cup, Perm	W80, kg/m^3	Thickness (m)	R-Value ($m^2 \cdot K/W$)
Exterior sheathing							
OSB	0.092		0.62	5.56	83.4	0.0125	0.136
Plywood (USA)	0.084		0.58	8.77	64.4	0.0150	0.179
Southern yellow pine	0.119		0.38	8.33	62.2	0.0200	0.168
Wood fiberboard	0.052		16.67	20.00	35.4	0.0125	0.240
Exterior gypsum	0.218		45.45	58.82	6.2	0.0127	0.058
Claddings							
Buff matt clay brick	0.43	0.00013	1.19	4.54	2.4	0.1040	0.242
Brick old	0.4	0.2083	2.56	4.35	3.3	0.1040	0.260
Vinyl siding		0	0.10	0.10	0		0.018
Stucco	0.399	0.0033	1.02	2.08	106.6	0.0200	0.050
Western red cedar	0.084	0.0011	0.21	1.02	33.7	0.0200	0.238
Fiber cement siding	0.245	0.03	2.08	20.00	185.8	0.0080	0.033
Vapor retarders							
PA membrane			0.88	16.67			
VB: 0.1 perm			0.1	0.10			
VR: 0.5 perm			0.5	0.50			
VR: 1.0 perm			1	1.00			
VR: 5.0 perm			5	5.00			
WRBs							
WRB: 0.5 perm			0.50	0.50			
WRB: 5.0 perm			5.00	5.00			
WRB: 10 perm			10.00	10.00			
WRB: 50 perm			50.00	50.00			
Cavity insulation			Permeance per inch				R-value per inch
Fiberglass			118				0.652
Closed-cell spray foam			1.46				1.057

Additionally, to evaluate the performance in a more refined location and weather conditions, the climatic information was converted to annual weather parameters to assess the ability of ML to correlate the hygrothermal performance to the weather parameters. In the ASHRAE Research Project 1325, "Environmental weather loads for hygrothermal analysis and design of buildings," [25] annual weather parameters were found to correlate with the durability performance of wall assemblies, and the method to select moisture reference years was created using annual average weather parameters. Therefore, it is reasonable to assume that the weather parameters could be used in the ML analyses instead of named weather locations as categorical inputs. Table 4 lists the weather locations and the annual average weather parameters for temperature (T, °C), relative humidity (RH, %), water vapor pressure (Pv, Pa), cloud index (Cloud, -), solar radiation on a wall facing north (RadN, W/m^2), solar radiation on a wall facing the orientation with the most wind-driven rain (RadWDR, W/m^2), rain on a wall facing north (RainN, mm/h), rain on a wall facing the orientation with the most wind-driven rain (RainWDR, mm/h), average daily minimum temperature (Av.Tmin, °C), and average daily maximum temperature (Av.Tmax, °C).

Table 4. Average annual weather parameters for weather locations. Solar radiation and wind-driven rain are shown for two orientations: north (N) and the orientation of the wall that receives most of the wind-driven rain (WDR). The locations in italics were used as testing locations and were not part of the training set.

City, State	T (°C)	RH (%)	Pv (Pa)	Cloud -	RadN (W/m^2)	RadWDR (W/m^2)	RainN (mm/h)	RainWDR (mm/h)	Av.Tmin (°C)	Av.Tmax (°C)
Miami, FL	24.5	73.3	2031	3.8	70	88	0.0651	0.0799	20.7	28.5
Houston, TX	18.7	78.9	1762	4.2	62	78	0.0651	0.0713	13.4	24.8
Mobile, AL	*18.7*	*74.6*	*1663*	*4.2*	*62*	*116*	*0.0734*	*0.0982*	*14.2*	*24.2*
Phoenix, AZ	23.3	43.2	1113	2.6	59	116	0.0080	0.0171	17.3	29.4
Tucson, AZ	18.9	37.0	987	2.5	59	158	0.0135	0.0227	12.0	25.9
Atlanta, GA	14.7	71.4	1357	4.4	63	104	0.0386	0.0713	10.1	20.0
Los Angeles, CA	17	75.8	1478	3.6	64	126	0.0070	0.0435	13.7	21.2
San Francisco, CA	13.9	78.0	1261	3.7	56	133	0.0034	0.0775	10.4	18.6
Baltimore, MD	12.4	68.7	1218	4.2	56	70	0.0599	0.0720	7.1	17.7
Knoxville, TN	14.9	76.2	1439	4.8	61	96	0.0360	0.0426	9.5	20.4
Albuquerque, NM	13.4	44.5	912	3.2	57	146	0.0121	0.0143	6.8	20.6
Seattle, WA	11.1	77.3	1086	5.5	48	99	0.0049	0.0812	7.4	15.3
Chicago, IL	9.7	70.0	1062	5.0	58	80	0.0612	0.0974	4.6	14.5
Madison, WI	7.7	76.3	1012	5.3	60	68	0.0461	0.0489	2.2	12.6
Syracuse, NY	8.5	75.9	1038	5.3	56	88	0.0244	0.0449	3.0	13.5
Grand Island, NE	*11.1*	*70.1*	*1114*	*4.2*	*55*	*55*	*0.0381*	*0.0381*	*5.7*	*17.7*
Flagstaff, AZ	6.9	52.8	765	3.0	56	56	0.0435	0.0435	−1.1	14.6
Minneapolis, MN	7.8	74.0	1010	4.9	55	94	0.0332	0.0630	3.2	12.3
Burlington, VT	*6.5*	*74.3*	*1006*	*5.4*	*56*	*113*	*0.0248*	*0.0317*	*1.7*	*17.7*
Boise, ID	11.1	60.2	859	4.5	54	72	0.0135	0.0154	4.8	17.4
Anchorage, AK	2	74.5	747	5.7	40	93	0.0090	0.0188	−2.6	6.1
Fairbanks, AK	−1.8	68.9	624	4.8	42	80	0.0029	0.0148	−6.9	3.3

4.5. Examples of Inputs for Machine Learning

The material properties for each layer in the wall assemblies and the annual average weather parameters were used as inputs for ML (Table 5).

Table 5. Descriptions of the inputs for training the ML models. The italicized parameters are not used as inputs in training the model but only to help identify the results afterward.

Parameter Assembly Inputs	Description	Parameter Weather Inputs	Description Annual Average of
Od_weather	Weather location index	T	Outdoor air temperature
Exterior_cladding_id	Exterior cladding index	RH	Outdoor air relative humidity
Air_gap_id	Air gap index indicating existence of air gap behind siding	Pv	Outdoor air vapor pressure
Ext_sheathing_id	Exterior sheathing index	Cloud	Cloud index
Wall_structure_category_id	Wall structure type index	Rad	Solar radiation on wall
A_clad	Liquid uptake of cladding	Rain	Wind-driven rain on wall
R_clad	R-value of cladding	Av_Tmin	Daily minimum temperature
Z_clad_dry/wet	Water vapor resistance of cladding, dry and wet cup test	Av_Tmax	Daily maximum temperature
R_exsh	R-value of exterior sheathing		
Z_exsh_dry/wet	Water vapor resistance of exterior sheathing, dry and wet cup test		
R_ci_ext	R-value of exterior continuous insulation	**Outputs**	

Table 5. *Cont.*

Parameter Assembly Inputs	Description	Parameter Weather Inputs	Description Annual Average of
Z_ci_ext	Water vapor resistance of exterior continuous insulation	Mold_index	Mold index
R_ci_int	R-value of interior continuous insulation	MaxMC	Maximum moisture content of the exterior sheathing
Z_ci_int	Water vapor resistance of interior continuous insulation		
R_cav	R-value of cavity insulation		
Z_cav	Water vapor resistance of cavity insulation		
Z_wrb	Water vapor resistance of water-resistive barrier		
Z_vb_dry/wet	Water vapor resistance of vapor barrier, dry and wet cup test		
S_clad	Moisture storage capacity of cladding		
S_exsh	Moisture storage capacity of exterior sheathing		
Airgap_Z	Water vapor resistance of air gap		

4.6. Preprocessing and Visualizing the Data

Figure 8 shows the histogram of different input features used for the ML model to predict the mold index. Most weather inputs are widely scattered between the minimum and the maximum values. Some of the inputs are concentrated on several values, such as the vapor resistance of the vapor barrier (Z_vb_wet) and the thermal resistance of the continuous insulation on the interior side of the wall (R_ci_int). The histograms help identify which inputs would require refinements to improve the ML predictions. Ideally, we would want the distribution of values to be evenly scattered across the range of the property values.

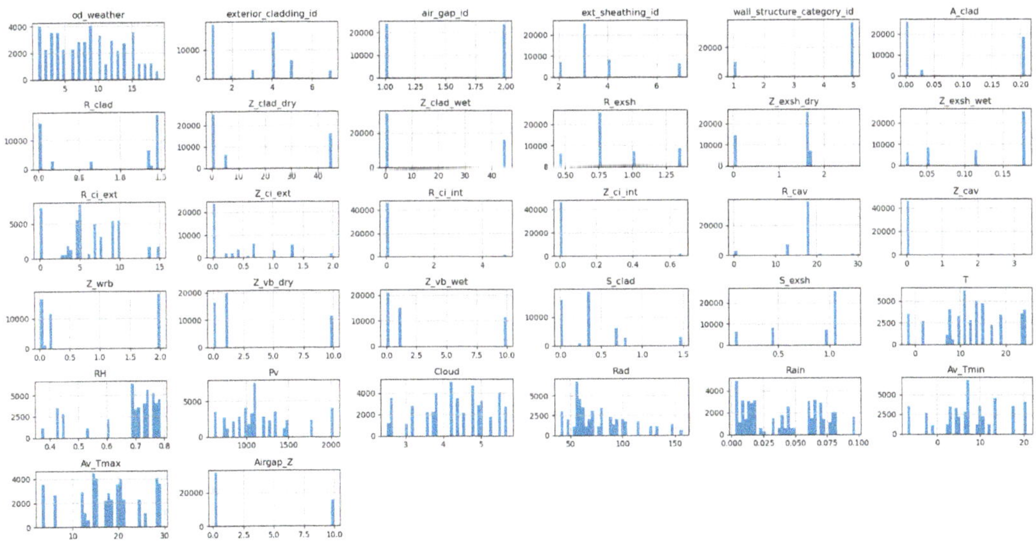

Figure 8. Distributions of the mold index and different input features (x-axis: input parameter value, y-axis: count per value).

4.7. Machine Learning Analyses for Mold Index and Maximum Moisture Content

A commercially available ML analysis tool (Google AutoML for Tables [26]) was used to evaluate the tabular data. Google Vertex AI and its AutoML require little effort and can be used to create a benchmark case for a data scientist's development [27]. The caveat is that the system is not free, which makes a case for testing and developing a local model if cost is an issue. Additionally, the system does many tasks without the user's knowledge, and the methods used in the specific model are not disclosed. The benefit of using AutoML is its ease of use, without setting up any local hardware or software environment. However, the user can evaluate the goodness of the performance with several parameters. Therefore, one can achieve good results by using a plug-and-play setup for the data. The authors will present their local model development in a follow-up paper. The AutoML tool conducts several ML tasks behind the scenes:

- Preprocessing the data;
- Performing automatic feature engineering;
- Model architecture searching;
- Model tuning;
- Cross-validating;
- Automatic model selection and ensembling.

Four steps were followed to train and test the ML models. First, the tool trained the ML models to predict the mold index and the maximum moisture content using a data set of simulated results. Second, we tested the performance of the models using the model to predict the performance of new materials that were not part of the training set. Third, the predictions for new climates were tested with materials that were already part of the training set. Fourth, the performance of the ML models was tested with both the new climates and the new materials.

4.7.1. Training the Machine Learning Model

A data set of 48,855 lines of simulated values was used to train the ML tool. The default 80%/10%/10% random split was used (training/validation/test). The prediction was optimized for residual mean squared error (RMSE) with mold index as the target. Table 6 lists the mean absolute error (MAE), RMSE, and coefficient of determination (R^2) for predicting the maximum mold index and the maximum moisture content.

Table 6. Calculated accuracies for the mold index and maximum moisture content prediction.

Prediction Target	MAE	RMSE	R^2
Maximum mold index	0.024	0.058	0.997
Maximum moisture content	0.154	0.568	0.995

The features in the input data are shown in the order of importance in Figure 9 as determined by the ML tool. The definitions of the parameters are listed in Table 5. The most important features in determining the mold index in the walls are the water vapor resistance (1/permeance) of the vapor barrier (Z_vb_wet and Z_vb_dry) and the thermal resistance of the continuous insulation (R_ci_ext), which are in the top three for predicting the maximum moisture content. However, the thermal resistance of the exterior sheathing is now the most important feature for determining maximum moisture content. Otherwise, the features are in a similar order for the mold index and maximum moisture content predictions, with slight changes in the order of importance.

The predicted values from ML models for the mold index and moisture content are depicted vs. the simulated values from hygrothermal simulation tools in Figure 10a,b, respectively. In contrast to the mold index, the moisture content has sections where some of the simulated and predicted values diverge at moisture contents greater than 15%. Despite the divergence, the correlation seems strong, as reflected by R^2 values of 0.997 and 0.995, respectively.

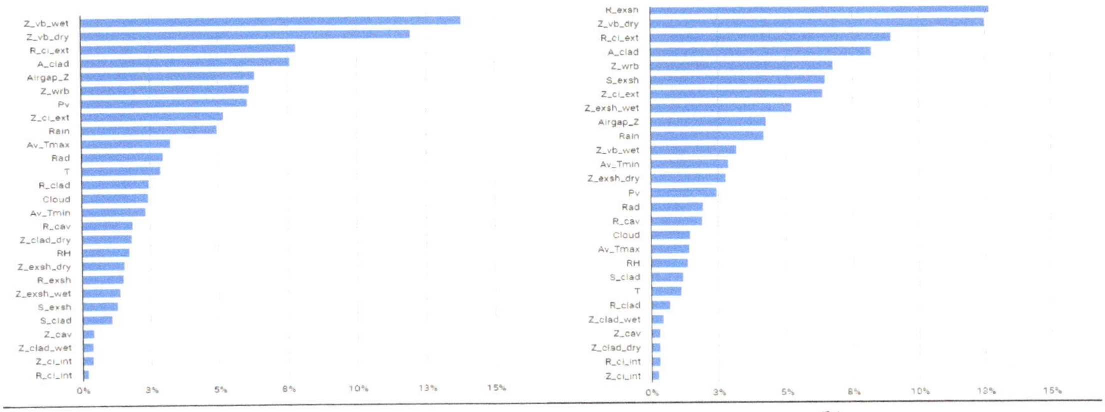

(**a**) (**b**)

Figure 9. Importance of feature per Google AutoML for the mold index and moisture content prediction. (**a**) Mold index prediction. (**b**) Maximum moisture content prediction.

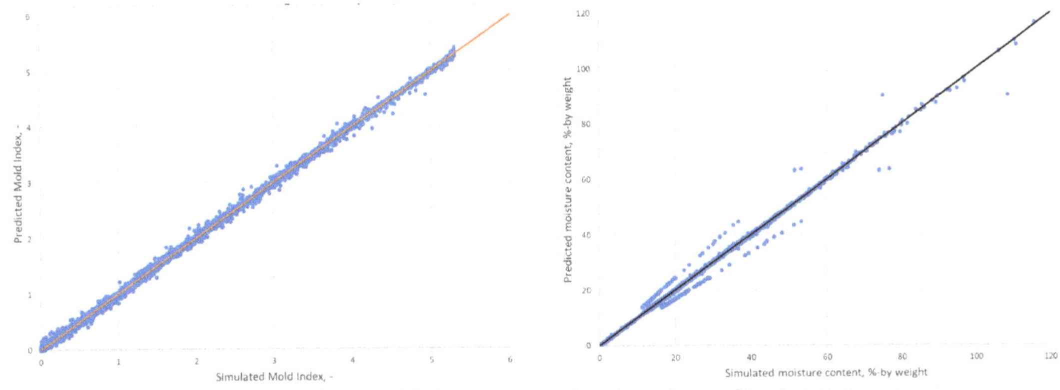

(**a**) (**b**)

Figure 10. Predicted results from ML models vs. simulated values from hygrothermal simulation tools for (**a**) the mold index and (**b**) the maximum moisture content. Charts include all input data in the training set. The line indicates the perfect fit.

4.7.2. Predicting the Performance Parameters with New Materials

Two new materials, chipboard exterior sheathing and wood fiber exterior insulation, were introduced, and simulation cases with those materials were carried out. Figure 11 shows the correlation between the ML-predicted and simulated mold index and maximum moisture content for the new materials. Table 7 lists the performance indicators. The method predicts these new materials well with no outliers. The results are well correlated with the input data with R-squared over 0.97 both for the mold index and maximum moisture content predictions and the MAE and RMSE representing less than 5% error from the maximum values. However, at high moisture contents >20% by weight, the predictions show consistently higher than simulated moisture contents, and the errors in the predictions are concentrated in this range. This is possibly due to fewer input data being available in ML training with high moisture contents and the highly nonlinear nature of sorption isotherms. The 20% by weight moisture content or higher for wooden materials is considered unacceptable, so both the simulations and the ML predictions predict unacceptable performance.

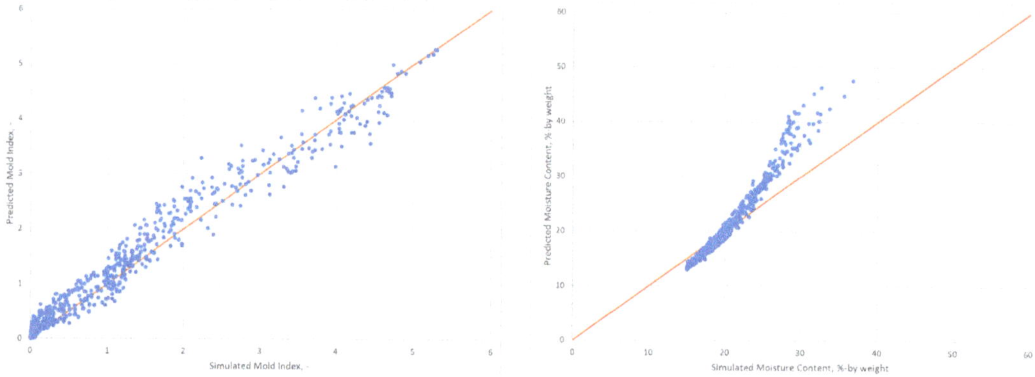

(a) (b)

Figure 11. Predicted vs. simulated mold index and maximum moisture content for new materials (i.e., chipboard exterior sheathing and wood fiber exterior insulation) with existing climates in training. (**a**) Mold index prediction. (**b**) Maximum moisture content prediction. The line indicates perfect fit.

Table 7. Calculated accuracies for the mold index and maximum moisture content prediction for new materials with existing climates in training.

Prediction Target	MAE	RMSE	R^2
Mold index	0.134	0.210	0.974
Maximum moisture content	1.940	2.716	0.977

4.7.3. Predicting the Performance with New Climates

Three new climates—Mobile, Alabama, Burlington, Vermont, and Grand Island, Nebraska—were introduced into the simulations with materials that were included in the training set for ML. Figure 12 shows the correlations for the predicted mold index and maximum moisture content as a function of the simulated values. Table 8 shows the performance indicators. The quality of the predictions is worse for the mold index, with both the MAE and RMSE higher and the R^2 lower for the new climate prediction than for testing with the new materials. The three climates also have different trends: the predictions for Mobile, Alabama, are generally on the high mold index and moisture content side, whereas in Burlington, Vermont, the results are mostly on the low side. Burlington is one of the coldest climates in the data set, and small changes in material and weather parameters can cause large differences in performance. Therefore, more focus should be placed on the extreme climates in the simulations that provide the input data. The ML model predicts maximum moisture content better than the mold index with the new climates.

The performance with weather parameters for Mobile, Alabama, and Grand Island, Nebraska, is conservative and provides safe guidance regarding mold growth by mainly predicting a higher mold index than simulated. However, for Burlington, Vermont, the predictions for the mold index are lower than those simulated, giving a false sense of moisture safety for the wall design. The predictions for the maximum moisture content are similarly low for Burlington, Vermont, but align much better with the simulated data. The poor performance in predicting the mold index in one of the climates indicates that more work is needed to improve the accuracy with new climate locations by adding new different climates into the training set and investigating potential additional weather parameters and their impact on predictions, although the general trend is acceptable.

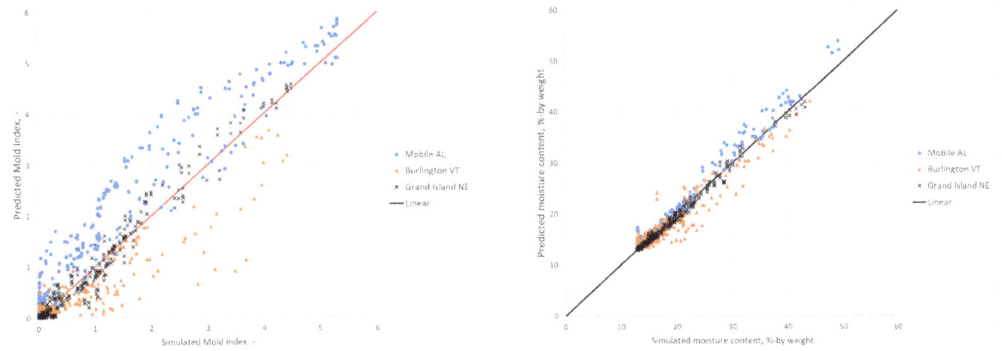

(a) (b)

Figure 12. Predicted vs. simulated mold index and maximum moisture content for new climates (i.e., Mobile, AL, Burlington, VT, Grand Island, NE) with existing materials in training. (**a**) Mold index prediction. (**b**) Maximum moisture content prediction. The line indicates perfect fit.

Table 8. Calculated accuracies for the mold index and maximum moisture content prediction for new climates with existing materials in training.

Prediction Target	MAE	RMSE	R^2
Mold index	0.238	0.420	0.901
Maximum moisture content	0.697	1.270	0.960

4.7.4. Predicting Performance with New Materials and New Climates

Finally, both the new materials and new climates were introduced together into the simulations, and the performance was predicted with the ML tool. Figure 13 shows the results for wall assemblies, including the two new materials in the three new climates, and Table 9 shows the performance indicators of the predictions. As expected, the predictions are now worse for MAE and RMSE. On the other hand, R^2 slightly improved for the mold index prediction compared with the predictions with new climates only. Most of the errors in the predictions are caused by the new climates, and closer investigations of weather parameters are needed.

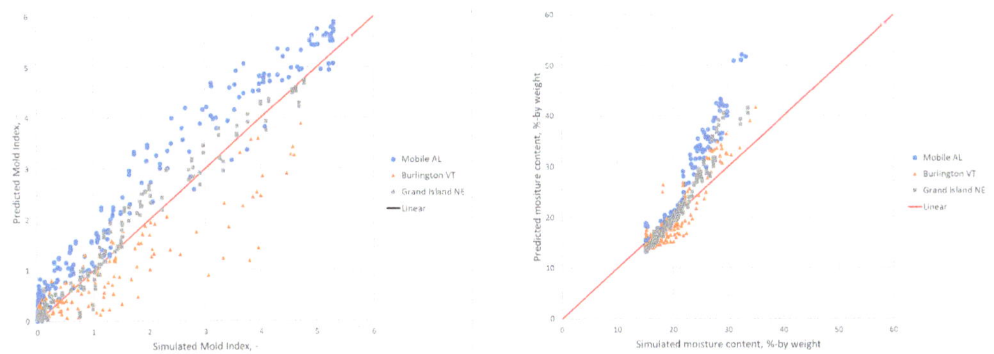

(a) (b)

Figure 13. Predicted vs. simulated mold index and maximum moisture content for new materials (i.e., chipboard exterior sheathing and wood fiber exterior insulation) and new climates (i.e., Mobile, AL, Burlington, VT, Grand Island, NE). (**a**) Mold index prediction. (**b**) Maximum moisture content prediction. The line indicates perfect fit.

Table 9. Calculated accuracy for the mold index and maximum moisture content prediction for new climates and materials.

Prediction Target	MAE	RMSE	R^2
Mold index	0.264	0.449	0.916
Maximum moisture content	2.400	3.413	0.909

5. Discussion

The benefits of this ML tool are, first, in the early design stages, to guide proper systems before running a detailed evaluation on many systems. Thereby, it can reduce the significant efforts in running hygrothermal analyses. Second, the tool will be the basis for optimizing the durability performance while assessing thermal performance and decarbonization possibilities. The ability to create granularity by interpolating performance as a function of material properties allows for guiding new material development and selections.

This study shows that ML can predict the hygrothermal performance of a building envelope design with reasonable accuracy ($R^2 > 0.90$). The results indicate that performance is strongly correlated to the vapor permeance of the vapor barrier and thermal resistance of the continuous insulation layer for the selected envelope design. The dependence of the accuracy on weather data is not as strong, which could explain why the model's predictive capability was not strong when the ML model was used to predict performance in other climate zones. Regardless, the model needs additional refinement in filling data gaps to address the performance of different assemblies and exposure to different climate zones, which will be the emphasis of future work.

ML is as good as the input data given in training. The current data set still includes large gaps between the high and low values of the input parameters, and many input values have only a few options. The ML methods cannot accurately learn the dependency of the output as a function of the input values unless there are input values in the region where the change in performance occurs. Finding these regions of input values is also part of future work.

The future development of the tool includes guiding the designer by helping to optimize the material layers in the wall assemblies to meet the user's requirements for thermal, energy, carbon, and moisture performance. The tool would guide the user to select materials and layers that result in a moisture-safe building envelope. Based on the developed ML tool, which uses material properties as inputs for training, we can run multi-objective optimization to select the optimal material that minimizes embodied carbon of the assembly while meeting all the insulation and moisture durability requirements. The tool would guide improving the assembly for better performance in terms of durability, energy losses, and carbon content.

To develop our own ML model, future work may consider testing various regression models to determine the most accurate algorithm instead of using a commercially available automated ML tool. For model training, more aggressive feature extraction will be applied to reduce the amount of redundant data in the input data set, which can help increase the accuracy and generalization of the model. We will also try different weight initialization and fine-tune the learning model, such as the activation function and a number of hidden layers. Commercially available automated ML tools allow quick benchmarking of our ML model.

Author Contributions: Conceptualization, M.S.; data curation, J.D. and G.A.; funding acquisition, S.M.; investigation, M.S.; methodology, M.S. and J.D.; project administration, M.S., D.H., E.W., and S.M.; validation, M.S., A.D., and B.L.; visualization, J.D.; writing—original draft, M.S., A.J.A.J. and P.B.; writing—review and editing, A.D., E.I., B.L., G.A. and D.H. All authors have read and agreed to the published version of the manuscript.

Funding: This manuscript has been authored in part by UT-Battelle, LLC, under contract DE-AC05-00OR22725 with the US Department of Energy (DOE). The publisher acknowledges the US government license to provide public access under the DOE Public Access Plan (http://energy.gov/downloads/doe-public-access-plan, accessed on 25 January 2023).

Data Availability Statement: Not applicable.

Acknowledgments: This research used resources at the Building Technologies Research and Integration Center, a DOE Office of Energy Efficiency and Renewable Energy User Facility operated by the Oak Ridge National Laboratory.

Conflicts of Interest: The authors declare no conflict of interest. The funders had no role in the design of the study; in the collection, analyses, or interpretation of data; in the writing of the manuscript; or in the decision to publish the results.

Nomenclature

Abbreviation	Meaning	Unit
A	water absorption coefficient	$kg/m^2 \cdot s^{0.5}$
AI	artificial intelligence	—
ASHRAE	American Society of Heating, Refrigerating and Air-Conditioning Engineers	—
Av.Tmax	annual average daily maximum temperature	°C
Av.Tmin	annual average daily minimum temperature	°C
BSA	Building Science Advisor	—
cav	cavity	—
ci	continuous insulation	—
clad	cladding	—
Cloud	cloud cover index	—
CMU	concrete masonry unit	—
dry	dry cup test value	—
EPD	Environmental Product Declaration	—
ext	exterior side	—
int	interior side	—
k-dry	thermal conductivity, dry material	W/m·K
MAE	mean absolute error	same as target
MaxMC	maximum moisture content of layer	wt %
ML	machine learning	—
Mold_index	mold growth index	—
OSB	oriented strand board	—
PA	polyamide	—
Permeance	permeance of material layer	US perm (1 US perm = 57 $ng/s \cdot Pa \cdot m^2$)
Pv	annual average partial water vapor pressure	Pa
R	thermal resistance of layer	$m^2 \cdot K/W$
RadN	annual average solar radiation on north facing wall	W/m^2
RadWDR	annual average solar radiation on wall with highest wind-driven rain	W/m^2
RainN	annual average rain on north facing wall	mm/h
RainWDR	annual average rain on wall with highest wind-driven rain	mm/h
RH	annual average relative humidity	%
RMSE	root mean square error	unit of target squared
S	moisture storage capacity of layer at 80% RH	kg/m^2
T	annual average temperature	°C
Thickness	thickness of material layer	m
VB	vapor barrier	—
VR	vapor retarder	—
W80	material moisture content at 80% RH	kg/m^3
wet	wet cup test value	—
WRB	water-resistive barrier	—
XPS	extruded polystyrene	—
Z	water vapor resistance (1/permeance)	1/US perm

References

1. Global Alliance for Buildings and Construction; International Energy Agency and the United Nations Environment Programme. *2019 Global Status Report for Buildings and Construction: Towards a Zero-Emission, Efficient and Resilient Buildings and Construction Sector*; United Nations Environment Programme: Nairobi, Kenya, 2019.
2. EC3 Tool. Available online: https://carbonleadershipforum.org/ec3-tool/ (accessed on 7 April 2022).
3. United Nations. *Paris Agreement to the United Nations Framework Convention on Climate Change*; United Nations: New York, NY, USA, 2015.
4. Cement. Available online: https://www.iea.org/reports/cement (accessed on 7 April 2022).
5. O.C.L. Ltd. Calculate and Optimize Building Carbon Footprint Easily. Available online: https://www.oneclicklca.com/construction/carbon-footprint/ (accessed on 4 July 2022).
6. Life Cycle Assessment. Available online: https://sphera.com/life-cycle-assessment-software-ppc/ (accessed on 7 April 2022).
7. Economy, C. Embodied Carbon—The ICE Database. Available online: https://circularecology.com/embodied-carbon-footprint-database.html (accessed on 4 July 2022).
8. *ISO 14025*; Environmental Labels and Dclarations—Type III Environmental Declarations—Principles and Procedures. International Organization for Standardization: Geneva, Switzerland, 2006.
9. EPD International. Environmental Product Declarations. Available online: https://www.environdec.com/all-about-epds/the-epd (accessed on 4 July 2022).
10. Buy Clean California Act (BCCA). Available online: https://leginfo.legislature.ca.gov/faces/codes_displayText.xhtml?division=2.&chapter=3.&part=1.&lawCode=PCC&article=5 (accessed on 4 July 2022).
11. Hong, T.Z.; Wang, Z.; Luo, X.; Zhang, W.N. State-of-the-art on research and applications of machine learning in the building life cycle. *Energ. Build.* **2020**, *212*, 10983. [CrossRef]
12. Tzuc, O.M.; Gamboa, O.R.; Rosel, R.A.; Poot, M.C.; Edelman, H.; Torres, M.J.; Bassam, A. Modeling of hygrothermal behavior for green facade's concrete wall exposed to nordic climate using artificial intelligence and global sensitivity analysis. *J. Build Eng.* **2021**, *33*, 101625. [CrossRef]
13. Tijskens, A.; Roels, S.; Janssen, H. Neural networks for metamodelling the hygrothermal behaviour of building components. *Build Environ.* **2019**, *162*, 106282. [CrossRef]
14. Tijskens, A.; Janssen, H.; Roels, S. Optimising Convolutional Neural Networks to Predict the Hygrothermal Performance of Building Components. *Energies* **2019**, *12*, 3966. [CrossRef]
15. Salonvaara, M.; Lee, S.; Iffa, E.; Boudreaux, P.; Pallin, S.; Desjarlais, A.; Aldykiewicz, A. Selecting durable building envelope systems with machine learning assisted hygrothermal simulations database. *J. Phys. Conf. Ser.* **2021**, *2069*, 012230. [CrossRef]
16. Kim, J.; Jung, J.-H.; Kim, S.-J.; Kim, S.-A. Multi-Factor Optimization Method through Machine Learning in Building Envelope Design: Focusing on Perforated Metal Façade. *Int. J. Archit. Environ. Eng.* **2018**, *11*, 1602–1609.
17. Bhamare, D.K.; Saikia, P.; Rathod, M.K.; Rakshit, D.; Banerjee, J. A machine learning and deep learning based approach to predict the thermal performance of phase change material integrated building envelope. *Build Environ.* **2021**, *199*, 107927. [CrossRef]
18. Solmaz, A.S. Machine learning based optimization approach for building energy performance. In *ASHRAE Topical Conference Proceedings*; American Society of Heating, Refrigeration and Air Conditioning Engineers, Inc.: Atlanta, GA, USA, 2020; pp. 69–76.
19. U.S. Department of Energy. *EnergyPlus™*; Version 22; U.S. Department of Energy: Washington, DC, USA, 2022.
20. Bansal, N.; Defo, M.; Lacasse, M.A. Application of Support Vector Regression to the Prediction of the Long-Term Impacts of Climate Change on the Moisture Performance of Wood Frame and Massive Timber Walls. *Buildings* **2021**, *11*, 188. [CrossRef]
21. ANSI/ASHRAE. *Criteria for Moisture Control Design Analysis in Buildings*; American Society of Heating, Refrigeration and Air Conditioning Engineers, Inc.: Atlanta, GA, USA, 2016.
22. Boudreaux, P.; Pallin, S.; Accawi, G.; Desjarlais, A.; Jackson, R.; Senecal, D. A rule-based expert system applied to moisture durability of building envelopes. *J. Build. Phys.* **2018**, *42*, 416–437. [CrossRef]
23. Ojanen, T.; Viitanen, H.; Peuhkuri, R.; Lähdesmäki, K.; Vinha, J.; Salminen, K. Mold growth modeling of building structures using sensitivity classes of materials. In Proceedings of the 11th International Conference on Thermal Performance of the Exterior Envelopes of Whole Buildings, Buildings XI, Clearwater Beach, FL, USA, 5–9 November 2010.
24. *WUFI®Pro*; Fraunhofer Institute for Building Physics: Holzkirchen, Germany, 2011.
25. Salonvaara, M.; Zhang, J.; Karagiozis, K. Environmental weather loads for hygrothermal analysis and design of buildings. In *Simulation Studies and Dataanalysis*; ASHRAE RP-1325; American Society of Heating, Refrigeration and Air Conditioning Engineers, Inc.: Atlanta, GA, USA, 2011.
26. Google. Vertex AI. Available online: https://cloud.google.com/vertex-ai (accessed on 4 July 2022).
27. Röhrich, G. Benchmark your Machine Learning Models using Cloud Based AutoML. Available online: https://towardsdatascience.com/benchmark-your-models-6ef942e2683f (accessed on 4 July 2022).

energies

MDPI

Article

Evaluation of Heated Window System to Enhance Indoor Thermal Comfort and Reduce Heating Demands Based on Simulation Analysis in South Korea

Hyomun Lee [1], Kyungwoo Lee [2], Eunho Kang [1], Dongsu Kim [1], Myunghwan Oh [3] and Jongho Yoon [1,*]

1 Department of Architectural Engineering, Hanbat National University, Daejeon 34158, Republic of Korea
2 Sustainable Design Team of R & D Center, Junglim Architecture, Seoul 04526, Republic of Korea
3 Center for Climatic Environment Real-Scale Testing, Energy Division, Korea Conformity Laboratories, Seosan 31900, Republic of Korea
* Correspondence: jhyoon@hanbat.ac.kr

Abstract: Heated glass can be applied to improve windows' condensation resistance and indoor thermal comfort in buildings. Although this applied technology has advantages, there are still some concerns in practical applications, such as additional energy consumption and control issues. This study evaluates the effectiveness of a heated window heating (HWH) system in terms of thermal comfort and heating energy performance (HEP). The simulation-based analysis is performed to evaluate the effectiveness of the HWH using a residential building model and to compare it with radiant floor heating (RFH) and hybrid heating (HH) systems (i.e., combined HWH and RFH). This study also investigates the peak and cumulative heating loads using HWH systems with various scenarios of control methods and setpoint temperature. The predicted mean vote (PMV) is used as an indoor thermal comfort index. The ratio of cumulative thermal comfort time to the entire heating period is calculated. The results show that HWH and HH can reduce the heating load by up to 65.60% and 50.95%, respectively, compared to RFH. In addition, the times of thermal comfort can be increased by 12.55% and 6.98% with HWH and HH, respectively. However, considering the social practices of South Korea, HH is more suitable than HWH. Further investigations for HH show that a surface setpoint of 26 °C is proper, considering both heating demands and thermal comfort. In addition, the setpoint temperature should be determined considering HEP and the thermal comfort for HWH, and the optimal setpoint temperature was suggested under specific conditions.

Keywords: heated window heating; radiant floor heating; thermal comfort; predicted mean vote; control method

Citation: Lee, H.; Lee, K.; Kang, E.; Kim, D.; Oh, M.; Yoon, J. Evaluation of Heated Window System to Enhance Indoor Thermal Comfort and Reduce Heating Demands Based on Simulation Analysis in South Korea. *Energies* **2023**, *16*, 1481. https://doi.org/10.3390/en16031481

Academic Editor: Paulo Santos

Received: 12 January 2023
Revised: 26 January 2023
Accepted: 30 January 2023
Published: 2 February 2023

1. Introduction

According to a 2022 UN report, approximately 34% of the global energy consumption in 2021 was attributed to buildings and construction, of which about 30% was consumed for heating, cooling, water heating, lighting, and cooking in buildings. In the buildings' part, 21% was consumed by residential buildings [1].

To improve the energy performance of buildings, the Republic of Korea has strengthened the thermal insulation performance of each part of the building. The country also implemented a zero-energy building certification system in 2020 to enhance the overall energy performance rather than that of specific building parts [2]. Currently, residential buildings in Seoul, Republic of Korea, must achieve a thermal insulation performance of at least 0.15 $W/m^2 \cdot K$ for exterior walls and 0.9 $W/m^2 \cdot K$ for windows. These performances are similar to the level for passive house standards (i.e., wall: 0.15 $W/m^2 \cdot K$, window: 0.8 $W/m^2 \cdot K$) [3]. Although the thermal insulation performance of windows has been substantially improved, they are still the thermally weakest part of the building envelope when compared with the walls, roofs (0.15 $W/m^2{\cdot}K$), and floors (0.15 $W/m^2{\cdot}K$), which also constitute the envelope.

Windows with relatively poor thermal insulation degrade buildings' energy performance [4]. They also cause lower surface temperatures than the walls, roof, and floors, which negatively affects the thermal comfort of occupants owing to cold drafts [5]. Raising the heating setpoint to improve this situation increases energy consumption [6]. Moreover, low surface temperatures increase the probability of condensation on the window surface [7], obstructing the occupant's view of the outside [8].

One method to prevent condensation is to control the window's surface temperature to an appropriate level [4] by applying heated glass [9]. Heated glass is implemented by inserting a heating wire into the glass or using a transparent conductive coating layer on the glass surface [10]. When current is applied to the heated glass implemented with a heating wire or transparent conductive coating (TCC), heat is generated through electrical resistance. In architecture, the transparent conductive coating layer is widely used as opposed to heating wire insertion owing to its superior light transmittance and visibility, and research on this is actively underway.

Kurnitski et al. [9] investigated a method for estimating the efficiency and thermal transmittance of heated glazing by using the heat transfer theory. For typical glazing, the thermal transmittance can be determined as a single value under specific conditions depending on the glass, coating, air layer thickness, and composition of gas forming the air layer. However, for heated glazing, the thermal transmittance varies because different amounts of heat pass through the glazing depending on the temperature of the heated glass. Thus, researchers proposed a method for estimating the thermal transmittance and efficiency that reflects this feature. Lee et al. [10] experimentally measured the internal and external surface temperatures and heat flux of vacuum-heated glazing applied to residential buildings. These are measured by dividing to the center and edge of a large-area heated glazing. They also analyzed the heat gain and overheating tendencies based on measured data. The results were examined to identify the considerations for heated glass. They stressed that to prevent overheating, the temperature must be appropriately set depending on the purpose of the heated glass (condensation, comfort, heating, etc.) and to solve the difference in operating temperature between the center and the edge of a large-area heated glass. Cakó et al. [11] measured the thermal comfort of heated glass using two devices and analyzed the results according to the surface temperature and distance. In conditions where the metabolic rate is 1.0 MET, and the clothing insulation is 1.0 clo, the surface temperature of heated glass for thermal comfort is at least 40 °C. The authors stated that the required setpoint temperature for thermal comfort might vary with the experimental conditions (e.g., distance to an occupant, etc.), and stressed that thermal comfort is improved by using the heated glass. Lee et al. [12] used a simulation program to evaluate the heating energy performance of heated windows according to the improvements in the thermal insulation performance of building envelopes, such as the outer walls, floors, and windows in South Korea. They compared the performance with those of air-based heating and radiant floor heating and found that as the thermal insulation performance improved, the heating methods showed similar annual heating energy consumption. Based on these results, they noted the usefulness of heated glass for the heating system. To investigate the efficiency curves of heated glazing, Lee et al. [13] conducted both experimental and analytical studies to examine the heat fluxes under heating conditions of heated glass. Based on the results, they derived the efficiency curves according to the setpoint of the heated glass and the difference in air temperature between the internal and external. Moreau et al. [6] used an analytical method to evaluate the heating energy characteristics of heated windows. They found that the heating energy consumption was lower than required for typical double-glazed windows. They also emphasized that the effect was greater when the heated windows were installed on the east or west side than on the south side. Borys et al. [14] proposed a numerical analysis model based on the heat transfer for heated windows' indoor and outdoor heat flow and experimentally validated the model. Krukovski et al. [15] performed an analytical method to evaluate the appropriate capacity for a heated window heating system in Ukraine. Mitsui and Sato [16] investigated the sheet resistance according to the materials and coating thickness used for the transparent conductive coating and experimentally derived the appropriate coating thicknesses of the materials.

The previous studies can be broadly divided into four categories. The first group of studies examined the sheet resistance of the transparent conductive coating layer, including the coating materials and thickness. The second group of studies researched the efficiency related to the internal and external heat flux of heated glazing. The subjects of the third and fourth groups are thermal comfort and the heating system as the research of architectural usefulness. Despite extensive research, few studies have comprehensively examined heating performance and thermal comfort for practical applications in buildings. Moreover, the perspectives of occupant comfort and energy use have not been adequately studied. To improve the thermal comfort of occupants close to the window, typically in the winter season [17], heating appliances underneath exterior windows are generally used. However, an inappropriate setpoint temperature of the heating appliances causes overheating [18], which is neither efficient for thermal comfort nor efficient for the energy consumption of buildings. It is necessary to evaluate the usefulness by comprehensively considering the thermal comfort and energy performance of heated windows.

Therefore, this study was conducted to comprehensively evaluate the usefulness of heated windows considering thermal comfort and energy performance. The evaluation was performed by modeling the living room of a residential building using a simulation program, and the region and weather conditions of Gangneung in South Korea. To evaluate the effectiveness of heated windows, the thermal comfort and energy performance were compared with the radiant floor heating. In this study, the usefulness of heated windows as a heating appliance was proven, and finally, a setpoint temperature was proposed by comprehensively considering thermal comfort and energy performance in the condition of this study.

2. Methodology

The effectiveness of the heated window system was evaluated by assessing heating loads and indoor thermal comfort behaviors based on the simulation-based analysis of residential zones. Three different heating systems (i.e., radiant floor heating, heated window heating, and hybrid heating systems) were considered to carry out the comparative analysis of each heating system. This section describes the features of the heated window system used in this study, details of the analysis model, and the evaluation method.

2.1. Properties of Heated Glass

Heated glass is implemented through three main methods. The first method involves inserting a nichrome heating wire into a typical glass, such as clear glass. This method is used in the rear windows of automobiles to ensure visibility by removing accumulated snow or frost. The second method involves attaching a heating film to the glass to generate heat [19], and the third method uses a TCC on a substrate. This method uses sputter-coating [12] or pyrolytic-coating to form TCC on the glass. The heated glass used in this study is manufactured using a atmospheric pressure chemical vapor deposition (AP-CVD) during glass manufacture. [20]. In the heated glass, glass is used as the substrate and is coated with a transparent conductive oxide to form the TCC layer. Suitable materials include indium tin oxide and fluorine-doped tin oxide (FTO). The heated glass used in this study was coated with FTO [20].

Figure 1 shows the transmittance and reflectance of heated glass [21] according to the wavelength measured with a solar spectrometer [22]. In the solar spectrometer, tungsten-halogen and deuterium lamps are used as a light source to measure reflectance and transmittance according to the wavelength. Table 1 shows the optical properties of the heated glass, calculated according to ISO 9050 [23] and KS L 2514 [24] based on the measurements in Figure 1. Light transmittance and reflectance were calculated using the measured transmittance and reflectance by the solar spectrometer and the daylight spectrum as D_{65} [25], provided by the International Commission on Illumination. As a result, the ratio of the total transmitted energy or the total reflected energy to the total energy of D_{65} becomes the visible light reflectance or visible light transmittance. The other optical properties were calculated using transmittance and reflectance of the solar spectrometer and weighting factors for each wavelength. The heated glass showed high light transmittance of 82.5%, whereas the emissivity of the coated side was 0.16, approximately 20% that of the glass side. The results in Table 1 were input to the simulation program. The heated window was composed by assembling one heated glass and two low-emissivity glasses in the simulation program.

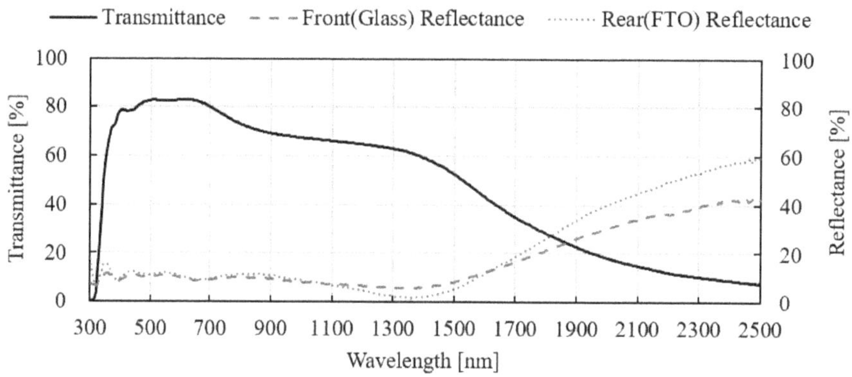

Figure 1. Transmittance and reflectance of heated glass at various wavelengths.

Table 1. Optical properties of heated glass.

Light Transmittance (%)	Light Reflectance (%)		Total Solar Transmittance (%)	Total Solar Reflectance (%)		UV Transmittance (%)	Emissivity (-)	
	Glass Side	FTO Side		Glass Side	FTO Side		Glass Side	FTO Side
82.5	10.4	11.3	69.7	10.6	11.6	52.5	0.84	0.16

2.2. *Simulation Model for Analysis of Heating Load and Thermal Comfort*

This study used ESR-r (i.e., Environmental System Performance-reference) [26], a dynamic simulation program for assessing building energy, to analyze the heating load and thermal comfort when a heated window was used for heating supply. A feature of ESP-r that defines the heating at a specific node was used to evaluate the application of heated glass to heating. Figure 2 shows the building model and the cross-section of the heated glazing in ESP-r. In Figure 2b, the thick lines represent the low-e coating and TCC, and the dots are the nodes that can be defined as the heating surface of the window in ESP-r. The red node is the heating surface of the model in this study. Since this method of modeling heated windows in EPS-r has already been used and validated in a previous study [13], an additional verification process was not performed in this study.

The building simulation model for the analysis included three zones: the glass zone located on the southernmost side, followed by the room zone and the common zone. There was no physical zone division between the glass zone and the room zone. Since the analysis program does not allow the use of multiple heating systems in a zone, this study assumed the zone (i.e., the domestic area) was divided into two zones, including room and glass zones. The building was modeled with the size of a living room in a 100 m^2 apartment house, the standard housing area in the laws for a place in South Korea [27]. The floor area of the building model, excluding the common zone, was 24 m^2, the window area on the south side was 9.6 m^2, and the window-to-floor ratio was 40%.

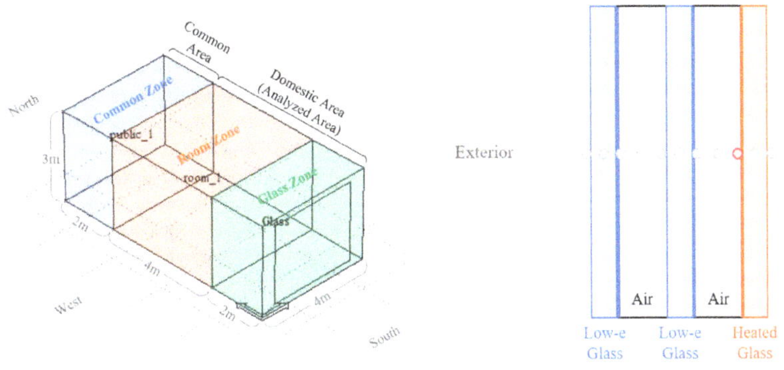

(**a**) (**b**)

Figure 2. Illustration of the model and heated glazing in the simulation program. (**a**) The building model in the simulation program; (**b**) The section layout of heated glazing.

The boundary conditions for the analysis model's side walls (east-facing and west-facing walls) were modeled under the adiabatic with no heat flow, assuming they are in contact with the adjacent households. The boundary condition for the north wall of the room zone was modeled to be in touch with common spaces (corridors, stairwells, etc.), thus creating a heat flow even though it was not directly exposed to the outdoor air. Two models with different floor boundary conditions were generated to evaluate the effect of heat loss through the ground. Model 01 used a boundary condition that assumed contact with the ground. Model 02 used the adiabatic boundary condition with no heat flow between households, assuming the household was located on an intermediate story. The thermal insulation performance of each component (i.e., walls, roofs, floors, etc.) was modeled to meet Korean regulations. The thermal transmittance of typical windows and windows with heated glass was identically defined. Since the optical properties of heated windows differ from those of typical windows, a heating function was added. Table 2 summarizes the main aspects of the analysis model.

Table 2. Details of the simulation model.

Category and Items		**Model 01 (Ground Floor)**			**Model 02 (Typical Floor)**		
Floor area (m^2)		24					
Window area (m^2)		96					
Window-to-floor ratio (%)		40					
Thermal transmittance ($W/m^2{\cdot}K$)	Exterior wall	0.129					
	Interior wall	0.184					
	Window	0.721					
	Heated window	0.721					
	Between floors	-			0.782		
	Ground floor	0.147			-		
Boundary condition	Zone	Glass	Room	Common	Glass	Room	Common
	Roof (ceiling)	Similar current			Adiabatic		
	Wall (east, west)	Adiabatic		Exterior	Adiabatic		Exterior
	Wall (south)	Exterior	Zone (Glass)	Zone (Room)	Exterior	Zone (Glass)	Zone (Room)
	Wall (north)	Zone (Room)	Zone (Common)	Exterior	Zone (Room)	Zone (Common)	Exterior
	Floor	Ground			Adiabatic		

To focus the evaluation on the effect of applying the heated glass to heating, internal heat gain elements such as occupants, lighting, and devices were not modeled. The heating setpoint was set to 22 °C, and the heating systems worked to control the indoor air temperature. The heating device capacity of each system was determined based on auto-sized values automatically calculated with the design-day condition. Radiant floor heating and heated window heating were applied to analyze the performance variations between the two methods. Table 3 and Figure 3 show each heating method's heat source locations and indoor air-temperature sensor locations.

Table 3. Sensor, actuator, and setpoint of each heating system.

	Radiant Floor	Heated Window	Hybrid 01	
			Radiant Floor	Heated Window
Sensor	Dry bulb temperature	Dry bulb temperature	Dry bulb temperature	Dry bulb temperature
Sensor location	Room zone	Glass zone	Room zone	Glass zone
Actuator	Room zone	Glass zone	Room zone	Glass zone
Setpoint (°C)	22			
Heating period	January to March, September to December			

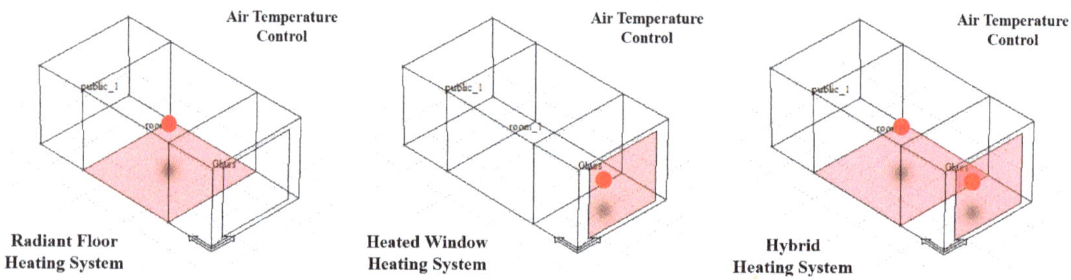

Figure 3. Heating surface and air temperature sensing point of each heating system.

The weather data for the Gangneung region in South Korea were used to investigate the heating load and thermal comfort, which were included in the simulation tool. The data from the International Weather for Energy Calculations 1.1, created and provided by the American Society of Heating, Refrigerating, and Air-conditioning Engineers (ASHRAE), were used as the weather data. Table 4 presents the monthly characteristics of each meteorological item. Since the ground temperature was not included in these meteorological data, the ground temperature was defined based on the average ground temperature from 1981 to 2010, which was provided by the Korea Meteorological Administration [28].

Figure 4 depicts the location and climate classification of Gangneung, South Korea, using the Koppen–Geiger climate classification map [29]. The climate classification of South Korea, including Gangneung, belongs to the humid continental climate. The northern states of the United States, and the border area between Europe and Russia, show the same climatic characteristics.

Table 4. Weather data used in the simulation model.

Month	Dry Bulb Temperature (°C)			Relative Humidity (%)	Wind Speed (m/s)	Ground Temperature (Depth: 1.5 m) (°C)
	Avg	Max	Min			
January	0.4	10.7	−9.8	46.8	3.0	9.1
February	2.2	13.6	−8.5	56.6	2.6	7.1
March	5.6	15.4	−3.5	54.4	2.7	7.2
April	12.4	27.0	1.7	58.6	2.9	9.3
May	18.0	34.0	6.9	62.5	2.7	12.6
June	20.6	29.9	14.3	73.9	1.6	16.0
July	24.0	34.1	17.2	77.4	2.2	19.3
August	24.9	37.1	16.5	78.3	1.5	21.7
September	20.1	28.5	14.2	72.9	1.9	22.0
October	15.2	25.7	5.3	66.6	2.3	19.9
November	9.0	19.6	−4.8	54.6	3.1	16.5
December	3.1	12.1	−6.2	50.6	3.3	12.6

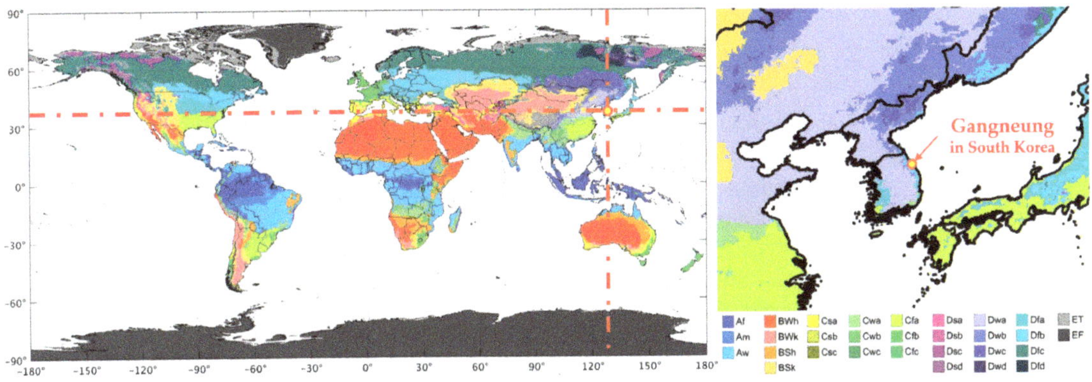

Figure 4. Location and climate classification of Gangneung in the Koppen–Geiger climate classification map [29].

2.3. Description of Indoor Thermal Comfort

The predicted mean vote (PMV), the thermal comfort index used in this study, was proposed in an experimental study by Fanger at the Technical University of Denmark. It expresses the average thermal sensation experienced by people in a given environment [30]. The PMV can be used as an index to evaluate the comfort level of indoor environments in residential buildings, offices, hospitals, etc., and to control the heating and cooling facilities. It has been introduced as a thermal comfort index in ISO 7730 [31] and ASHRAE 55 [32], which are international standards. PMV is calculated by considering the heat transfer between the body and surrounding environment using two parameters of the occupant (i.e., metabolic rate and insulating of clothing) and four parameters of the indoor environment (i.e., air temperature, mean radiant temperature (MRT), air velocity, and humidity). Additionally, as shown in Table 5, PMV is calculated as a value ranging from +3 to −3, which is the range of thermal sensation from hot to cold. Values closer to −3 indicate a thermal comfort level where the occupants feel colder, whereas values closer to +3 indicate an environment where occupants feel hotter. Hence, a PMV of 0 represents the optimal comfort level; statistically, 95% of the occupants are satisfied with the indoor environment. ISO 7730 [31] and ASHRAE 55 [32] recommend a PMV range of −0.5 to 0.5 to provide indoor comfort.

Table 5. Seven-point thermal sensation scale.

Hot	Warm	Slightly Warm	Neutral	Slightly Cool	Cool	Cold
+3	+2	+1	0	−1	−2	−3

The input variables were set to evaluate the thermal comfort according to each heating method, as shown in Table 6. The input meteorological data and model characteristics were reflected to calculate the air temperature, relative humidity, and MRT in the analysis program. In addition, the air velocity, metabolic rate, and insulation of clothing values were direct inputs into the simulation program. A value of 0.1 m/s was input for the air velocity considering the indoor conditions, whereas those for metabolic rate and clothing insulation were 1.5 MET and 0.7 clo, respectively. According to ISO 7730, insulation of clothing of 0.7 clo signifies wearing undergarments, shirts, trousers, socks, and shoes, and a metabolic rate of 1.5 MET designates standing and light activity (shopping, laboratory work, light industry). The PMV of the glass zone and room zone were calculated every hour using the analysis program, based on which the cumulative hours in each PMV range were analyzed to evaluate thermal comfort behaviors.

Table 6. Input values to calculate the PMV.

Air Temperature (°C)	Relative Humidity (%)	Mean Radiant Temperature (°C)	Air Velocity (m/s)	Metabolic Rate (-)	Cloth (-)
Value calculated by the simulation model			0.1	1.5	0.7

3. Heating Load and Thermal Comfort of Heating Systems

3.1. Heating Load Comparison for Each Heating System

As shown in Figure 3, in radiant floor heating, the room zone and glass zone are heated via room zone heating, whereas in heated window heating, they are heated via glass zone heating. Therefore, the indoor air temperature may differ between the heating zone and adjacent zones, and the accuracy of the analysis model used in this study will deteriorate if the temperature difference is large.

Figure 5 shows each model's indoor air temperatures of the glass zone and room zone according to the floor boundary condition and heating system. The top graph presents the indoor air temperature over time for each model, and the scatter plots below this present the indoor temperature distribution of the room zone to the glass zone for each model. According to the graphs of radiant floor heating and hybrid heating, the air temperature of the room zone with respect to the glass zone was controlled at a low level. The times were observed in which the air temperature of the glass zone was high owing to daytime solar radiation, which caused a difference in the air temperature between the two zones. Since these times are different from the times needed for heating in the model, it is expected to have no substantial impact on the results of this study.

Figure 6 presents the heating load concerning the outdoor air temperature for each heating method. The load is depicted only for outdoor air temperatures lower than the heating setpoint (22 °C). In Figure 6a, the floor boundary condition is the ground, whereas Figure 6b shows the adiabatic, which relates to the floor between two stories. Radiant floor heating showed the highest peak heating load under both floor boundary conditions, followed by hybrid heating and window heating. Table 7 presents the peak heating load of each heating method.

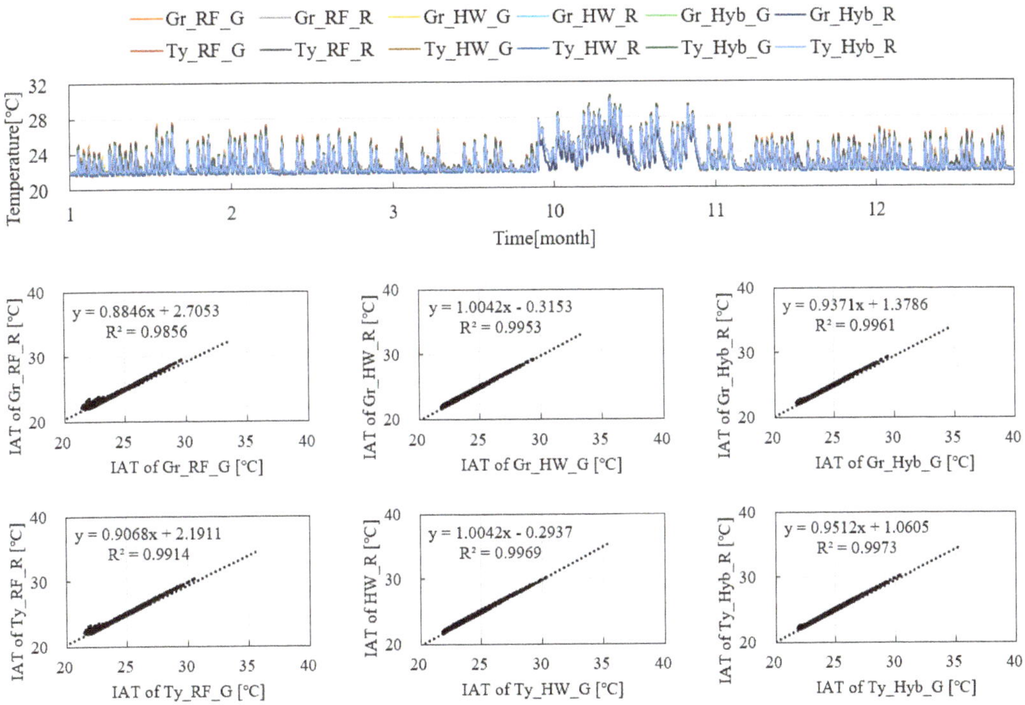

Figure 5. Outdoor air temperature (OAT) and indoor air temperature (IAT) of the glass zone (G) and room zone (R) of simulation models according to the floor boundary condition (ground floor: Gr, typical floor: Ty) and heating system.

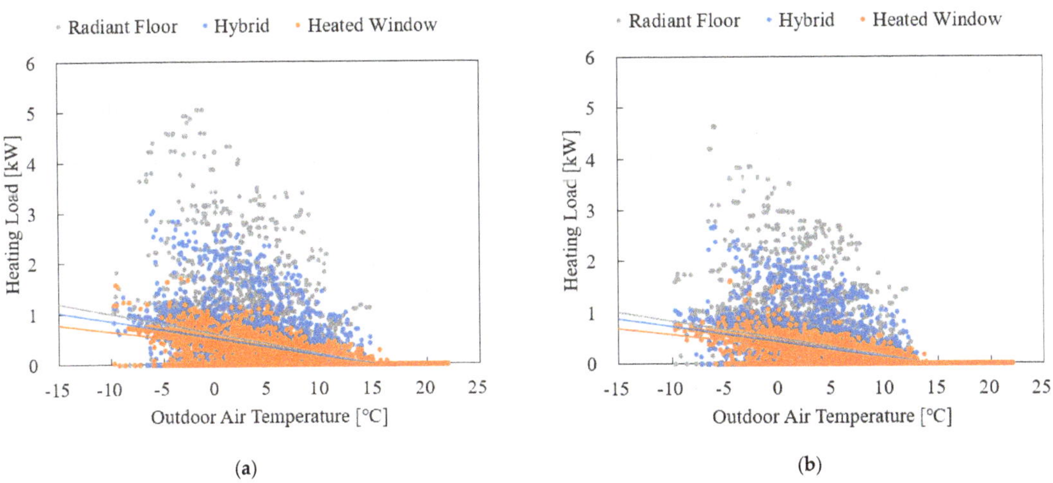

Figure 6. Variation in heating load with outdoor air temperature for each simulation model. (**a**) Ground floor; (**b**) Typical floor.

Table 7. Peak heating load and cumulative heating load of each simulation model.

Boundary Condition		Ground Floor				Typical Floor			
Heating System		Radiant Floor	Heated Window	Hybrid		Radiant Floor	Heated Window	Hybrid	
				Radiant Floor	Heated Window			Radiant Floor	Heated Window
Peak Load (kW)		5.04	1.70	2.47	0.85	4.62	1.62	2.21	0.95
Cumulative heating load (kWh)	January	395.78	266.83	231.45	114.81	327.78	233.5	186.11	103.65
	February	318.90	213.66	188.58	92.51	249.03	178.34	144.24	78.27
	March	327.06	223.02	196.55	92.09	253.96	184.14	152.17	76.98
	April	-	-	-	-	-	-	-	-
	May	-	-	-	-	-	-	-	-
	June	-	-	-	-	-	-	-	-
	July	-	-	-	-	-	-	-	-
	August	-	-	-	-	-	-	-	-
	September	-	-	-	-	-	-	-	-
	October	23.62	11.93	12.68	5.5	5.37	2.62	2.88	1.59
	November	188.52	122.17	109.82	53.91	147.83	97.11	85.8	40.15
	December	320.15	210.05	192.31	86.92	259.03	180.78	156.09	74.03
	Annual	1574.03	1047.66	931.39	445.74	1242.99	876.49	727.29	374.67

Compared with the ground floor model (ground floor boundary condition), the peak heating load of the typical floor model (adiabatic floor boundary condition) was small. Under radiant floor heating, heated window heating, and hybrid heating, the peak heating load was reduced by approximately 8.33%, 4.71%, and 12.17%, respectively. This trend is because the typical floor model has no heat loss through the floor.

The peak heating load ratio of each heating model to the radiant floor heating model was analyzed among the same boundary conditions of the floor. First, for the model with the ground floor, the peak heating load ratio relative to radiant floor heating was 33.73% for heated window heating and 60.32% for hybrid heating, whereas for the model with the typical floor, the ratios were 35.06% and 57.79%, respectively.

Table 7 presents each model's peak heating load and monthly and annual cumulative heating loads. The incremental heating loads exhibited the same trend observed in the above peak heating load analysis. Compared with the ground floor results, the typical floor showed small monthly and annual cumulative heating loads, and window and hybrid heating showed lower cumulative heating loads than radiant floor heating. However, compared with the reduction in the peak heating load by applying heated windows, the reduction in the cumulative heating load was small. For the ground floor model, the annual cumulative heating load ratio relative to radiant floor heating was 66.56% for heated window heating and 87.49% for hybrid heating, whereas the ratios were 70.51% and 88.65%, respectively, for the typical model.

According to the above analysis, the peak and cumulative heating loads were reduced when heated window heating was applied. This was because heating is provided via the heated window, which has the lowest thermal insulation performance (0.721 W/m^2) among the envelope components where heat is exchanged with the outdoor air or a common zone. The inner surface temperature of the heated window on heating is higher than the outdoor air and indoor temperatures, thus preventing heat loss through the window and reducing the heating load. Figure 7 shows the window's outdoor air temperature, indoor air temperature, and inner surface temperature in December for radiant floor heating and heated window heating in the typical floor model. As explained, under heated window heating, the window's inner surface temperature was higher than the indoor air temperature. In contrast, under radiant floor heating, the window's inner surface temperature was lower than the indoor air temperature at night but higher during the daytime. This trend was also observed in the ground floor model.

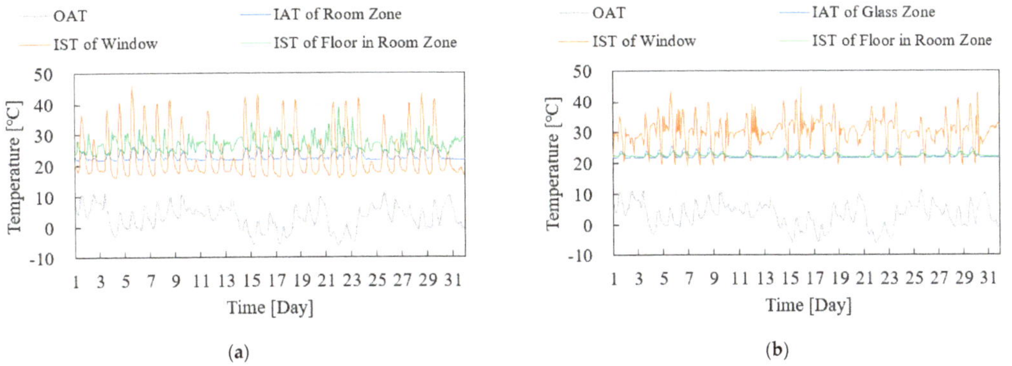

Figure 7. Outdoor air temperature (OAT), indoor air temperature (IAT), and interior surface temperature (IST) of window and floor for the ground floor model. (**a**) Radiant floor heating system; (**b**) Heated window heating system.

Under hybrid heating, the peak and cumulative heating loads were lower than those of radiant floor heating, although the heating load was higher than in heated window heating. The annual cumulative heating load was 31.9% and 25.8% higher than in heated window heating for the ground floor and typical floor, respectively. Figure 8 shows the analysis results for the typical floor. Figure 7a shows the cumulative heat fluxes for the window, south wall (glass zone), and north wall (room zone) of each model at night (00:00–04:00), that is, without the influence of solar radiation. Positive values indicate heat gain, whereas negative values indicate heat loss. Figure 7b shows the difference in indoor air temperature between each model's room zone and common zone for one week in December.

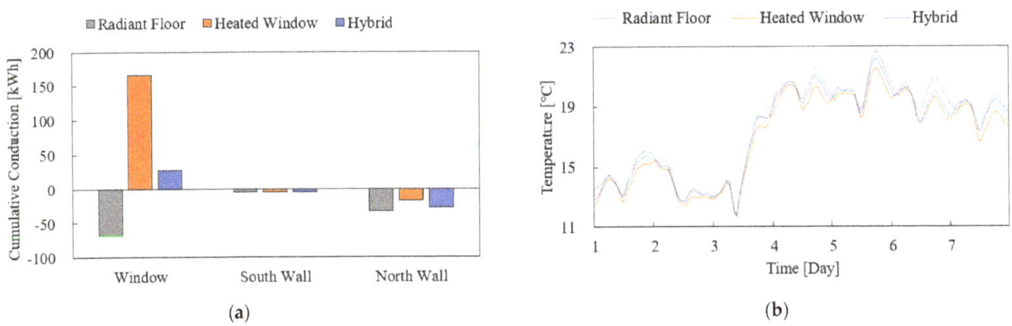

Figure 8. Inner surface conduction and difference in air temperature between room and common zones. (**a**) Cumulative conduction on inner surfaces of window, south wall (glass zone), and north wall (room zone) for each heating system at night (00:00–04:00); (**b**) Difference in air temperature between room zone and common zone for each heating system for one week (1–7 December).

According to Figure 7a, the heat fluxes of the window varied with the heating method, whereas the heat flux of the south wall did not greatly vary with the heating method. However, the heat flux of the north wall greatly varied with the heating method. As the thermal transmittance and boundary condition of the north wall were identical for each model, they should not differ according to the heating method; however, this study observed a difference. As shown in Figure 3, the temperature sensor and heating surface location varied for each model according to the heating method. In the hybrid heating model, because temperature sensing and heating were applied in the room zone and glass zone, the air temperature of the room zone was formed according to the setpoint of 22 °C.

Conversely, in the heated window heating model, temperature sensing and heating were applied to the glass zone; hence, the glass zone and room zone had similar air temperatures, whereas the air temperature in the room zone was a little lower than 22 °C. This caused a difference in the air temperature variation and, consequently, a heat flux difference between the room zone and the common zone for each model. Figure 7b shows the air temperature difference between each model's room zone and common zone, which was larger in the hybrid heating model than in the heated window heating model.

3.2. *Thermal Comfort Comparison for Each Heating System*

Figure 9 and Table 8 show the ratio according to each PMV range for each model over the entire analysis period. The glass zone and room zone results for each model under both floor boundary conditions are presented. Since heating was applied, the PMV was never lower than −0.5 and was occasionally higher than 1.0. The hours when the PMV was outside the thermal comfort range and evaluated as "warm" or "hot" were attributed to irradiance during the daytime.

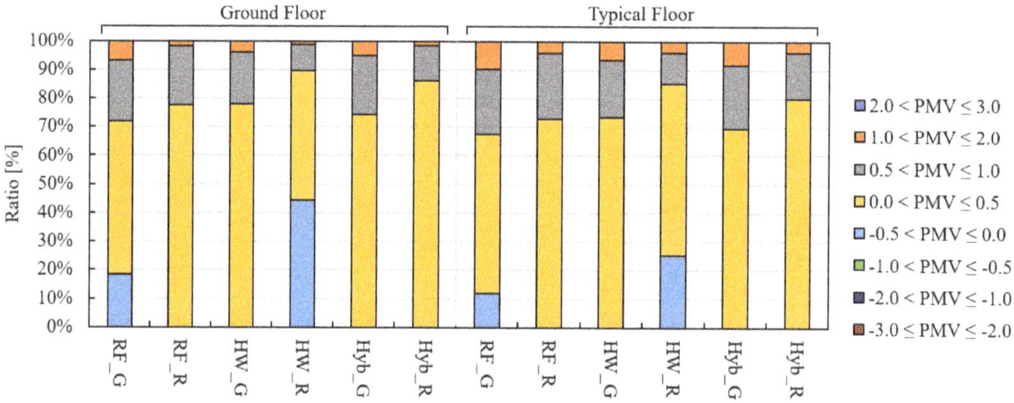

Figure 9. Ratio of comfort time for each model.

Table 8. Cumulative hours for various ranges of PMV and comfort ratio ($-0.5 \leq PMV \leq 0.5$).

PMV Range	Ground Floor						Typical Floor					
	Radiant Floor		Heated Window		Hybrid		Radiant Floor		Heated Window		Hybrid	
	G	R	G	R	G	R	G	R	G	R	G	R
$2.0 < PMV \leq 3.0$	-	-	-	-	-	-	2	-	2	-	2	-
$1.0 < PMV \leq 2.0$	293	80	171	60	227	70	426	179	286	170	359	177
$0.5 < PMV \leq 1.0$	934	897	793	392	891	534	1000	1003	8872	479	971	699
$0.0 < PMV \leq 0.5$	2336	3391	3403	1984	3247	3764	2423	3186	3208	2605	3036	3492
$-0.5 < PMV \leq 0.0$	805	-	1	1932	3	-	517	-	-	1114	-	-
Comfort Ratio	71.91	77.63	77.93	89.65	74.40	86.19	67.31	72.94	73.44	85.14	69.51	79.95

First, differences were observed in the PMV according to the floor boundary condition. Compared with the ground floor, the ratio of time that PMV was lower than 0 was higher on the specific model with the typical floor because heat loss through the floor did not occur during heating. Additionally, the ratio of time that PMV was higher than 0.5 was higher in all models with the typical floor because heat loss through the floor did not occur under overheating caused by irradiance.

Heated window heating and hybrid heating exhibited better thermal comfort than radiant floor heating. According to the PMV analysis results for the typical floor, 70.12% (glass zone: 67.31%, room zone: 72.94%) of the total 4368 h of radiant floor heating was in the thermal comfort range in terms of the PMV. Under heated window heating and hybrid heating, 79.29% (glass zone: 73.44%, room zone: 85.14%) and 74.73% (glass zone: 69.51%, room zone: 79.95%) were in the thermal comfort range, respectively. Hence, the heating methods, including heated windows, were more advantageous than radiant floor heating in terms of thermal comfort.

For the typical ground condition, the ratio of time of thermal comfort according to the heating method was compared based on the room zone. The ratio showed that heated window heating and hybrid heating were approximately 16.73% and 9.60% higher than radiant floor heating. This indicates that heated window heating improves the occupants' thermal comfort.

Among the factors determining the PMV, the MRT can be influential depending on the heating method. Except for radiant floor heating, other heating methods have heated window heating; therefore, when heat is generated from windows with low surface temperatures in winter, relatively high window surface temperatures are formed. This results in a higher MRT than that achieved in radiant floor heating. Figure 10 presents the MRT of the glass zone for each model under the typical floor condition from 1 to 14 December. The MRT in the daytime was increased because of irradiance. At night, when there is no influence of irradiance, the MRT varied with the heating device, and it was approximately 22.5 °C under radiant floor heating and about 24 °C under heated window heating, which is a difference of roughly 1.5 °C. The MRT of hybrid heating was also below that of heated window heating, although it was higher than that of radiant floor heating.

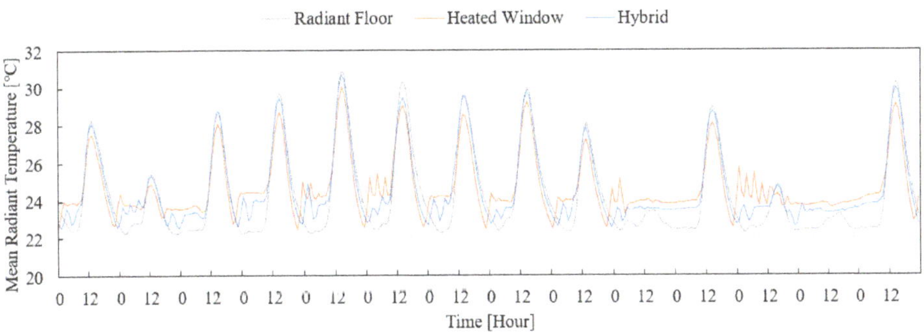

Figure 10. Mean radiant temperature of the glass zone of each model under the typical floor condition.

Radiant floor heating is the traditional and typical heating method for residential buildings in South Korea. This is related to the Korean custom of not wearing shoes in residential spaces. Hence, the floor surface temperature is a major factor determining the occupants' thermal comfort. Figure 11 shows the floor surface temperature of the room zone for the period of 1–7 December for each model. Under heated window heating, the floor surface temperature was distributed from approximately 22 to 23 °C. In contrast, the surface temperature was higher than 25 °C during the heating period under radiant floor heating and hybrid heating. In the area shaded in red, the temperature difference between heated window heating and hybrid heating was approximately 5 °C, which is greater than that with radiant floor heating.

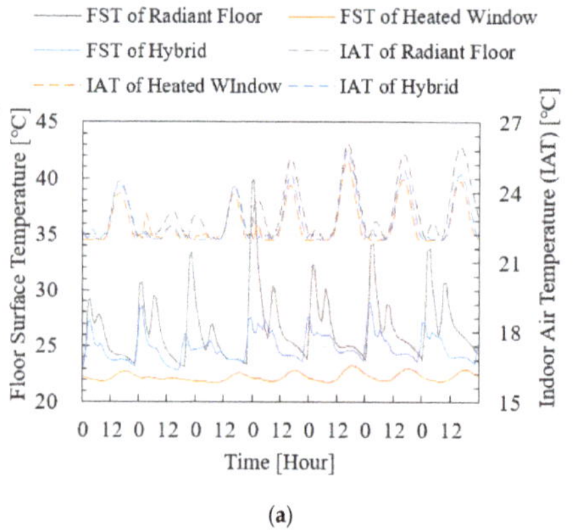

Figure 11. Each model's floor surface temperature (FST) and the indoor air temperature (IAT) of each model. (**a**) Ground Floor; (**b**) Typical Floor.

Considering the results of the heating load and thermal comfort from the previous analysis, heated window heating appears to be the most appropriate heating method. However, as explained above, the floor surface temperature is important in Korea. In a previous study [2] that applied a low-temperature radiant floor heating system using an air-source heat pump, the indoor air setpoint was set to 26–28 °C to increase the floor surface temperature. Therefore, hybrid heating combining heated window heating with radiant floor heating is considered suitable for such a case.

4. Comparative Study Based on Control Method of the Hybrid Heating System

The heating load and thermal comfort according to the control method of hybrid heating, which combines radiant floor heating with heated window heating, were evaluated. Figure 12 shows the models of two hybrid heating systems. Figure 12a shows the air-temperature sensing model for heated window systems in the hybrid heating system. It is the same hybrid heating system as that used in Section 3. Figure 12b shows the model of surface-temperature sensing for the hybrid heating system's heated window heating system. The difference between the two models is only the temperature sensing location to control the heated window heating system. In addition to the temperature sensing location for controlling the operation of the heated window, the setpoint for controlling the surface-temperature sensing was changed from 22 to 36 °C, and the heating load and thermal comfort characteristics were comparatively analyzed. The floor boundary condition applied here was the same as that of the typical floor.

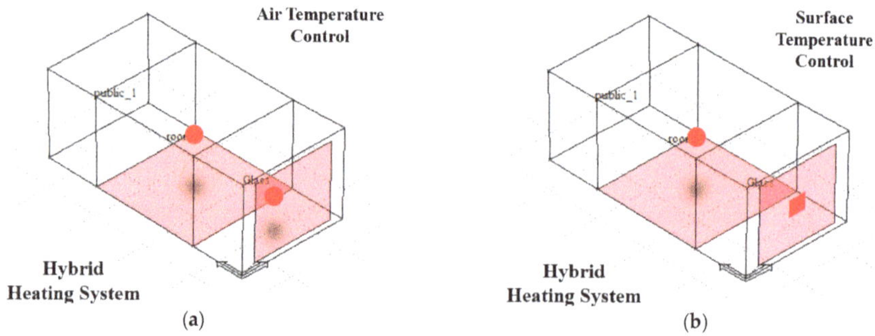

Figure 12. Heating surface and temperature sensing point of hybrid heating systems. (**a**) Air-temperature sensing model; (**b**) Surface-temperature sensing model.

4.1. Heating Load Comparison According to Control Method and Setpoint

Figure 13 shows the variation in the heating load under hybrid heating according to the control method of the heated window and the setpoint. "Air" indicates the air temperature control method, as in Figure 12a, and "Sur" indicates the surface temperature control method, as in Figure 12b. First, the hybrid heating model using the air temperature control method with a setpoint of 22 °C showed an annual cumulative heating load of 1101.96 kWh. When the temperature sensing location was changed to the glass surface while maintaining the setpoint, the annual cumulative heating load increased by approximately 6.60% to 1174.65 kWh. The annual cumulative heating load changed as the setpoint of the heated window surface increased, with the lowest cumulative heating load observed at a setpoint of 30 °C. The annual cumulative heating load was 1063.17 kWh, approximately 5.52% lower than that in the air temperature control model and 9.49% lower than that in the surface temperature control model with a setpoint of 22 °C. The load tended to increase at setpoints higher than 30 °C.

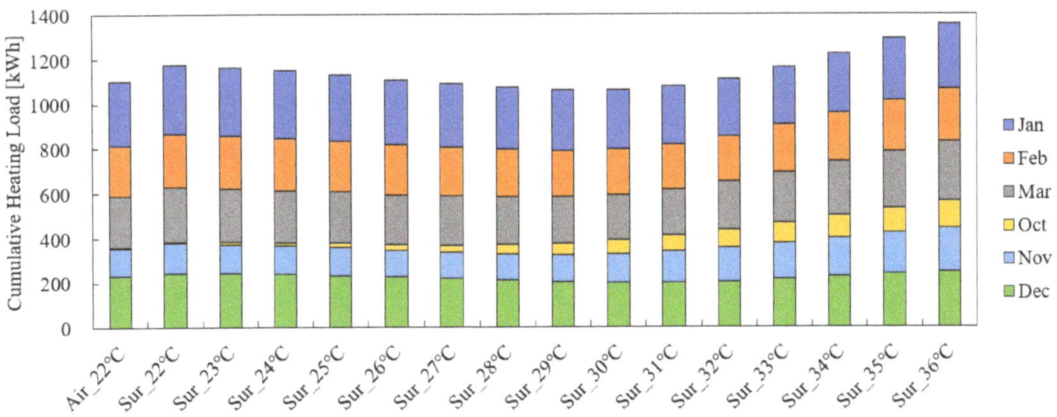

Figure 13. Variation in monthly heating load according to the control method and setpoint.

Figure 14 shows the ratio of the monthly heating load to the annual cumulative heating load of each model. By month, from January to March and October to December, the hybrid heating model using air temperature control with a setpoint of 22 °C showed ratios of 26.29%, 20.19%, 20.79%, 0.41%, 11.43%, and 20.88%, respectively. The total annual cumulative heating load increased when the temperature sensing location was changed to the glass surface while maintaining the setpoint. However, there was no significant difference in the ratio. As the setpoint of the surface temperature control method increased, the monthly ratio changed; in January, February, and December, it tended to decrease as the setpoint increased. In contrast, as the setpoint increased in October, the ratio continuously increased from 0.61% to 8.93%. In March and November, it tended to increase after showing the lowest ratio at a specific setpoint. In March, the annual cumulative heating load was approximately 19.12% at the lowest setpoint of 30 °C; in November, it was about 10.70% at 27 °C.

Figure 15 shows the ratio of the heating load to the annual cumulative heating load of each heating system for each model. First, in the air temperature control model with a setpoint of 22 °C, radiant floor heating showed a ratio of approximately 66.0%, whereas that for heated window heating was about 34.0%. In the surface temperature control model with the same setpoint, radiant floor heating and heated window heating showed ratios of 82.94% and 17.06%, respectively. As the heated window setpoint increased, the heating load of radiant floor heating tended to decrease, and the heating load ratio was approximately 0.14% at 34 °C.

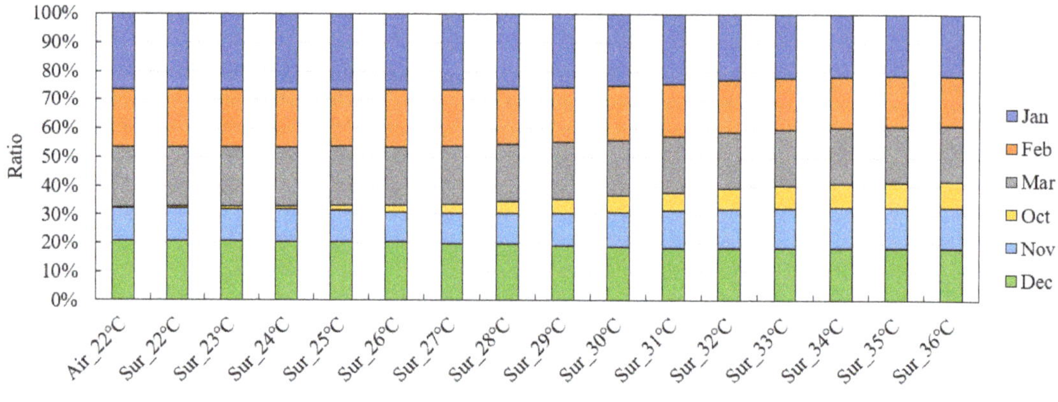

Figure 14. Ratio of monthly heating load to annual heating load.

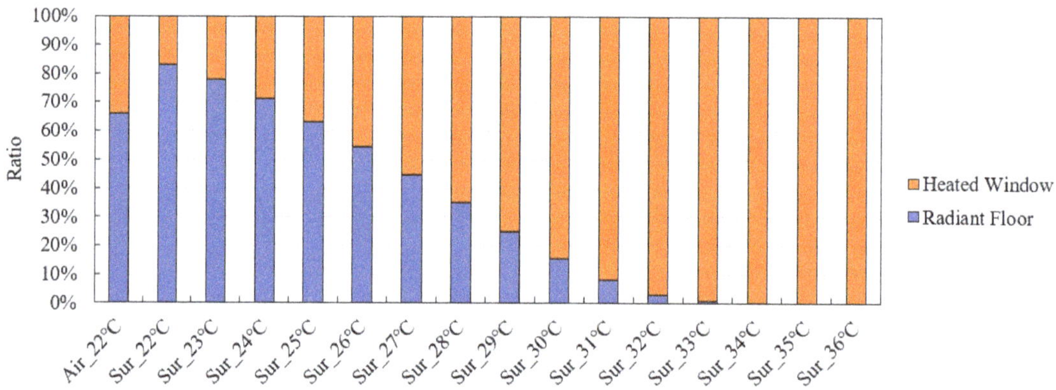

Figure 15. Ratio of heating load to annual heating load for each heating system.

The surface setpoint at which the heating load ratios to the heating systems' total heating load were the most similar to "Air_22 °C" was 25 °C; at this setting, the load ratios of floor heating and window heating were 63.26% and 36.74%, respectively. This was because the air temperature control method with a setpoint of 22 °C and the window surface temperature control method with a setpoint of 25 °C had a similar inner surface temperature of the heated window. Figure 16 presents the inner surface temperatures of the heated window in the air temperature control model with a setpoint of 22 °C and the surface temperature control model with setpoints of 22 °C, 25 °C, and 28 °C, from 1 to 14 December. During the daytime when the window temperature increased owing to irradiance, the same temperature was observed regardless of the control method and setpoint. During the nighttime, each model showed different inner surface temperatures of the heated window, and the model of the air temperature control and the model of the surface temperature control with a setpoint of 25 °C exhibited the most similar surface temperature trends.

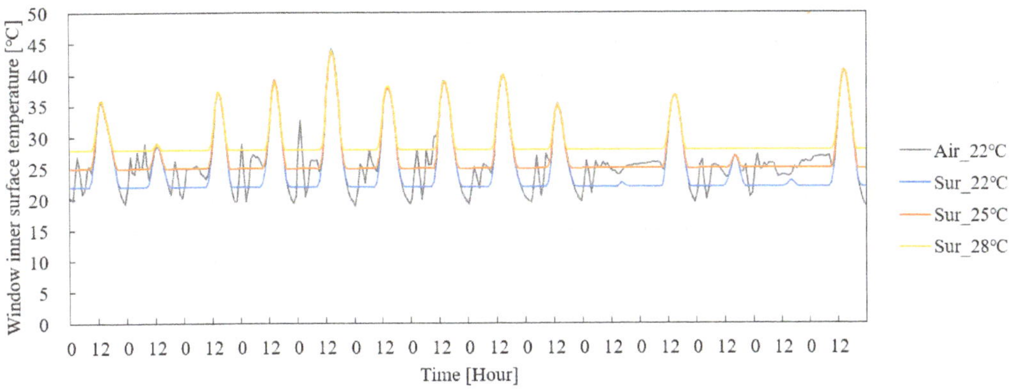

Figure 16. Inner surface temperature of the window.

4.2. Thermal Comfort Comparison According to Control Method and Setpoint

The thermal comfort of the hybrid heating models was analyzed. The thermal comfort in the glass and room zones was analyzed, and the ratio of the time for cumulative analysis for each PMV range to all periods was examined.

Figure 17 shows the thermal comfort analysis results in the glass zone of each model. As heated window heating was applied to the glass zone, the changes in the PMV were sensitive to the heated window setpoint. In the air temperature control model, the cumulative time for thermal comfort ($-0.5 \leq \text{PMV} \leq 0.5$) was 3036 h, approximately 69.51% of the total time (4368 h). The time for thermal comfort tended to decrease as the setpoint of the heating window increased, and it sharply decreased from a specific temperature.

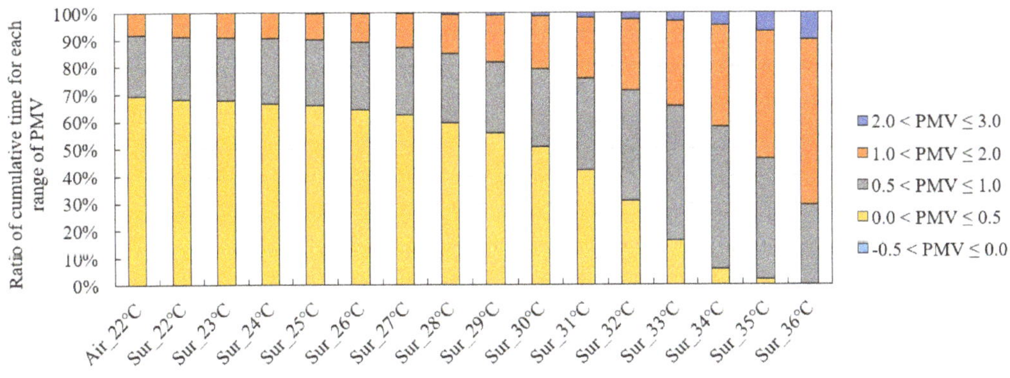

Figure 17. Ratio of cumulative time for each range of PMV for the glass zone.

Figure 18 presents the glass zone's air temperature and heated window inner surface temperature for the air temperature control model with a setpoint of 22 °C and the surface temperature control model with various setpoints. The figure shows the temperature distribution for one week from 1 to 7 December. The air temperature of the glass zone for the air temperature control model was similar to the air temperature control model with a surface setpoint of 22 °C and 25 °C. However, suppose that the surface setpoint is higher than 30 °C. In that case, the glass zone's air temperature for the surface temperature control model is higher than the air temperature control model during daytime and nighttime. It means that the air temperature is higher than the air setpoint of the room zone.

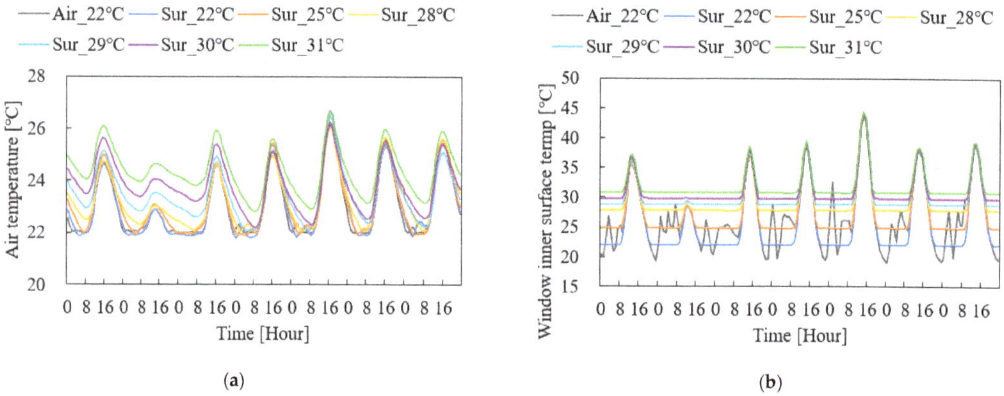

Figure 18. Air temperature of the glass zone and the inner surface temperature of the window. (**a**) Air temperature of glass zone; (**b**) Window inner surface temperature.

Figure 19 shows the cumulative time ratio according to the thermal comfort range in the room zone. Overall, the cumulative time ratio of the thermal comfort range ($-0.5 \leq$ PMV ≤ 0.5) was higher than that of the glass zone. In the air temperature control method, the time for thermal comfort was 3492 h, approximately 79.95% of the total time. In contrast, the surface temperature control method with the same setpoint was about 3% lower, at 76.85%.

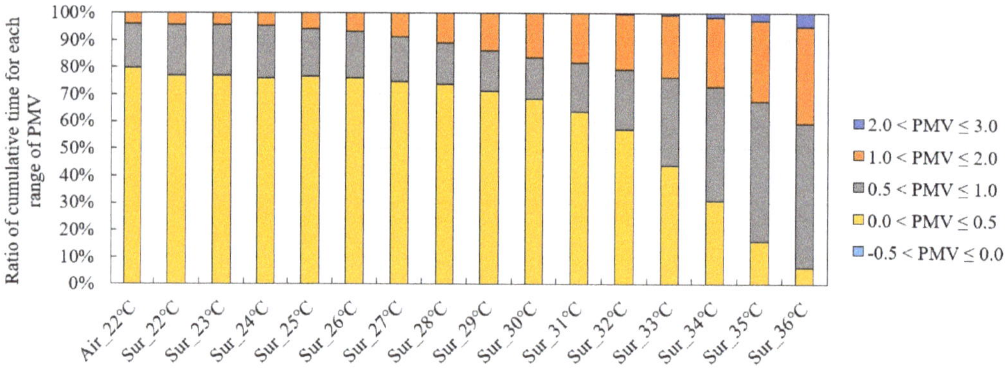

Figure 19. Ratio of cumulative time for each range of PMV for the room zone.

Figure 20 shows the combined annual cumulative heating load and the ratio of time for thermal comfort of the air temperature control and surface temperature control models. As analyzed above, when the surface setpoint was 30 °C, the heating load was the lowest at 1063.17 kWh. Therefore, a surface setpoint of 30 °C was found to be the most suitable for the heating load. In contrast, the time for thermal comfort was the highest in both the glass and room zones when the air setpoint was 22 °C. Table 9 summarizes the optimal setpoints and differences in the parameters according to the control method.

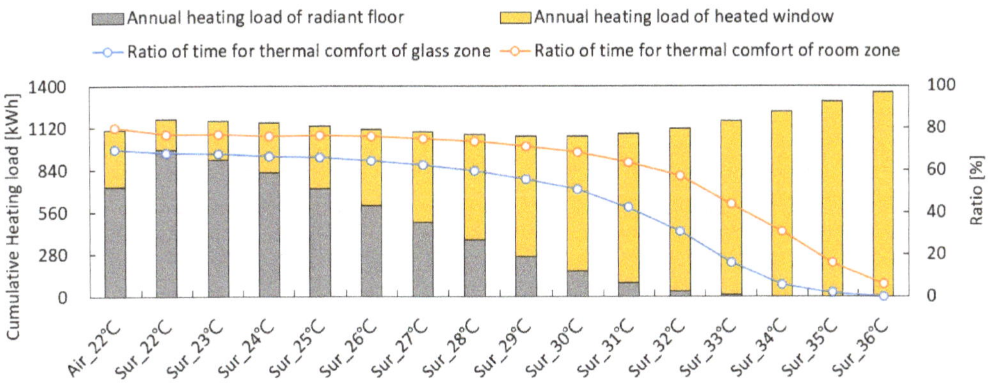

Figure 20. Annual heating load and ratio of time for thermal comfort.

Table 9. Setpoint, heating load, and thermal comfort for each control method.

		Optimal Setpoint (°C)	Annual Heating Load (kWh)	Ratio of Time for Thermal Comfort (%)	
				Glass Zone	Room Zone
Air temperature control		22	1101.96	69.51	79.95
Surface temperature control	Heating load	30	1063.17	50.80	68.59
	Thermal comfort	22	1174.65	68.22	76.85
	Optimal control	26	1109.26	64.4	76.08

In surface temperature control, the heating load of the model optimized for heating load was 9.05% lower than the model optimized for thermal comfort, and the ratio of time of the thermal comfort was also 25.54% and 10.75% lower in the glass zone and room zone, respectively. In contrast, for the model for thermal comfort, the heating load increased by approximately 10.49%, and the ratio of time for thermal comfort improved by 33.29% and 12.04% in the glass and room zones, respectively.

To propose the optimal surface setpoint by synthesizing the heating load and thermal comfort, it is necessary to find the surface setpoint that maximizes the decrease in heating load while minimizing the reduction in the ratio of the time of thermal comfort. Based on Figure 19, the setpoint can be estimated by comprehensively considering the heating load and thermal comfort. Up to a setpoint of 26 °C, the reduction in the ratio of time for thermal comfort in the glass zone and room zone was not large, and the heating load was improved.

5. Conclusions and Discussion

This study evaluated the effectiveness of the heated window heating system in a typical residential building in South Korea. This study compared the heating load and thermal comfort of three different heating system types (i.e., radiant floor, heated window, and hybrid heating systems) in the Gangneung location, South Korea. The surface temperature variations of the heated window were investigated to determine appropriate setpoint temperatures regarding indoor thermal comfort and energy consumption.

The compared results of heating loads for each heating system indicated that applying the heated window heating method decreased the peak and cumulative heating load. The window surface was heated by the heated window system, which caused an increase in the surface temperature of the glass. Theoretically, if the indoor surface temperature of the heated window system is higher than the indoor air temperature, the heat loss through the window is zero, which is advantageous regarding the heating load.

Compared to the heating load for radiant floor heating, the heating reductions were about 34.40% (ground floor: 3.73%, typical floor: 35.06%) and 59.05% (ground floor: 60.32%, typical floor: 57.79%) for the heated window and hybrid heating systems, respectively. In addition, the ratios of time for thermal comfort in the glass zone and room zone were 72.45% (ground floor: 74.77%, typical floor: 70.12%) for radiant floor heating, 81.54% (ground floor: 83.79%, typical floor: 79.29%) for heated window heating, and 77.51% (ground floor: 80.29%, typical floor: 74.73%) for hybrid heating.

Such results demonstrated that the heated window heating system could present better performance compared to the existing radiant floor heating method regarding the heating load and thermal comfort. However, because the floor surface temperature was lower under the heated window heating system than in the radiant floor heating system, the hybrid heating system is more suitable for residential household types in South Korea.

A suitable control method and setpoint were estimated by evaluating the heating load and thermal comfort according to the control method in the hybrid heating system. The annual heating load was the lowest when the heated window surface temperature was set to 30 °C. However, operating at a surface temperature of 30 °C caused the indoor space to overheat, which was disadvantageous in terms of indoor thermal comfort levels. In the case of the air temperature being set to 22 °C, the time ratio for the indoor thermal comfort increased when compared with the surface temperature control. However, by comprehensively considering the heating load and thermal comfort to estimate the control method and setpoint, the heating load and thermal comfort levels were enhanced when the heated window surface temperature was controlled to 26 °C.

This study has proved the effectiveness of the heated window systems as a building heating appliance. Although the thermal insulation performance of windows has substantially improved, the windows continue to be a thermally weak component of the building envelope. Heat loss and condensation occur in thermally inefficient windows, and cold drafts can affect the occupant's thermal comfort. Hence, adding a heating functionality to these windows should not only prevent condensation and improve thermal comfort, but can also make them applicable to indoor heating. It was confirmed that the control of the heated windows for the thermal comfort of occupants close to the heated windows could adversely affect the thermal comfort of occupants far from the heated windows. In addition, the optimal setpoint temperature for the thermal comfort and energy performance under the conditions of this study (i.e., type of building, the region and the climate, internal gain, thermal insulation performance of building element) was also proposed. However, since the building model and area for evaluation of usefulness was limited to one, additional research is needed to generalize the optimal setpoint temperature.

This study used a heating load to assess the effectiveness of heated windows as an index. The primary energy consumption and carbon emissions can vary according to the energy source of the heating system (electricity, gas, etc.). Furthermore, because this study only utilized simulation-based analysis, future work should include experimental analysis and evaluation through full-scale mock-ups and their application in a practical manner.

Author Contributions: Conceptualization, J.Y., D.K., H.L. and K.L.; methodology, H.L., K.L. and M.O.; software, H.L. and K.L.; validation, H.L. and K.L. and D.K.; formal analysis, H.L. and K.L.; investigation, H.L., K.L. and M.O.; resources, H.L. and K.L.; data curation, H.L., K.L. and E.K.; writing—original draft preparation, H.L. and K.L.; writing—review and editing, H.L. and D.K.; visualization, H.L. and K.L.; supervision, J.Y. and D.K.; project administration, J.Y. and D.K.; funding acquisition, J.Y. All authors have read and agreed to the published version of the manuscript.

Funding: This work is supported by the Korea Agency for Infrastructure Technology Advancemen (KAIA) grant funded by the Ministry of Land, Infrastructure and Transport (Grant 21CTAP-C163698-01).

Data Availability Statement: Not applicable.

Conflicts of Interest: The authors declare no conflict of interest.

References

1. UN Environment Programme, 2022 Global Status Report for Buildings and Construction. Available online: https://www.unep.org/resources/publication/2022-global-status-report-buildings-and-construction (accessed on 22 December 2022).
2. Lee, H.; Choi, M.; Lee, R.; Kim, D.; Yoon, J. Energy performance evaluation of a plus energy house based on operational data for two years: A case study of an all-electric plus energy house in Korea. *Energy Build.* **2021**, *252*, 111394. [CrossRef]
3. International PASSIVE HOUSE Association. Passive House Guidelines. Available online: https://passivehouse-international.org/index.php?page_id=80 (accessed on 22 December 2022).
4. Ihm, P.; Park, L.; Krarti, M.; Seo, D. Impact of window selection on the energy performance of residential buildings in South Korea. *Energy Policy* **2012**, *44*, 1–9. [CrossRef]
5. Myhren, J.A.; Holmberg, S. Flow patterns and thermal comfort in a room with panel, floor and wall heating. *Energy Build.* **2008**, *40*, 524–536. [CrossRef]
6. Moreau, A.; Sansregret, S.; Fournier, F. Modeling and study of the impacts of electrically heated windows on the energy needs of buildings. In Proceedings of the WSEAS International Conference on Heat Transfer, Thermal Engineering and Environment, Rhodes, Greece, 20–22 August 2008; pp. 76–83.
7. Werner, A.; Roos, A. Condensation tests on glass samples for energy efficient windows. *Sol. Energy Mater. Sol. Cells* **2007**, *91*, 609–615. [CrossRef]
8. Werner, A.; Roos, A. Simulations of coatings to avoid external condensation on low U-value windows. *Opt. Mater.* **2008**, *30*, 968–978. [CrossRef]
9. Kurnitski, J.; Jokisalo, J.; Palonen, J.; Jokiranta, K.; Seppänen, O. Efficiency of electrically heated windows. *Energy Build.* **2004**, *36*, 1003–1010. [CrossRef]
10. Lee, R.; Kang, E.; Lee, H.; Yoon, J. Heat flux and thermal characteristics of electrically heated windows: A case study. *Sustainability* **2022**, *14*, 481. [CrossRef]
11. Cakó, B.; Lovig, D.; Ózdi, A. Measuring the effects of heated windows on thermal comfort. *Pollack Periodica* **2021**, *16*, 114–119. [CrossRef]
12. Lee, D.-H.; Yoon, J.-H.; Lee, K.-W.; Seo, M.-G. Heating energy performance of heated glass according to insulation level of the residential building. *J. KSES* **2021**, *41*, 73–84. [CrossRef]
13. Lee, D.-H.; Yoon, J.-H.; Oh, M.-H. Analysis of temperature and total heat of heated glass through experimental measurement and three-dimensional steady-state heat transfer analysis. *KIEAE J.* **2015**, *15*, 111–116. [CrossRef]
14. Basok, B.; Davydenko, B.; Goncharuk, S.; Pavlenko, A.; Lysenko, O.; Novikov, V. Features of heat transfer through a window with electric heating. In Proceedings of the 2022 IEEE 8th International Conference on Energy Smart Systems (ESS), Kyiv, Ukraine, 12–14 October 2022; pp. 195–198. [CrossRef]
15. Krukovskyi, P.; Smolchenko, D.; Krukovskyi, G.; Deineko, A. Analysis of the heating capacity of electrically heated windows. *Thermophys. Therm. Power Eng.* **2021**, *43*, 62–67. [CrossRef]
16. Mitsui, A.; Sato, K. Thermal stability of electrical resistance of (ZnO:Ga,Y)/(ZnO:Ga)/(ZnO:Ga,Y) multilayers for electrically heated windows. *Vacuum* **2004**, *74*, 747–751. [CrossRef]
17. Rueegg, T.; Dorer, V.; Steinemann, U. Must cold air down draughts be compensated when using highly insulating windows? *Energy Build.* **2001**, *33*, 489–493. [CrossRef]
18. Wu, Q.; Wang, Z.; Dong, J.; Liu, J. A method for judging the overheating of the radiator in the compensation of window downdraught based on thermal image velocimetry. *Build. Environ.* **2021**, *197*, 107858. [CrossRef]
19. Kim, S.-J.; Choi, H.; Kim, K.-H.; Hong, Y.-J.; Hwang, I.-S.; Lee, D.-W.; Chun, S.-K. Heating Glass and Manufacturing Method Thereof. U.S. Patent 8916805B2, 12 December 2014.
20. Pilkington. NSG TEC Overview. Available online: https://www.pilkington.com/en/global/knowledge-base/types-of-glass/powered-by-nsg-tec/overview (accessed on 20 January 2023).
21. Pilkington. NSG TEC Technical Data Sheet. Available online: https://www.pilkington.com/ (accessed on 20 January 2023).
22. Perkin Elemer. Available online: https://www.perkinelmer.com/ (accessed on 22 December 2022).
23. *ISO 9050*; Glass in building—Determination of Light Transmittance, Solar Direct Transmittance, Total Solar Energy Transmittance, Ultraviolet Transmittance and Related Glazing Factor. International Organization for Standardization: Geneva, Switzerland, 2003.
24. *KS L 2514*; Testing Method on Transmittance and Emittance of Heat Glasses and Evaluation of Solar Heat Gain Coefficient. Korea Industrial Standards: Seoul, Republic of Korea, 2011.
25. International Commission on Illumination. CIE Standard Illuminant D65. Available online: https://cie.co.at/datatable/cie-standard-illuminant-d65 (accessed on 20 January 2023).
26. Energy System Research Unit of University of Strathclyde Glasgow. Available online: https://www.strath.ac.uk/research/energysystemsresearchunit/applications/esp-r/ (accessed on 22 December 2022).
27. The Center of Information for National Law. Available online: https://www.law.go.kr (accessed on 20 January 2023).
28. Korea Meteorological Administration. Available online: https://data.kma.go.kr/ (accessed on 22 December 2022).
29. Beck, H.; Zimmermann, N.; McVicar, T.; Vergopolan, N.; Berg, A.; Wood, E.F. Present and future Köppen-Geiger climate classification maps at 1 km resolution. *Sci. Data* **2018**, *5*, 180214. [CrossRef]
30. Fanger, P. *Thermal Comfort—Analysis and Applications in Environmental Engineering*; Danish Technical Press: Copenhagen, Denmark, 1982; pp. 128–133.

31. *ISO 7730:2005*; Ergonomics of the Thermal Environment—Analytical Determination and Interpretation of Thermal Comfort Using Calculation of the PMV and PPD Indices and Local Thermal Comfort Criteria. International Organization for Standardization: Geneva, Switzerland, 2005.
32. *Standard 55-2010*; Thermal Environmental Conditions for Human Occupancy. ANSI/ASHRAE: Peachtree Corners, GA, USA, 2010.

energies

MDPI

Article

Embodied vs. Operational Energy and Carbon in Retail Building Shells: A Case Study in Portugal

Ana Ferreira [1,*], Manuel Duarte Pinheiro [1], Jorge de Brito [1] and Ricardo Mateus [2]

1 CERIS, Instituto Superior Técnico, Universidade de Lisboa, Av. Rovisco Pais, 1, 1049-001 Lisboa, Portugal
2 ISISE, School of Engineering, University of Minho, 4800-058 Guimarães, Portugal
* Correspondence: anaferreiraleonardo@tecnico.ulisboa.pt

Abstract: (1) Background: The embodied energy of building materials is a significant contributor to climate change, in tandem with the energy use intensity (EUI). Yet, studies on the material impacts of European retail buildings, namely with relation to EUI, are missing. Hence, this study set out to: (i) evaluate the embodied energy and carbon emissions for a European retail building; (ii) quantify the material flow in terms of mass; (iii) compare the embodied aspects to the operational EUI and carbon use intensity (CUI); (iv) assess building materials with higher impacts; and (v) investigate strategies to mitigate materials' impacts. (2) Methods: A Portuguese retail building was selected as a case study. A simplified LCA method was followed (cradle-to-gate), analysing the shell building materials in terms of primary energy demand and global warming potential. (3) Results: the embodied energy represented 32% of total lifecycle energy while the embodied carbon represented 94%. EUI was 1×kWh/m^2/y while CUI was 21 kg CO_2eq/m^2/y. The embodied energy was 4248 kWh/m^2, and the embodied carbon was 1689 kg CO_2eq/m^2. Cement mortar, steel, concrete, and extruded polystyrene were the most intensive materials. (4) Conclusions: The embodied impacts of the analysed store could decrease by choosing stone wool sandwich panels for the facades instead of extruded polystyrene panels and roof systems with metal sheet coverings instead of bitumen materials.

Keywords: retail buildings; building shell; environmental impact; embodied energy; primary energy; GHG emissions; GWP; LCA

check for updates

Citation: Ferreira, A.; Pinheiro, M.D.; de Brito, J.; Mateus, R. Embodied vs. Operational Energy and Carbon in Retail Building Shells: A Case Study in Portugal. *Energies* **2023**, *16*, 378. https://doi.org/10.3390/en16010378

Academic Editor: Paulo Santos

Received: 24 November 2022
Revised: 21 December 2022
Accepted: 23 December 2022
Published: 29 December 2022

1. Introduction

Buildings are the largest materials consumers in Europe, representing 50% of all extracted materials, 42% of the final energy consumption, 35% of greenhouse gas (GHG) emissions and 32% of waste flow [1]. GHG emissions are transversal to all aspects of the construction industry [2]. There is now robust evidence that the embodied energy of building materials is a significant contributor to climate change [3], in tandem with the consumption of energy in buildings during the operational stage [4,5]. In this context, the choice of building materials is of the utmost importance, not only because of the embodied impacts they entail, but also because they can determine lifelong profiles of energy use intensities (EUI) in buildings [3], especially regarding building envelopes. The life cycle assessment (LCA) is a preferable tool that assists in assessing the environmental impacts of building materials. It is encouraged to promote sustainable design, namely in building sustainability assessment (BSA) methods, such as LEED or BREEAM [6,7]. The European Commission has also acknowledged the importance of LCA to estimate potential environmental impacts in buildings, and as a result, CEN TC 350 was mandated for the development of standards for the sustainability assessment of construction works [1]. Subsequently, the European Commission has launched research projects to analyse the performance of residential and office buildings according to resource use, benchmarking it against the standard and best practices, to develop an LCA model for these types of buildings [1,8].

Nonetheless, retail buildings are often disregarded from such studies. However, retail buildings in particular have one of the highest EUI [9], which makes them an important sector to target towards reducing carbon emissions and supporting the transition to a competitive low-carbon economy [10]. Furthermore, retail is a particularly intensive sector in terms of direct emissions, mostly due to refrigeration and HVAC systems. Since the operational energy of buildings is being pushed down by increasingly tighter regulations, such as the Energy Performance of Buildings Directive [11], the embodied aspects of building materials are increasingly important to reduce environmental impacts in the construction industry [1]. It is therefore necessary to systematically assess the environmental impacts of retail buildings through LCA methods, as to support decision making towards building material choices that mitigate impacts, while delivering equivalent performance.

The literature review has retrieved several LCA studies that assess the environmental impact of buildings, mostly office buildings [12–15], residential buildings [16–21], or a mix of both [5]. These studies aimed mainly at identifying the impacts of building materials and understanding the relationship between embodied energy, and operational energy and quantified them with different results. Even though these studies do not relate to retail buildings, they nonetheless offer a methodology that can be transposed to this building typology, and results, namely by impact per group of material, that can be used to assess differences and similarities between several building types. Most authors agree on typical material groups where impacts are higher, as well as on the growing importance of the embodied aspects of building materials in terms of energy and carbon emissions. In that sense, Ramesh et al. [5] pointed out that operating effects were responsible for 80–90% of the impacts, against 10–20% for embodied effects. Kofoworola and Gheewala [22] suggested that the operational energy was 52% of total life-cycle impacts, despite the importance of steel and concrete regarding most of the embodied environmental impacts; Karimpour et al. [20] proposed that in milder regions, embodied energy could represent up to 25–35% of the total life cycle energy. Thormark [19] demonstrated that, for low energy buildings, 40–60% of total life cycle energy was embodied energy, and Gustavsson and Joelsson [23] argued that with increased efficiency in operational energy and effectiveness of insulation materials, embodied energy was becoming a significant factor in life cycle energy, representing 45% in conventional residential buildings and up to 60% in low energy buildings. Other authors corroborate the opinion that embodied energy is increasingly more significant [20,24–26], reaffirming the need to balance the performance of buildings in terms of both embodied and operational aspects to mitigate global environmental impacts.

Inversely, studies referring to these aspects is that retail buildings are scarce. Very few studies have addressed the embodied and operational energy in commercial buildings, and in particular, in retail buildings: Ooteghem and Xu [27] studied structural and envelope building materials in a single-story retail unit in Canada from an LCA perspective. Chau et al. [28] investigated the environmental impact of building materials for commercial buildings in Hong Kong (in which retail buildings were included). Cinneli et al. [29] examined the embodied energy contents of materials in a commercial building in Canada. Khoa et al. [30] assessed the life cycle greenhouse gas emissions for several typical commercial building fabrics in Australia, and Luo et al. [31] quantified the embodied carbon emissions of building materials in residential, office, and commercial buildings in China. These studies are too limited in number to provide approximate threshold levels that could be used to compare, for instance, the percentage of embodied aspects of building materials in terms of energy and carbon emissions vs. the percentage of energy used during the operational stage, throughout the building life cycle. They represent a small fraction of LCA knowledge concerning the impacts of retail buildings, since they vary in location, scope, and systems boundaries. In effect, the literature review confirmed that current knowledge about the impacts of embodied and operational energy in retail buildings is extremely limited, which is evidence of the gap in knowledge this study intends to address. Hence, this research intends to assess the embodied aspects of building materials in terms of energy and carbon emissions for a retail building in Portugal. In addition, it intends to compare the embodied energy and carbon emission aspects to the energy and carbon emissions consumed and produced during store operation. This quantification is in line with the studies previously mentioned for residential and office buildings and is essential to discern mitigation strategies that can be of use to reduce the environmental impact of retail buildings. The need for applied research on the subject of material impacts in retail buildings via the investigation of case studies has been identified by Omer and Noguchi, Cabeza et al. [4,24] and van Ooteghem and Xu [27]. On this matter, van Ooteghem and Xu [27] stated that the life cycle environmental impacts of retail buildings have been largely neglected and called for more investigation in this area since retail buildings have unique characteristics when compared to office and residential buildings that need special attention, namely their higher EUI in the operational stage.

Out of the few studies that addressed retail buildings, none addressed aspects related to the shell of these buildings, namely sandwich panels in facades, which is one of the most common wall solutions of recent standalone retail buildings in Europe. This is another gap in knowledge that this study will address. Van Ooteghem and Xu [27] studied the embodied impacts of steel building systems and of roofs in single-story commercial buildings in Canada. Wall systems, such as those identified in the present research, were not addressed. Khoa et al. [30] assessed four typical commercial building fabrics in Australia, leaving out sandwich panels in facades. Likewise, Luo et al. [31] and Chau et al. [28] calculated the carbon emissions of the most used materials in retail buildings in China and Hong Kong. However, these studies considered building solutions based on their geographic location, which may not be similar to those often used in European countries. Concrete blocks and bricks, for instance, were the most used construction materials for walls in the literature review. Contemporary retail buildings in west European countries tend to have an envelope made of a cladding wall system or simply of sandwich panels juxtaposed to an auxiliary structure. Thermal performance is currently of the utmost importance in Europe, due to climate change regulation. For that matter, studies on the thermal performance of retail buildings were also researched, particularly of extruded polystyrene as a material frequently used in retail building envelopes. Some of the retrieved studies explored the embodied energy of this type of material [32–34], but not the role it played in retail buildings' life cycle energy.

Hence, this study's major contributions can be summarized as follows: (i) to evaluate the embodied energy and carbon emissions for a typical contemporary European retail building, namely regarding its shell; (ii) to quantify the material flow of the shell materials in terms of mass (kg/m^2 of sales area); (iii) to compare the embodied impacts to the energy and carbon emissions consumed and produced during store operation, over the building life cycle (estimated in 50 years); (iv) to assess which typical building material impacts are higher; and (v) to investigate strategies that can be used to mitigate materials' impacts in retail buildings, while delivering equivalent technical performance.

As the choice of building solutions, as well as the performance of envelope materials, depends on the climatic conditions where buildings are located, a European retail building in a Mediterranean Climate was selected as a case study. To evaluate the building's environmental impact in a rapid way, a simplified LCA method with a defined system boundary is suggested. For the present study, impacts were assessed during the material production stage (cradle-to-gate), excluding the transportation and construction stage.

2. Materials and Methods

2.1. Case Study Description

A retail store located in Loulé, Algarve, Portugal (Figure 1), was selected as a case study for the analysis of the impacts of shell building material elements in terms of embodied energy and GHG emissions. The selected store is part of a top global do-it-yourself (DIY) retail group in terms of revenue [35], with stores in Europe, America, and Asia.

Figure 1. Approximate location of the retail store in Algarve, Portugal.

This store was selected as a case study because it is one of the group's most recent stores in Portugal (opened in 2017), with enough years in operation to collect data from energy bills.

The case study consists of a building located on a southeast slope, with two ventilated underground parking floors, plus the sales area floor, a mezzanine for staff facilities, and a roof with limited access, where photovoltaic (PV) panels are installed to produce green electricity (Figure 2).

Figure 2. Front facade and roof with photovoltaic panels of the case study store in Algarve, Portugal.

The retail building has a total gross floor area of 16,473 m^2 and a sales area of 9556 m^2 (Figure 3). The main facade of the building is opaque, and the window-to-wall ratio is about 5%, with southeast exposure (Figure 4).

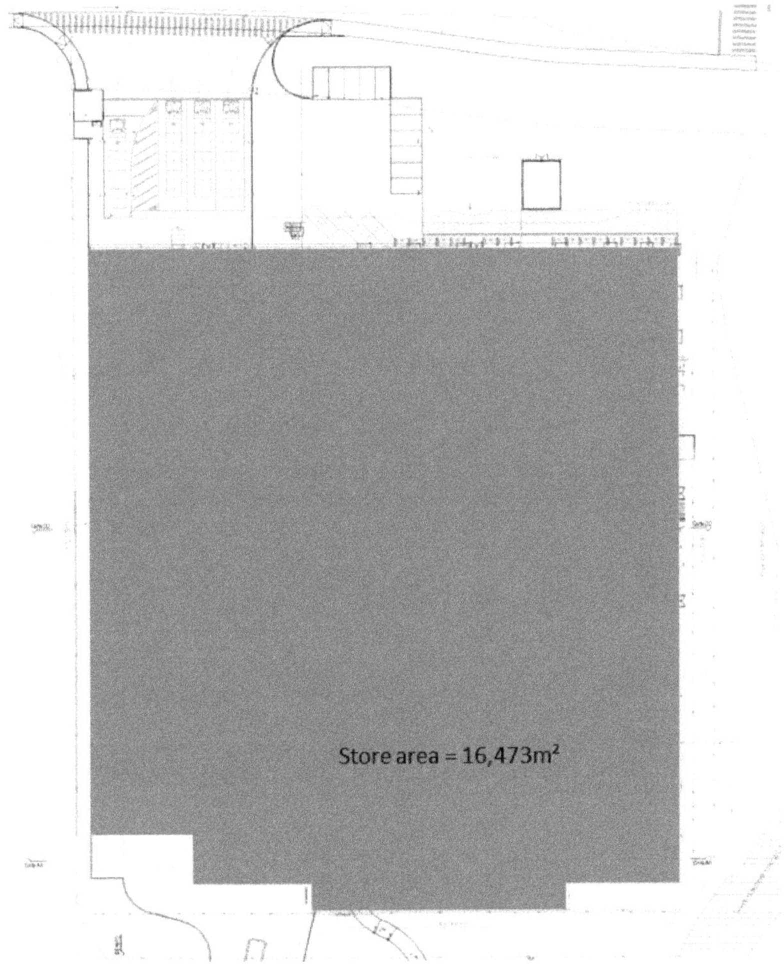

Figure 3. Sales level plan of the retail non-food unit in Loulé, Algarve [36].

The store is open 12 h per day, seven days per week. According to the Köppen Climate Classification, subtype for this climate is "Csa" or Mediterranean Climate. The warmest month, on average, is July, with an average temperature of 29 °C (minimum temperature of 21 °C and maximum temperature of 32 °C). The coolest month on average is January, with an average temperature of 16 °C (minimum temperature of 13 °C and maximum temperature of 22 °C). The selected case study is a typical big-box nonfood store. It is representative of a large nonfood store from an international retailer in Portugal due to its peripheral urban location, store size (<9000 m^2 in sales area), and store layout. The choice of building solutions and building materials is also typical of retail buildings: the superstructural elements of the underground floors are made of concrete, while those of the sales floor are made of a conventional hot-rolled steel structure. The facades comprise sandwich panels, while the roof comprises ceramic blocks covered by rock wool and a flexible waterproofing bitumen membrane (Table 1).

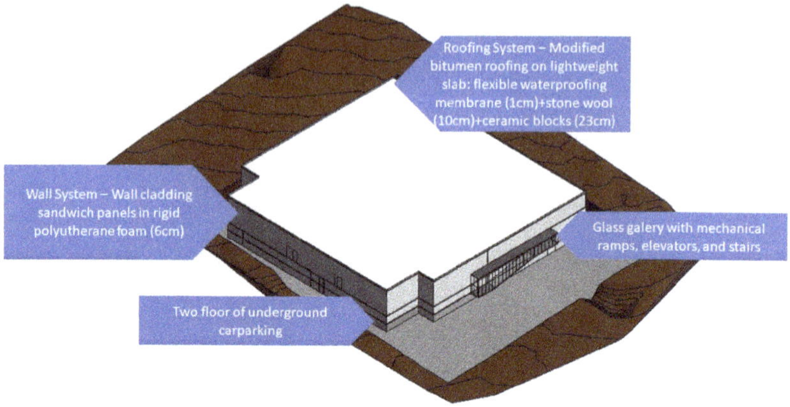

Figure 4. Case study's building modelling in software REVIT® with shell building material specifications.

Table 1. Building materials used in the envelope of the case study.

	Energy Consumption per Building System
HVAC	17%
Lighting	41%
Equipment (elevators and escalators)	42%
Total annual energy consumption	

In terms of equipment, the store has LED lighting, two elevators, and two escalators. The HVAC systems comprise three heat pump packaged rooftop units for the sales area and a VRF system for the staff rooms. The breakdown of the annual energy consumption of the main building systems is indicated in Table 2.

Table 2. Breakdown of the annual energy consumption of the case study's main building systems.

Envelope Element	**Building Element Detailed Composition**	**Building Element Abbreviation**
Walls	Exterior wall in white colour composed by rigid polyurethane foam (PUR) sandwich panels (6 cm thickness and a density of 35–50 kg/m^3)	6 cm thick sandwich panels in rigid polyurethane foam (PUR)
Roofs	The roof is 34 cm thick, the external surface is white and has the following composition (from the interior to the exterior: (i) lightweight ceramic pot (2 rows of holes) and beam slab (total thickness of 23′cm and thermal resistance = 0.23 m^2°C/ W); (ii) rock wool (density= 35–100 kg/m^3, 10 cm thickness, and thermal resistance=2.50 m^2°C/W); and (iii) flexible waterproofing membrane impregnated with bitumen (1 cm thickness)	flexible waterproofing bitumen membrane 1 cm + rock wool 10 cm + lightweight pot and beam slab 23 cm

2.2. Methodology

The data required to conduct this study were obtained from different sources: (i) the retail group's technical department provided the building's projects, bills of quantities for all building services, winning contractor's bid, and energy bills of the store; (ii) the worksheet, where the weights of building materials were inserted to calculate the total mass of building elements, was downloaded from Level(s), an European Union voluntary BSA framework [37]; (iii) the weights of building materials were estimated based on published trade literature, product technical datasheets and catalogues; and (iv) the impacts of building materials were calculated in SimaPro version 9.0.0.48 software, based on Ecoinvent version 3 database, according to the Cumulative Energy Demand (LHV) V1.00 methodology.

Hence, six main steps were considered in this study to calculate the impacts of shell building materials of a retail store (Figure 5).

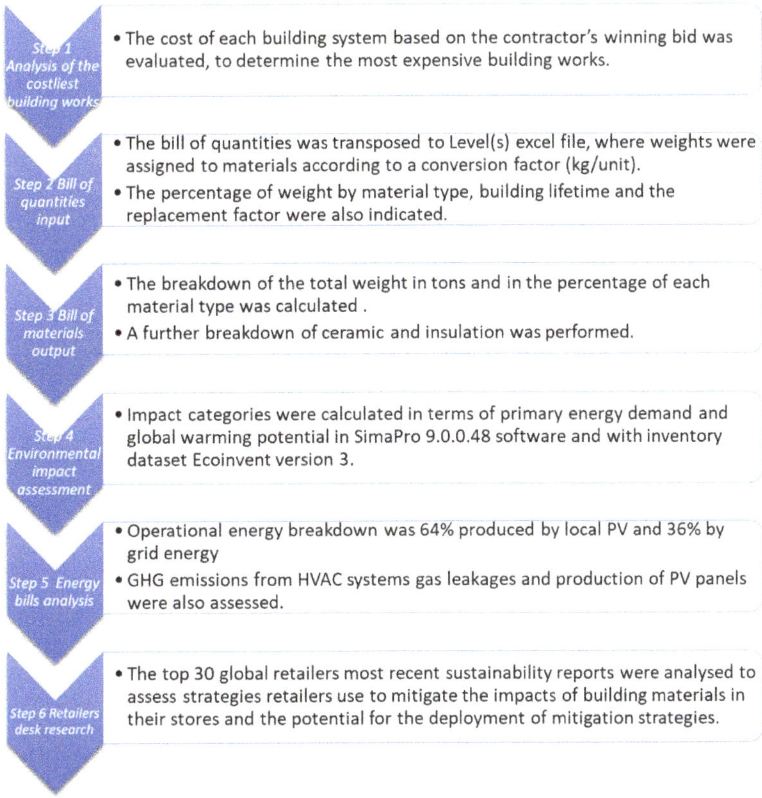

Figure 5. Methodology flow chart.

In step 1, analysis of the costliest building works, the cost of each building system based on the contractor's winning bid was evaluated (Figure 6 and Table S1), to determine the most expensive building works. As retailers value cost structure in their investments and prioritize decision making according to potential financial gain [38], shell elements were selected to conduct the present assessment, representing more than half of the total construction cost, and therefore were chosen to perform the present LCA. Shell elements were also selected for having a similar lifetime in years to that of the building (50), with low replacement factors.

In step 2, bill of quantities (BoQ) input, the bill of quantities for shell elements based on the contractor's winning bid was transposed to a Level(s) excel file and organized according to the main building parts proposed by the Level(s) template (Table A1 in Appendix A). Weights were assigned to building elements' materials according to a conversion factor (kg/unit). The percentage of weight split by material type was also indicated in this excel file, as well as the assumed building lifetime (considered 50 years in the present case study) and the replacement factor of each building component (Table A2 in Appendix B).

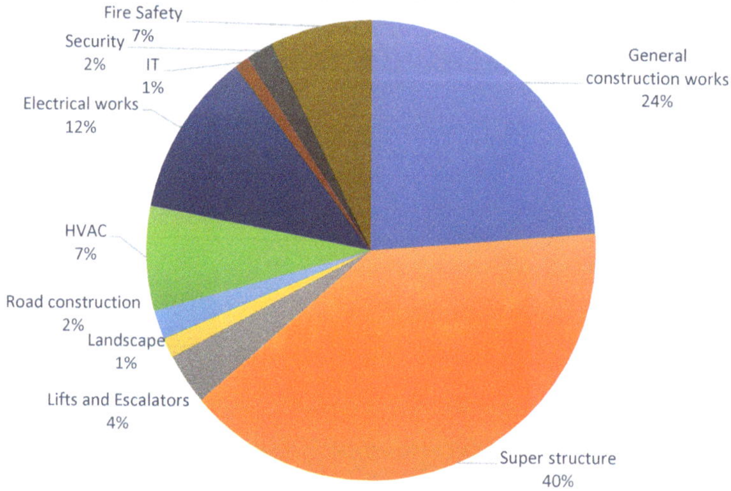

Figure 6. Cost disaggregation by the percentage of the cost of building system for the case study store.

In step 3, bill of materials (BoM) output, the breakdown of the total weight in tons and in the percentage of each material type was calculated (Table 3). Since Level(s) bundles insulation materials into one category and ceramic materials into the category "concrete, brick, tile, natural stone, ceramic", a further breakdown of these two categories was performed. Hence, insulation materials were split into extruded polystyrene and stone wool, and ceramic materials were split into round gravel, concrete block, cement mortar, screed, crushed limestone, ceramic tile, and concrete.

Table 3. Bill of materials output regarding shell building elements by material type in terms of total weight and percentage of weight.

Material Type Breakdown	Material Total (t)	Material Total (%)
Glass	**0.37**	**0.00%**
Plastic	**2.19**	**0.01%**
Bituminous mixtures	**86.02**	**0.20%**
Metals	**2320.69**	**5.50%**
Electrical and electronic equipment	**70**	**0.17%**
Concrete, brick, tile, natural stone, ceramic	**39,415.43**	**93.44%**
Round gravel	3.74	0.01%
Concrete block	1509.3	3.58%
Cement mortar	56.08	0.13%
Screed	326.98	0.78%
Crushed limestone	7154.82	16.96%
Ceramic tile	1.38	0.00%
Reinforced concrete	30,363.13	71.98%
Insulation materials	**289.61**	**0.69%**
Rock wool	167.93	0.40%
Extruded polystyrene	121.67	0.29%
Combined total *	**42,184.31**	**100.00%**

* Material flow: 4414,43 Kg/m^2 (building floor area of 9556 m^2).

In step 4, environmental impact assessment, the impact categories of shell materials in terms of primary energy demand (PE) and global warming potential (GWP) in kg CO_2 eq were calculated in SimaPro 9.0.0.48 software, as a preferable life-cycle assessment tool [30]. The inventory dataset chosen to evaluate the life cycle impacts of the selected case study was Ecoinvent version 3. European averages of the Ecoinvent database inventory were selected for materials' impact calculation, and where these were not available, rest-of-the-world values were considered. System boundaries were delimited by the production and transportation of materials to the marketplace (cradle-to-gate). All other lifecycle stages were excluded from this study, namely the construction and transportation stages. One kg of material was selected as a declared unit. Primary energy demand was calculated according to the cumulative energy demand (LHV) V1.00 methodology and expressed in MJ, and later converted to kWh. GWP was calculated according to the IPCC 2013 GWP 100a V1.03 methodology and expressed in kg CO_2 eq. These impact categories were selected to provide comparison to familiar metrics for retailers in terms of energy use during the operation stage, namely, to compare material impacts results to EUI values and carbon use intensity (CUI) values (Table 4). Hence, to analyse further total impacts' results, they were normalized per m^2 of sales area of the building, which gave rise to the indicators embodied energy intensity (EEI) kWh/m^2 and embodied carbon intensity (ECI) in kg CO_2 eq /m^2.

Table 4. Impact categories of shell building elements by material type in terms of primary energy demand (in MJ and kWh) and GWP (in kg CO_2 eq).

Impact Category *	Energy (MJ)	Energy (kWh)	%	IPCC GWP 100a (kg CO_2 eq)	%
Concrete, brick, tile, natural stone, ceramic					
Gravel, round {RoW} | market for gravel, round | Cut-off, S	656	182	0%	44	0%
Concrete block {GLO} | market for | Cut-off, S	1,222,540	339,594	1%	135,314	1%
Cement mortar {RoW} | market for cement mortar | Cut-off, S	61,511,953	17,086,654	42%	8,178,828	51%
Screed total amount	485,943	134,984	0%	62,547	0%
Limestone, crushed, washed {RoW} | market for limestone, crushed, washed | Cut-off, S	582,959	161,933	0%	39,466	0%
Ceramic tile {GLO} | market for | Cut-off, S	16,250	4514	0%	1130	0%
Concrete	20,014,386	5,559,552	14%	2,774,402	17%
Glass			0%		0%
Plastic			0%		0%
Acrylic varnish, without water, in 87.5% solution state {RoW} | market for acrylic varnish, without water, in 87.5% solution state | Cut-off, S	66,874	18,576	0%	4249	0%
Polycarbonate {GLO} | market for | Cut-off, S	79,777	22,160	0%	6558	0%
Bituminous mixtures			0%		0%
Bitumen adhesive compound, hot {GLO} | market for | Cut-off, S	4,519,716	1,255,477	3%	43,842	0%
Metals			0%		0%
Reinforcing steel {GLO} | market for | Cut-off, S	57,631,652	16,008,792	39%	4,893,783	30%
Insulation materials *			0%		0%
Stone wool, packed {GLO} | market for stone wool, packed | Cut-off, S	3,021,457	839,294	2%	226,984	1%
Polystyrene, extruded {GLO} | market for | Cut-off, S	11,559,108	3,210,863	8%	1,176,143	7%
Total	146,132,705	40,592,419		16,140,162.98	

* SimaPro 9.0.0.48 software, EcoInvent database, Method: cumulative energy demand (LHV) V1.00 and IPCC 2013 GWP 100a V1.03.

In step 5, energy bills analysis, the energy bills of the store were collected from the year 2021, and data regarding the annual values of energy consumption and correspondent GHG emissions were gathered from the bills (Table 5). Normalizing annual energy consumption and GHG emissions per m^2 of sales area obtained EUI and CUI values for the case study store. As the store produces solar energy locally (64% of the total energy consumption), the store's EUI was calculated assuming that the energy bills represented 36% of the total operational energy. GHG emissions related to the production and installation of the PV panels were accounted for at 38 g CO_2eq/kWh [39]. Annual fugitive GHG emissions from HVAC systems were considered null in the present study since up to 2022, and there were no gas leakages in this store. This may be due to proper maintenance, or because the HVAC system is still relatively new. Nonetheless, it is possible that gas leakages may occur in the future. The gas load of HVAC equipment is described in Table 5. To compare the percentage of lifecycle energy and GHG emissions in terms of EEI/EUI and ECI/CUI, a static approach was considered within the defined system boundaries, in which values are set without analysing their variation over time [40].

Table 5. Annual energy consumption and related GHG emissions according to the store's energy bills.

Energy Bill Month	Energy (kWh)	GHG Emissions (kg CO_2 eq)
Jan/21	9736	66,619
Feb/21	15,121	50,634
Mar/21	10,955	36,684
Apr/21	11,802	39,518
May/21	9101	33,723
Jun/21	11,727	43,456
Jul/21	11,909	44,130
Aug/21	14,755	54,674
Sep/21	14,832	51,815
Oct/21	15,415	53,850
Nov/21	14,470	50,551
Dec/21	17,106	59,758
		43,183 *
Total	**156,930**	**628,595** **

* GHG emissions from production stage of the installed PV panels (38 g CO_2 eq/kWh according to the PV supplier)

** HVAC systems did not have any leakages thus far. Their gas load is the following:

Heat pump packaged rooftop unit 1	79,340
Heat pump packaged rooftop unit 2	79,340
Heat pump packaged rooftop unit 3	22,970
VRF system	41,550

In step 6, retailers desk research, the top 30 global retailers most recent sustainability reports [41] were analysed to assess strategies they use to mitigate the impacts of building materials in their stores.

In the Conclusion section, the main findings of the environmental impacts of shell elements for retail buildings in terms of embodied energy and GHG emissions are described, as well as suggestions for future studies.

3. Results

The case study's bill of quantities based on the winning contractor's bid was analysed to assess which building system weighted the most in terms of cost (Figure 6). The costliest building systems were superstructure, general construction works, and electrical works.

Shell elements (general construction works, superstructure, and lifts and escalators) represented 67% of the total construction cost, while core elements represented 29% and external works 3%. When looking at shell elements in detail, to identify the costliest building components, concrete superstructure (31%), steel superstructure (19%) and locksmiths (15%) were the most expensive elements (Table S1).

In Table 3, the total weight of materials by material is presented. Overall, reinforced concrete was the material with the highest weight (72%), followed by crushed limestone (17%) and steel (5.5%).

In Table 4, the results of the SimaPro analysis are presented in terms of the materials' primary energy (in MJ and kWh) and GWP (in CO_2 eq). Cement mortar is the material with the highest embodied energy (42%) and GWP (51%), followed by steel (39% and 30%, respectively) and concrete (14% and 17%, respectively). Extruded polystyrene, which is the main material of façade sandwich panels, accounted for 8% of primary energy and 7% of GWP, whereas its contribution in weight was 0.29%. Stone wool and extruded polystyrene, as insulation materials, have important differences in their impacts, as the contribution of roofing stone wool in terms of primary energy was 2% and of GWP was 1%.

In Table 5, the store's annual energy consumption and GHG emissions are presented. The EEI of the analysed store is 4248 kWh/m^2, and the ECI is 1689 CO_2 eq/m^2, whereas the annual EUI is 180 kWh//m^2/y and annual CUI is 21 kg CO_2 eq/m^2/y.

Correspondingly, the operational energy represents 68% of the projected life cycle energy, whereas the embodied energy represents around 32%. Moreso, the GHG emissions in the operation stage represent 6% of the projected life cycle emissions, whereas those embodied in building materials represent 94%.

4. Discussion

By assessing the impacts of materials in retails stores, it is possible to determine the contribution of design choices in terms of environmental impact over the building life cycle, which would remain undisclosed if not for LCA accounting methods. Such information would support designers, project teams, retail owners and developers in the decision-making process of the design and refurbishment of existing retail stores. Therefore, in the Discussion section, we intend to address the results obtained for each of our research goals, which are presented in the following subsections.

4.1. Embodied vs. Operational Energy and Carbon Emissions

The results of the present study indicate that according to the system boundaries defined for the analysed case study, the impacts of shell building materials in terms of primary energy demand (thereby referred to as embodied energy) represent 32% of total life cycle energy, whereas the impacts of shell building materials in terms of GWP (thereby referred to as embodied carbon) represent 94%. Inversely, 68% of life cycle energy and 6% of GHG emissions occur in the operation stage of the building. This corroborates the growing importance of embodied energy in building materials pointed out by other authors [20,23–26], as energy efficiency measures and more efficient building envelopes minimize the annual energy consumption during the operational stage. With relation to the literature review, the findings to this study corroborate the findings of Karimpour et al. [20] that argued that in milder regions, embodied energy could represent up to 35% of the total life cycle energy, of Bribián et al. [40], according to which the proportion of embodied energy in conventional buildings is of up to 38%, and of Kofoworola and Gheewala [22] that placed operational energy at 52% of total life cycle impacts, despite the importance of steel and concrete in embodied impacts. It is expected that with regulations on energy efficiency, namely the European Energy Performance of Building Directive (EPBD) [11] and the Energy Efficiency Directive [42], Member States will push towards highly energy-efficient and decarbonized buildings, as measures to mitigate climate change by 2050 [10]. Thus, the amount of embodied energy in buildings is likely to increase, especially in low energy buildings, at about 40–60% of total embodied life cycle energy, as defended by Thormark [19] and Gustavsson and Joelsson [23]. In the case of retail buildings, the ratio of operational energy versus embodied energy tends to be higher than in other building typologies, as retail EUI values in the operation stage are three to five times higher than those of office and residential buildings. Nonetheless, in retail buildings, the ratio of embodied energy is likely to augment as EUI average values have decreased in the retail sector over the past decade [43], and thus is the ratio of GHG emissions, with the increasingly more popular local production of renewable energy in retail stores.

In the analysed store, EUI was found at 180 kWh/m^2/y, which is 20% below the average for non-food retailers mentioned in the literature [44]. The CUI was 21 kg CO_2 eq/m^2/y, which is a top performance result based on the literature benchmarks for retail buildings [44]. The CIU of the analysed store is very low since 64% of the store's energy consumption is produced by PV panels installed onsite, which means there are no carbon emissions involved for all the energy produced on site. In addition, about half of the purchased grid energy was produced from renewable sources. Thus, with the increasing penetration of renewable energy in the market, minimizing the total energy consumption in retail buildings during the operation stage is more challenging than minimizing carbon emissions from energy consumption. From the analysed case study, with locally produced PV energy and with proper maintenance of HVAC systems, annual GHG emissions derived from the operation stage can be drastically reduced. Hence, it is apparently easier for retailers to find strategies to reduce energy consumption and carbon emissions during the operational stage than to mitigate the embodied energy and carbon emissions in buildings, mostly because of the lack of knowledge regarding the environmental impacts of building materials, a gap in knowledge which this study addressed as a research goal.

For instance, most of the energy consumed in non-food retail buildings is for HVAC systems [9], which makes the choice to use envelope building materials of retail buildings of great importance. Envelope materials with high thermal resistance enable the reduction of energy losses, as they function as powerful thermal insulators and therefore can reduce internal heat gains, in turn leading to lower energy consumption in HVAC systems [30,45]. Nonetheless, different insulation materials with similar thermal resistance can have different environmental impacts. In this sense, extruded polystyrene, the main material of the sandwich panels used in this store, accounted for 8% of embodied energy and 7% of embodied carbon, whereas rock wool used for roofing had significantly lower embodied energy (2%) and embodied carbon (1%). These findings are similar to those of Bríbian et al. [40], according to whom stone wool has a primary energy demand 4 times lower and a carbon footprint 4.7 times lower than that of rigid polyurethane foam. Hence, when using low impact materials, attention must be paid to their thermal characteristics and to their expected lifetime to effectively reduce the total building's life cycle impacts [24]. In the case of the analysed store, material impacts could be reduced by using stone wool sandwich panels for facades instead of extruded polystyrene panels, with similar thermal performance.

4.2. *Quantification of Material Flow and Materials Intensity*

The material flow of the analysed building was 4414 kg/m^2. In terms of the total mass, reinforced concrete (72%), crushed limestone (17%) and steel (5.5%) were found to be the most significant materials. In terms of primary energy demand in kWh, cement mortar (42%), steel (39%), concrete (14%), and extruded polystyrene (8%) were found to be the most intensive materials. In terms of GWP in kg CO_2 eq, these materials were also the most intensive ones, accounting for 51%, 30%, 17% and 7% of GHG emissions, respectively. These findings are in line with those of Cabeza et al. [24] for commercial buildings that placed steel, cement, and sand as the materials with the largest contributors to embodied energy due to their mass. Likewise, Chau et al. [28] placed concrete and rebar as the largest contributors to total life cycle environmental impacts in a commercial building.

In the case of insulation materials, the findings of the present study also confirm those of Chau et al. [28], in the sense that materials with little mass may have important lifecycle impacts.

In relation to the total sales area, the EEI and ECI of shell elements were found to be 4248 kWh/m^2 and 1689 kg CO_2 eq/m^2, respectively. The breakdown of the most relevant building materials' impacts was the following: cement mortar 6437 kWh/m^2

and 856 kg CO_2 eq/m^2, steel 6031 kWh/m^2 and 512 CO_2 eq/m^2, concrete 2094 kWh/m^2 and 290 CO_2 eq/m^2, and extruded polystyrene 1210 kWh/m^2 and 123 kg CO_2 eq/m^2. Comparison of these values with other case studies found in the literature is difficult, as studies vary according to building type, methodology, scope, and localization, as stated by van Ooteghem and Xu [27]. In China, Luo et al. [31] obtained a material flow of 494 kg CO_2 eq/m^2, ranking steel, concrete, and mortar as the most carbon-intensive materials, which is in line with the findings of the present study. Nevertheless, the impact of steel was 195.13 CO_2 eq/m^2, the impact of concrete was 105.05 CO_2 eq/m^2, and the impact of mortar 58.05 CO_2 eq/m^2, which is significantly less than the results obtained in the present case study. Nevertheless, for the analysed case study, the amount of concrete is higher, as the store has two underground car parking floors. As for steel elements, facades and roofing, the material quantities found in this store are similar to those of similar standalone stores.

In Sri Lanka, Kumanayake et al. [46] obtained a material flow of 2318.27 kg/m^2 for a commercial office building, and the embodied carbon in the material production phase was 629.60 kg CO_2 eq/m^2. Reinforced steel, concrete and clay bricks were major carbon-emitting materials. In Hong Kong, Chau et al. [28] concluded that concrete, steel, plaster and render and screed were the main contributors to the total impact of commercial buildings, in which concrete accounted for up to 28% of the total impact, rebar for up to 22%, and plaster, render and screed for up to 15%.

Overall, there is a consensus in these studies that steel, cement, and concrete account for the majority of environmental impacts at the manufacturing stage [24,30,40], which is also in line with the findings of the present study.

4.3. *Strategies to Mitigate Materials' Impacts in Retail*

Some of the solutions more frequently mentioned in the literature to mitigate life cycle impacts of building materials are [20,46]: (i) to reduce quantities of materials, (ii) to incorporate lower carbon intensity materials in concrete, (iii) to promote the use of low-carbon materials, (iv) to use recycled materials and (v) to favour materials with environmental labelling (e.g., Environmental Product Declarations (EPD)). Nonetheless, not all solutions can be performed in the retail sector. For instance, some strategies to mitigate environmental impacts must be operated during material production processes. The embodied impacts of concrete could decrease by a more eco-efficient production of clinker, the use of alternative fuels in the cement industry, the use of different types of cement waste, and the use of lime mortars instead of cement mortars [40]. The mitigation strategies that can be more sustainable for retail buildings are described in the next subsections. Some retailers already mentioned them in their sustainability reports, with varying degrees of implementation.

4.3.1. Use of Low Environmental Impact Materials

Wood is acclaimed as a viable construction material for superstructures by many authors, as it is almost carbon neutral [3,31,40,47]. According to Le et al. [30], structures that combine timber with other materials have fewer impacts than those using metal, brick, or concrete. Wooden structures have been used sparsely in retail, creating a building visually lighter, adaptable, and recyclable. Some retailers' green building concepts include the use of organic building materials, namely wood for structural frames, front facades and, in many cases, the roof shell. Light steel framing has also been identified as a promising structure due to it being lightweight and for allowing for increased floor area [31]. The replacement of limestone-based clinker in Portland cement by supplementary cementitious materials (namely fly ash, granulated blast-furnace slag, and calcined clay) is suggested by Rissman et al. [47] as a way to reduce the amount of cement needed in concrete. The replacement of conventional insulation with natural materials such as cork, wood fibre and sheep's wool, or with recycled materials such as cellulose fibre, is suggested by Bribián et al. [40], which further points to the difficulty in recycling polystyrene or polyurethane at their current assemblage process, thus stressing the need to design for disassembly. Other solutions that minimize the impact of building materials mentioned in the literature include the substitution of ceramic tiles by light clay or silica-calcareous bricks [40], or the replacement of concrete by unfired bricks and stone [26].

In a general way, the use of environmentally friendly and recyclable materials is commonly highlighted by retailers in their sustainability reports, namely the use of recycled asphalt with recycled aggregate as a base material used in parking lots, the use of resources sparingly and the increased amount of recycled and sustainable materials employed in the construction of stores, or the use of prefabricated structural steel systems in new stores made of up to 80% of recycled content. The diversion of construction waste from landfill by recycling eligible roof membranes and metal fixtures was also reported by retailers, as well as the reuse of shelving fixtures and chiller cases, and of steel frames and sashes, to reduce construction waste. These actions support the findings of Bribián et al. [40] in identifying the recycling of building materials as essential to reduce the embodied energy in buildings. According to Bribián et al. [40], the use of recycled metal in structural elements can provide savings of more than 50% in terms of embodied energy.

4.3.2. Optimized Design

As superstructure materials represent most of the embodied energy and carbon in buildings, largely due to their weight, the design optimization of structural elements is of the utmost importance to minimize embodied impacts. The use of bolted connections in conventional hot-rolled steel structures is also a good practice to allow for design for disassembly [27]. Nevertheless, despite the importance of steel and concrete in the total lifecycle impacts of buildings, the optimization of other shell elements such as window area, insulation level and concrete flooring, is also important [20]. Material efficiency in design also includes the reduction of material waste, the improvement of the durability of buildings, their reusability, ease of refurbishment, and towards the end of their lifecycle, their recyclability [47]. The promotion of optimized design principles is apparent in retail buildings, including modular design, reversible attachments, material standardization, and the use of prefabricated elements. These strategies are mentioned in some retailers' sustainability reports (dataset 1), although the extension of its use in retail stores remains undisclosed.

4.3.3. Use of BSA Tools

BSA methods help stakeholders quantify the environmental impact of buildings, namely in terms of material choice. The growth of guidelines in BSA methods addressing waste efficiency, as well as the quantification of the embodied impacts of materials by LCA approaches, corroborates the growing importance of building materials in sustainability assessment [3]. The two most internationally used BSA methods, LEED and BREEAM [48], have integrated several of the above-mentioned strategies to reduce the impacts of buildings under their assessment processes. LEED [6] encourages the use of materials for which life cycle information is available and that have environmentally, economically, and socially preferable lifecycle impacts. These include the sourcing of raw materials (namely bio-based materials, certified wood, salvaged or refurbished materials, and of recycled content and locally sourced materials) and the careful choice of material ingredients (awarding material ingredient reporting or optimization); BREEAM [7] encourages the reduction of environmental and social impacts of buildings under a lifecycle approach, namely through the conduction of an LCA study to assist in the selection of products with a low environmental impact, ensuring that all lifecycle impacts are taken into account in the design stage, and allowing optimal solutions to be identified and adopted early on. For that purpose, BREEAM awards points to EPDs for the responsible sourcing of construction products (with lower environmental, economic, and social impacts) and the design for durability and material efficiency (encouraging the reuse of existing materials and the use of materials with recycled content). The management of waste is also considered by BREEAM, in promoting the reduction and diversion of waste to landfills during the construction and operational stage and the design for disassembly and adaptability. Nevertheless, BSA methods in Europe and throughout the world are market oriented and national policy driven, and in this sense, there is lack of an international standardized approach to measure building sustainability [49]. This context has led the European Commission to launch Level(s), as a voluntary sustainability reporting framework [48,50], with a set of indicators and metrics that consider the full life cycle of the building. Level(s) can be used directly or indirectly with other BSA methods, and therefore, the use of its bill of quantities template file is encouraged (see Appendices A and B), as it facilitates the estimation of the weight of materials used in new construction or refurbishing projects, which is fundamental for LCA studies.

4.3.4. Applicability of Mitigation Strategies in Retail Buildings

The degree to which the described mitigation strategies varies within retailers may be related to several different factors, mostly related to knowledge on the environmental impact of building materials, maturity of alternative solutions in the construction industry, and cost. Many retailers are owner-occupiers, and they are in the best position to make long-term investment decisions about their buildings. They will tend to have a longer-term perspective and stand to benefit directly from their material choices. This applies both to new buildings and the refurbishment of existing ones. However, the limited mandate time of a board of administration in a retail company (usually 3 to 4 years) contributes to a greater focus and attraction on the short-term payback of the investments especially for refurbishments, thus limiting the range of alternative solutions to be considered for the buildings they use.

Developers are the primary players in commercial construction and can be speculative, which frequently results in a short-term perspective on the buildings' financial value. Speculative developers will typically be interested in material and energy efficiency only if it is a significant factor in the buying decision. This weakens the incentive for investments in materials impact mitigation strategies. Whichever is the circumstance, too much importance is placed on the initial investment required, rather than on life-cycle cost assessments and return-on-investment calculations.

The main reasons for using mitigation strategies in retail buildings are perceived to be long-term economic benefits, the availability of subsidies, image benefits, the desire to reduce environmental impacts, and because of corporate social responsibility. On the other hand, the most common barriers hindering retailers from deploying more mitigation strategies in material choices are perceived high capital costs, long payback times, ignorance and lack of knowledge regarding embodied impacts of building elements, a perception of risk and that alternative solutions are unproven, incoherent policy, and planning constraints.

Some of the suggested mitigation strategies are easy to use and do not pose a risk on retailers. If the use of wood in structural elements and light steel framing is too daring for some building owners, increasing the quantity of recycled materials in buildings is more feasible, namely in that of steel in structural elements, or that of recycled aggregates in concrete and in paving solutions. Optimizing the design of structural elements and designing for disassembly are also straightforward strategies to mitigate construction impacts. Choosing materials with equivalent performance, but with minor impacts, could also be easily achieved in retail. For instance, the embodied impacts of the materials of the analysed store could be minimized by choosing stone wool sandwich panels for the facades instead of extruded polystyrene panels and by choosing metal sheet coverings for the roof system instead of bitumen materials.

Lastly, the use of BSA tools to support project design in retail buildings will assuredly lead to better design choices, as most building sustainability aspects are covered by these methods, including different levels of LCA studies.

4.4. Implications

LCA studies can effectively assess the environmental performance of buildings and identify improvement opportunities; this was evident in the assessment of the case study, which identified materials that could be replaced for technically equivalent materials in terms of thermal performance (e.g., rock wool instead of extruded polystyrene or coated steel sheets instead of asphalt-based materials), at a fraction of the environmental impacts, in turn validating LCA methods as useful tools for sustainable business management and further encouraging its use.

Retail is a diverse but highly concentrated industry in terms of ownership and sales and is composed of a very large number of participants. This level of concentration allows for easier deployment of sustainable practices across the sector. In addition, the retail sector is strategically positioned in the construction industry and can influence the supply stream of materials used in this sector. Given that each studied retailer operates hundreds of stores, results show a key potential to reduce the impacts of building materials in the retail sector, in either new stores or refurbishing processes. Regarding the latter, there is a great opportunity for material impact mitigation in the upkeep of existing retail stores, as the embodied energy and carbon impacts of new structural elements could be avoided.

In addition, if core and external works' elements were included in the present study, the amount of embodied energy and embodied carbon would augment in proportion to operational energy, which further reinforces the effort that should be made in material choice at early design stages, to minimize total life cycle impacts. Accordingly, Chau et al. [28] argue that the total impact of non-structural elements is 1.4–1.6 times that of structural elements, since structural elements have no replacement factor, and core elements in retail buildings can be renovated as often as every five years, depending on market demands [51]. As the frequent refurbishment cycles of retail stores, motivated mostly by competition circumstances, lead to increased life cycle impacts in retail buildings [28], more attention should be given to strategies that save material quantities, reduce the use of materials with high energy and carbon intensities and promote the use of environmentally friendly materials [46].

4.5. Limitations

This study analysed the impacts of shell building materials in a retail building. The results of this study are approximate rather than precise since without EPDs, and impacts can only be estimated using existing inventories that, on occasion, are difficult to adapt to individual projects [40]. Inventory databases provide general values rather than specific, regional context-adapted values.

In further studies, it is also necessary to analyse the impact of core and external materials in the life cycle of retail buildings and to enlarge system boundaries to other life cycle stages. Nonetheless, as pointed out by Hertwich et al. [52] and Pomponi and Moncaster [3], an incomplete assessment is better than no assessment, and in the case of retail buildings, preliminary studies are necessary, as current knowledge about the impacts of materials is extremely limited [4,24].

Impacts in SimaPro software were calculated in terms of primary energy demand, whereas the EUI of the store was calculated in terms of final energy. The current default conversion coefficient is 2.1 in the European Union [53], which implies that for each unit of electricity, 2.1 units of primary energy are required. This study did not provide direct conversion of final energy into primary energy. Nevertheless, the values presented in this study provide an original understanding of the need to reduce the embodied impacts of materials at their initial lifecycle stages, since the tendency is for operational energy to decrease.

5. Conclusions

This study set out to assess the embodied energy and carbon emissions for a typical contemporary European retail building, while comparing it to the energy and carbon emissions consumed and produced during store operation, over the building life cycle, which represents a significant contribution to the existing body of knowledge, as no such studies were identified in the literature for European retail stores. In addition, this research also suggests strategies that can be used to mitigate materials' impacts in retail buildings in a straightforward way, enhancing the study's applicability.

The present study results place the embodied energy of a retail store in Portugal at 32% of total life cycle energy and the embodied carbon at 94% of total life cycle carbon emissions. EUI was found at 180 kWh/m^2/y, which is 20% below the average for non-food retailers found in previous research work. Likewise, the CUI was 21 kg CO_2 eq/m^2/y, which is a top performance result. The CIU of the analysed store is very low since most of the store's energy consumption is produced by PV panels installed onsite. In addition, about half of the purchased grid energy is produced from renewable sources. Thus, with the increasing penetration of renewable energy in the market, the CUI derived from energy consumption is likely to decrease, which reinforces the importance of building materials choice for the overall life-cycle impact of the building.

In relation to the store's sales area, the embodied energy was found to be 4248 kWh/m^2 and the embodied carbon 1689 kg CO_2 eq/m^2. The most intensive energy and carbon materials were cement mortar, steel, concrete, and extruded polystyrene. Embodied impacts for the present case study could be minimized by choosing stone wool sandwich panels for the facades instead of extruded polystyrene panels and by choosing metal sheet coverings for the roof system instead of bitumen materials.

Easy solutions to reduce material impacts in retail buildings include increasing the quantity of recycled materials in steel structural elements, in concrete and in paving solutions. Optimizing the design of structural elements and designing for disassembly are also straightforward strategies to mitigate construction impacts, as well as choosing materials with equivalent performance but with minor impacts.

There is a great opportunity for material impact mitigation in the upkeep of existing retail stores, especially in developed countries where the building stock is extensive, including the extension of the lifetime of buildings. The embodied energy and carbon in buildings could be drastically minimized in refurbishment processes, as the impacts of new structural elements could be avoided. In addition, with tighter regulations on the energy consumption of buildings, as set up by the EPBD, and in the emerging low-energy building era, the embodied impacts of materials are increasingly important, as the results suggest. Nevertheless, becoming carbon zero is easier to achieve in retail buildings, namely through the onsite production of renewable energy and energy offsetting methods, than becoming near zero energy, mainly due to the high EUI of the retail sector and extensive operating hours.

Recommendations and Future Research

In further studies, it is necessary to analyse the impact of core and external works' materials in the life cycle of retail buildings and to enlarge system boundaries to other life-cycle stages, including demolition or refurbishment options. A pluralistic approach is a key to the transition to a low-carbon built environment. In this sense, mitigation strategies that include strong policy and regulation at a governmental level, namely carbon mitigation offsets, emissions trading, carbon tax, carbon sequestration and decarbonization of the energy grid should also be under debate. Tangible benefits of integrating LCA studies in future building codes should also be considered, to accelerate the transition towards a sustainable, low-carbon building sector.

Supplementary Materials: The following supporting information can be downloaded at: https://www.mdpi.com/article/10.3390/en16010378/s1, Table S1: Cost disaggregation by building component for the case study store, based on the winning contractor's bid.

Author Contributions: Conceptualization, A.F. and M.D.P.; methodology, A.F.; validation, M.D.P., J.d.B. and R.M.; formal analysis, A.F. and M.D.P.; investigation, A.F.; resources, A.F.; writing—original draft preparation, A.F.; writing—review and editing, M.D.P., J.d.B. and R.M. All authors have read and agreed to the published version of the manuscript.

Funding: This research was funded by FCT—Fundação para a Ciência e Tecnologia (grant number PD/BD/127852/2016) under the Doctoral Program EcoCoRe—Eco-Construction and Rehabilitation. Support from CERIS and Instituto Superior Técnico is also acknowledged.

Data Availability Statement: The data presented in this study are openly available in: Santos Ferreira, Ana Sofia (2022), "Decarbonizing strategies of the retail sector following the Paris Agreement—27 highest revenue retailers' CSR reports", Mendeley Data, V2, https://doi.org/10.17632/rjtrcrps5p.2, accessed on 23 November 2022.

Acknowledgments: The authors would like to thank Vera Durão for the support in impacts calculation in SimaPro software.

Conflicts of Interest: The authors declare no conflict of interest. The funders had no role in the design of the study; in the collection, analyses, or interpretation of data; in the writing of the manuscript; or in the decision to publish the results.

Appendix A

Table A1. Bill of quantities input of shell building elements for the analysed retail store, according to level(s) template file.

Bill of Quantities Organised by the Main Building Parts and Elements								Floor Area (m^2)		9556.00
Tier 1 Building Element	Tier 2 Building Element	Tier 3 Building Element	Description of the Product/Material Being Purchased	Bill of Quantities	Unit	Conversion Factor (kg/Unit)	Total (kg)	Cost €/Unit	Cost €/kg	Total Cost (EUR)
Shell	roof	Weatherproofing	roofing system: 0.75 mm metal sheet, 150 kg/m^3 rock wool plates, waterproofing in 3 layers of 4 kg/m^2 asphalt membrane	9119	m^2	25	227,975.00	34.45	1.38	314,149.55
Shell	Roof	Weatherproofing	waterproofing membrane in polymer bitumen 3 kg/m^2 and fiberglass reinforcement, polymer bitumen 4 kg/m^2, extruded polystyrene sheets 50 mm, synthetic fibre geotextile blanket 150 g/m^2, pebble	78	m^2	56.65	4418.70	35.25	0.62	2749.50
Shell	Roof	Weatherproofing	waterproofing membrane in polymer bitumen 3 kg/m^2 and fiberglass reinforcement, polymer bitumen 4 kg/m^2, 50 mm extruded polystyrene plates, 150 g/m^2 geotextile synthetic fibre mat; slabs, 35 mm screed and 30 mm insulation	146	m^2	83.65	12,212.90	52.30	0.63	7635.80
Shell	Roof	Weatherproofing	steel sheet 0.75 mm	767	m^2	5	3835.00	16.25	3.25	12,463.75
Shell	Facades	External wall systems, cladding and shading devices	50 mm sandwich panels, with rock wool interior, and fastening structures	3903	m^2	19	74,157.00	33.15	1.74	129,384.45
Shell	Facades	External wall systems, cladding and shading devices	0.5 mm steel sheet, including secondary metal frame	1288	m^2	9	11,592.00	23.18	2.58	29,855.84
Shell	Facades	External wall systems, cladding and shading devices	3 mm perforated aluminium panels, including secondary aluminium frame	416	m^2	17	7072.00	101.14	5.95	42,074.24
Shell	Facades	External wall systems, cladding and shading devices	2 mm steel plate including fastening frame	200.15	m^2	17	3402.55	62.22	3.66	12,453.33
Shell	Facades	External wall systems, cladding and shading devices	2 mm metal grid including fastening frame	40	m^2	17	680.00	266.91	15.70	10,676.40
Shell	Facades	External wall systems, cladding and shading devices	reinforced masonry of concrete blocks, cement mortar and sand	5284	m^2	270	1,426,680.00	20.51	0.08	108,374.84
Shell	Facades	External wall systems, cladding and shading devices	masonry of concrete blocks 50 × 20 × 15 cm, cement and sand mortar	306	m^2	270	82,620.00	11.06	0.04	3384.36
Shell	Facades	External wall systems, cladding and shading devices	plastering with cement mortar, hydraulic lime and sand	674	m^2	83.2	56,076.80	7.97	0.10	5371.78
Shell	Facades	External paints, coatings and renders	Interior paintings on concrete block walls and plasterboard walls	5241	m^2	0.26	1362.66	3.06	11.77	16,037.46
Shell	Facades	External paints, coatings and renders	Interior paintings on concrete block walls and plasterboard walls	742	m^2	0.26	192.92	3.42	13.15	2537.64
Shell	Non_loadbea-ring_elements	Internal walls, partitions and doors	Steel door, metal frame, 80 cm	14	Unit	10.64	148.96	1114.00	104.70	15,596.00
Shell	Facades	Façade openings (including windows and external doors)	Fire door 60 min, 90 cm	1	Unit	11.97	11.97	1960.00	163.74	1960.00

Table A1. *Cont.*

Bill of Quantities Organised by the Main Building Parts and Elements								Floor Area (m^2)		9556.00
Tier 1 Building Element	**Tier 2 Building Element**	**Tier 3 Building Element**	**Description of the Product/Material Being Purchased**	**Bill of Quantities**	**Unit**	**Conversion Factor (kg/Unit)**	**Total (kg)**	**Cost €/Unit**	**Cost €/kg**	**Total Cost (EUR)**
Shell	Non_loadbearing_elements	Internal walls, partitions and doors	Fire door 60 min, 90 cm	9	Unit	11.97	107.73	1960.00	163.74	17,640.00
Shell	Non_loadbearing_elements	Internal walls, partitions and doors	Fire door 60 min, 90 cm	16	Unit	11.97	191.52	5932.00	495.57	94,912.00
Shell	Facades	Façade openings (including windows and external doors)	Automatic sliding doors, 2.00 m. Double glazing 8.8.6 (laminated colourless glass on the outside and tempered glass on the inside)	3	Unit	25	75.00	6973.00	278.92	20,919.00
Shell	Facades	Façade openings (including windows and external doors)	Thermo-lacquered aluminium frame with 6 mm tempered crystal glass	12	Unit	33.6	403.20	336.00	10.00	4032.00
Shell	Non_loadbearing_elements	Internal walls, partitions and doors	Fireproof sliding gate in iron frame, 60 min	16	Unit	200	3200.00	13,166.00	65.83	210,656.00
Shell	Facades	Façade openings (including windows and external doors)	Steel door, metal frame, 90 cm	6	Unit	11.97	71.82	1968.00	164.41	11,808.00
Shell	Facades	Façade openings (including windows and external doors)	Steel door, metal frame, 140 cm	7	Unit	18.62	130.34	6134.00	329.43	42,938.00
Shell	Facades	Façade openings (including windows and external doors)	Steel door, metal frame, 140 cm	3	Unit	14	42.00	1167.00	83.36	3501.00
Shell	Roof	Weatherproofing	Smoke exhaustion skylight 2000 × 2000 mm in transparent honeycomb polycarbonate	49	Unit	7.08	346.92	1487.00	210.03	72,863.00
Shell	Roof	Weatherproofing	Skylight 2000 × 2000 mm	89	Unit	4.58	407.62	668.00	145.85	59,452.00
Shell	Roof	Weatherproofing	Day light and smoke-exhaustion skylight 1400 × 1400 mm, 16 mm transparent honeycomb polycarbonate.	2	Unit	5.92	11.84	1166.00	196.96	2332.00
Shell	Roof	Weatherproofing	Natural tubular lighting system SOLATUBE®, 35 cm diameter	17	Unit	1.75	29.75	703.00	401.71	11,951.00
Shell	Facades	External wall systems, cladding and shading devices	Double-skinned sectional door filled with rigid polyurethane foam	1	unit	200	200.00	4742.00	23.71	4742.00
Shell	Facades	External wall systems, cladding and shading devices	Micro-perforated metal rolling grille	4	unit	22	88.00	9725.00	442.05	38,900.00
Shell	Facades	External wall systems, cladding and shading devices	Stapled glass facade composed of double-glazing: 12 mm tempered glass + 16 mm air chamber + 10+10.4 tempered, transparent laminated glass. Stainless steel fasteners.	1	unit	10.5	10.50	117,390.00	11,180.00	117,390.00
Shell	Parking_facilities	Above ground and underground (within the curtilage of the building and servicing the building occupiers)	Foam insulation 50 mm	73596	m^2	1.65	121,433.40	5.17	3.13	380,491.32
Shell	Parking_facilities	Above ground and underground (within the curtilage of the building and servicing the building occupiers)	soundproofing mineral wool 45 mm	565	m^2	4.5	2542.50	3.79	0.84	2141.35
Shell	Facades	External paints, coatings and renders	Bituminous emulsion; Waterproofing membrane in polymer bitumen 4 kg/m^2 and polyester reinforcement protected with polyethylene on both sides; High density polyethylene granular sheet with embedded geotextile	2613	m^2	4.4	11,497.20	15.00	3.41	39,195.00
Shell	Foundations_substructure	Basements	C12/15 concrete	2266	m^2	144.3	326,983.80	6.39	0.04	14,479.74
Shell	Foundations_substructure	Basements	Rockfill in limestone quarry gravel, Ø40/70 mm	8684	m^2	640	5,557,760.00	11.61	0.02	100,821.24
Shell	Foundations_substructure	Retaining walls	formwork with modular metal panels for reinforced concrete walls	5226	m^2	13	67,938.00	13.67	1.05	71,439.42
Shell	Foundations_substructure	Basements	C30/37 reinforced concrete wall A500 NR steel, 86.6 kg/m^3	676.32	m^3	2586.6	1,749,369.31	135.29	0.05	91,499.33
Shell	Foundations_substructure	Retaining walls	Gabion wall: 2.70 mm diameter galvanized steel wire mesh box, 80 × 100 mm^2 hexagonal mesh, and limestone, on tires	1050.695	m^3	1600	1,681,112.00	50.86	0.03	53,438.35
Shell	Foundations_substructure	Piles	metal panel formwork for foundations	2005.92	m^2	13	26,076.96	11.39	0.88	22,847.43
Shell	Foundations_substructure	Piles	Foundation in reinforced concrete C30/37, steel A500 NR, 42.6 kg/m^3	1430.316	m^3	2542.6	3,636,721.46	104.53	0.04	149,510.93
Shell	Foundations_substructure	Retaining walls	Reinforced concrete foundation and concrete walls C30/37, A500 NR steel, 67.9 kg/m^3	182.83	m^3	2567.9	469,489.16	122.22	0.05	22,345.48
Shell	Non_loadbearing_elements	Ground floor slab	Slab in reinforced concrete C30/37 and steel A500 NR, 50 kg/m^3	173.04	m^3	2550	441,252.00	109.70	0.04	18,982.49
Shell	Loadbearing_structural_frame	Frame (beams, columns and slabs)	Steel S275JR (Fe430) in metallic structure, HEA	450959	kg	1	450,959.00	2.18	2.18	983090.62
Shell	Loadbearing_structural_frame	Frame (beams, columns and slabs)	C30/37 reinforced concrete column, A500 NR steel, 235.6 kg/m^3; sheet metal formwork	317	m^3	2736	867,312.00	345.00	0.13	109,365.00
Shell	Loadbearing_structural_frame	Frame (beams, columns and slabs)	Reinforced concrete beam, C30/37, A500 NR steel, 122.8 kg/m^3; wooden formwork	1400	m^3	2623	3,672,200.00	176.00	0.07	246,400.00
Shell	Non_loadbearing_elements	Ground floor slab	Slab foundation in reinforced concrete C30/37, steel A500 NR, 48 kg/m^2; wooden formwork; quartz hardener powder 7 kg/m^2	8284.31	m^2	1305	10,811,024.55	89.40	0.07	740,617.31

Table A1. *Cont.*

Bill of Quantities Organised by the Main Building Parts and Elements								Floor Area (m^2)		9556.00
Tier 1 Building Element	**Tier 2 Building Element**	**Tier 3 Building Element**	**Description of the Product/Material Being Purchased**	**Bill of Quantities**	**Unit**	**Conversion Factor (kg/Unit)**	**Total (kg)**	**Cost €/Unit**	**Cost €/kg**	**Total Cost (EUR)**
Shell	Non_loadbearing_elements	Ground floor slab	Slab of reinforced concrete C30/37, steel A500 NR, 48 kg/m^2; wooden formwork; quartz hardener powder 7 kg/m^2	63.2	m^2	555	35,076.00	44.55	0.08	2815.56
Shell	Non_loadbearing_elements	Ground floor slab	Slab of reinforced concrete C30/37, steel A500 NR, 48 kg/m^2; quartz hardener powder 7 kg/m^2	317	m^2	805	255,185.00	57.30	0.07	18,164.10
Shell	Loadbearing_structural_frame	Upper floors	Fungiform slab of reinforced concrete C30/37, volume 0.269 m^3/m^2, steel A500 NR, 13.7 kg/m^2; lightweight concrete block with expanded clay FB65/40; electrowelded mesh AR42 of A500 EL steel and quartz powder hardener 7 kg/m^2	3295	m^2	1631	5,374,145.00	59.00	0.04	194,405.00
Shell	Loadbearing_structural_frame	External walls	C30/37 reinforced concrete wall, A500 NR steel, 183.4 kg/m^3	61	m^3	670.75	40,915.75	337.00	0.50	20,557.00
Shell	Non_loadbearing_elements	Ground floor slab	C30/37 concrete flooring, steel fibres, polyethylene mesh, quartz hardener powder 7 kg/m^2	8624	m^2	501.4	4,324,073.60	21.60	0.04	186,278.40
Shell	Loadbearing_structural_frame	Upper floors	Slab with galvanized steel plate and reinforced concrete C30/37, total volume of concrete 0.082 m^3/m^2, steel A500 NR, 1 kg/m^2	716	m^2	300.84	215,401.44	30.78	0.10	22,038.48
Shell	Non_loadbearing_elements	Stairs and ramps	Concrete staircase C30/37, A500 NR steel, 22 kg/m^2, olive leaf plate sill	114	m^2	504.4	57,501.60	79.88	0.16	9106.32
Shell	Non_loadbearing_elements	Stairs and ramps	Escalators, with a load capacity of 9000 kg/120 people	4	Unit	15000	60,000.00	66823.00	4.45	267,292.00
Shell	Non_loadbearing_elements	Stairs and ramps	Lift, with a load capacity of 2000 kg/27 people	2	unit	5000	10,000.00	34778.00	6.96	69,556.00
Shell	Non_loadbearing_elements	Internal walls, partitions and doors	Ceramic tile, mortar of cement, hydraulic lime and sand	153	m^2	9	1377.00	29.51	3.28	4515.03

Appendix B

Table A2. Bill of materials input of shell building elements according to material type for the analysed retail store, in terms of percentage of weight for each building element, assumed lifetime and replacement factor, according to Level(s) template file.

Bill of Materials by Material Type (% Weight).											
Concrete, Brick, Tile, Ceramic, etc.	**Glass**	**Plastic**	**Bituminous Mixtures**	**Metals**	**Insulation Materials**	**Electrical and Electronic Equipment**	**Total % (Should Be 100%)**	**Assumed Lifetime of Product/Material (Years)**	**Normalised Requirement Factor over Building Lifetime**	**Normalised Weight of Materials Needed over Lifetime**	**Normalised Cost of Materials Needed over Lifetime**
			32.0%	20.0%	48.0%		100.0%	50	1.5	341,962.50	471,224.32
84.7%			12.4%		2.9%		100.0%	50	1.5	6628.05	4124.25
89.7%			8.4%		2.0%		100.0%	50	1.5	18,319.35	11,453.70
				100.0%			100.0%	50	2	7670.00	24,927.50
				25.0%	75.0%		100.0%	50	2	148,314.00	258,768.90
				100.0%			100.0%	50	2	23,184.00	59,711.68
				100.0%			100.0%	50	1	7072.00	42,074.24
				100.0%			100.0%	50	1.5	5103.82	18,679.99
				100.0%			100.0%	50	2	1360.00	21,352.80
100.0%							100.0%	50	1	1,426,680.00	108,374.84
100.0%							100.0%	50	1	82,620.00	3384.36
100.0%							100.0%	50	1	56,076.80	5371.78
		100.0%					100.0%	50	3	4087.98	48,112.38
		100.0%					100.0%	50	3	578.76	7612.92
				75.0%	25.0%		100.0%	50	2	297.92	31,192.00
				75.0%	25.0%		100.0%	50	2	23.94	3920.00
				75.0%	25.0%		100.0%	50	2	215.46	35,280.00
				75.0%	25.0%		100.0%	50	2	383.04	189,824.00
	75.0%			25.0%			100.0%	50	2	150.00	41,838.00
	75.0%			25.0%			100.0%	50	2	806.40	8064.00
				75.0%	25.0%		100.0%	50	2	6400.00	421,312.00
				75.0%	25.0%		100.0%	50	2	143.64	23,616.00
				75.0%	25.0%		100.0%	50	2	260.68	85,876.00
				100.0%			100.0%	50	2	84.00	7002.00
		75.0%		25.0%			100.0%	50	2	693.84	145,726.00
		85.0%		15.0%			100.0%	50	2	815.24	118,904.00
		75.0%		25.0%			100.0%	50	2	23.68	4664.00
		75.0%		25.0%			100.0%	50	1.5	44.62	17,926.50
				75.0%	25.0%		100.0%	50	3	600.00	14,226.00
				100.0%			100.0%	50	2	176.00	77,800.00
	75.0%			25.0%			100.0%	50	1	10.50	117,390.00
					100.0%		100.0%	50	1	121,433.40	380,491.32
					100.0%		100.0%	50	1	2542.50	2141.35
			100.0%				100.0%	50	1	11,497.20	39,195.00
100.0%							100.0%	50	1	326,983.80	14,479.74
100.0%							100.0%	50	1	5,557,760.00	100,821.24
				100.0%			100.0%	50	1	67,938.00	71,439.42
95.0%				5.0%			100.0%	50	1	1,749,369.31	91,499.33
95.0%				5.0%			100.0%	50	1	1,681,112.00	53,438.34
				100.0%			100.0%	50	1	26,076.96	22,847.42
95.0%				5.0%			100.0%	50	1	3,636,721.46	149,510.93

Table A2. *Cont.*

Bill of Materials by Material Type (% Weight).											Normalised Weight of Materials Needed over Lifetime	Normalised Cost of Materials Needed over Lifetime
Concrete, Brick, Tile, Ceramic, etc.	Glass	Plastic	Bituminous Mixtures	Metals	Insulation Materials	Electrical and Electronic Equipment	Total % (Should Be 100%)	Assumed Lifetime of Product/Material (Years)	Normalised Requirement Factor over Building Lifetime			
95.0%				5.0%			100.0%	50	1		469,489.15	22,345.48
95.0%				5.0%			100.0%	50	1		441,252.00	18,982.48
				100.0%			100.0%	50	1		450,959.00	983,090.62
95.0%				5.0%			100.0%	50	1		867,312.00	109,365.00
95.0%				5.0%			100.0%	50	1		3,672,200.00	246,400.00
95.0%				5.0%			100.0%	50	1		10,811,024.55	740,617.31
95.0%				5.0%			100.0%	50	1		35,076.00	2815.56
95.0%				5.0%			100.0%	50	1		255,185.00	18,164.10
95.0%				5.0%			100.0%	50	1		5,374,145.00	194,405.00
95.0%				5.0%			100.0%	50	1		40,915.75	20,557.00
95.0%				5.0%			100.0%	50	1		4,324,073.60	186,278.40
95.0%				5.0%			100.0%	50	1		215,401.44	22,038.48
95.0%				5.0%			100.0%	50	1		57,501.60	9106.32
						100.0%	100.0%	50	1.5		90,000.00	400,938.00
						100.0%	100.0%	50	1.5		15,000.00	104,334.00
100.0%							100.0%	50	1		1377.00	4515.03

References

1. Gervasio, H.; Dimova, S. *Model for Life Cycle Assessment (LCA) of Buildings, EUR 29123 EN, Publications Office of the European Union, 2018, JRC110082*; Joint Research Center: Brussels, Belgium, 2018; ISBN 9789279799730. [CrossRef]
2. Yin, S.; Dong, T.; Li, B.; Gao, S. Developing a Conceptual Partner Selection Framework: Digital Green Innovation Management of Prefabricated Construction Enterprises for Sustainable Urban Development. *Buildings* **2022**, *12*, 721. [CrossRef]
3. Pomponi, F.; Moncaster, A. Embodied Carbon Mitigation and Reduction in the Built Environment—What Does the Evidence Say? *J. Environ. Manag.* **2016**, *181*, 687–700. [CrossRef] [PubMed]
4. Omer, M.A.B.; Noguchi, T. A Conceptual Framework for Understanding the Contribution of Building Materials in the Achievement of Sustainable Development Goals (SDGs). *Sustain. Cities Soc.* **2020**, *52*, 101869. [CrossRef]
5. Ramesh, T.; Prakash, R.; Shukla, K.K. Life Cycle Energy Analysis of Buildings: An Overview. *ENERGY Build.* **2010**, *42*, 1592–1600. [CrossRef]
6. Leadership in Energy and Environmental Design Home | LEED Lookbook. Available online: https://leed.usgbc.org/ (accessed on 14 October 2021).
7. BREEAM Sustainability Assessment Method. Available online: https://www.breeam.com/ (accessed on 14 October 2021).
8. Gervasio, H. Resource Efficient Construction towards Sustainable Design | EFIResources Project | Fact Sheet | H2020 | CORDIS | European Commission. Available online: https://cordis.europa.eu/project/id/707532 (accessed on 20 April 2022).
9. Galvez-Martos, J.-L.; Styles, D.; Schoenberger, H. Identified Best Environmental Management Practices to Improve the Energy Performance of the Retail Trade Sector in Europe. *Energy Policy* **2013**, *63*, 982–994. [CrossRef]
10. European Commission. *A Roadmap for Moving to a Competitive Low Carbon Economy in 2050*; European Commission: Brussels, Belgium, 2011.
11. European Parliament. *Directive 2018/844/EU Amending Directive 2010/31/EU on the Energy Performance of Buildings and Directive 2012/27/EU on Energy Efficiency*; European Parliament: Luxembourg, 2018.
12. Cole, R.J.; Kernan, P.C. Life-Cycle Energy Use in Office Buildings. *Build. Environ.* **1996**, *31*, 307–317. [CrossRef]
13. Junnila, S.; Horvath, A.; Guggemos, A.A. Life-Cycle Assessment of Office Buildings in Europe and the United States. *J. Infrastruct. Syst.* **2006**, *12*, 10–17. [CrossRef]
14. Chau, C.K.; Leung, T.M.; Ng, W.Y. A Review on Life Cycle Assessment, Life Cycle Energy Assessment and Life Cycle Carbon Emissions Assessment on Buildings. *Appl. Energy* **2015**, *143*, 395–413. [CrossRef]
15. Wallhagen, M.; Glaumann, M.; Malmqvist, T. Basic Building Life Cycle Calculations to Decrease Contribution to Climate Change - Case Study on an Office Building in Sweden. *Build. Environ.* **2011**, *46*, 1863–1871. [CrossRef]
16. Zabalza Bribian, I.; Aranda Uson, A.; Scarpellini, S. Life Cycle Assessment in Buildings: State-of-the-Art and Simplified LCA Methodology as a Complement for Building Certification. *Build. Environ.* **2009**, *44*, 2510–2520. [CrossRef]
17. Verbeeck, G.; Hens, H. Life Cycle Inventory of Buildings: A Contribution Analysis. *Build. Environ.* **2010**, *45*, 964–967. [CrossRef]
18. Ortiz, O.; Bonnet, C.; Bruno, J.C.; Castells, F. Sustainability Based on LCM of Residential Dwellings: A Case Study in Catalonia, Spain. *Build. Environ.* **2009**, *44*, 584–594. [CrossRef]
19. Thormark, C. The Effect of Material Choice on the Total Energy Need and Recycling Potential of a Building. *Build. Environ.* **2006**, *41*, 1019–1026. [CrossRef]
20. Karimpour, M.; Belusko, M.; Xing, K.; Bruno, F. Minimising the Life Cycle Energy of Buildings: Review and Analysis. *Build. Environ.* **2014**, *73*, 106–114. [CrossRef]
21. Blengini, G.A.; Di Carlo, T. The Changing Role of Life Cycle Phases, Subsystems and Materials in the LCA of Low Energy Buildings. *Energy Build.* **2010**, *42*, 869–880. [CrossRef]
22. Kofoworola, O.F.; Gheewala, S.H. Environmental Life Cycle Assessment of a Commercial Office Building in Thailand. *Int. J. Life Cycle Assess.* **2008**, *13*, 498. [CrossRef]
23. Gustavsson, L.; Joelsson, A. Life Cycle Primary Energy Analysis of Residential Buildings. *Energy Build.* **2010**, *42*, 210–220. [CrossRef]

24. Cabeza, L.F.; Rincón, L.; Vilariño, V.; Pérez, G.; Castell, A. Life Cycle Assessment (LCA) and Life Cycle Energy Analysis (LCEA) of Buildings and the Building Sector: A Review. *Renew. Sustain. Energy Rev.* **2014**, *29*, 394–416. [CrossRef]
25. Ding, G.K.C. Sustainable Construction—The Role of Environmental Assessment Tools. *J. Environ. Manag.* **2008**, *86*, 451–464. [CrossRef]
26. Dutil, Y.; Rousse, D.; Quesada, G. Sustainable Buildings: An Ever Evolving Target. *Sustainability* **2011**, *3*, 443–464. [CrossRef]
27. van Ooteghem, K.; Xu, L. The Life-Cycle Assessment of a Single-Storey Retail Building in Canada. *Build. Environ.* **2012**, *49*, 212–226. [CrossRef]
28. Chau, C.K.; Yik, F.W.H.; Hui, W.K.; Liu, H.C.; Yu, H.K. Environmental Impacts of Building Materials and Building Services Components for Commercial Buildings in Hong Kong. *J. Clean. Prod.* **2007**, *15*, 1840–1851. [CrossRef]
29. Cinelli, M.; Coles, S.R.; Kirwan, K. Analysis of the Potentials of Multi Criteria Decision Analysis Methods to Conduct Sustainability Assessment. *Ecol. Indic.* **2014**, *46*, 138–148. [CrossRef]
30. Le, K.N.; Tran, C.N.N.; Tam, V.W.Y. Life-Cycle Greenhouse-Gas Emissions Assessment: An Australian Commercial Building Perspective. *J. Clean. Prod.* **2018**, *199*, 236–247. [CrossRef]
31. Luo, Z.; Yujie, C.; Zhang, N.; Liu, Y.; Liu, J. A Quantitative Process-Based Inventory Study on Material Embodied Carbon Emissions of Residential, Office, and Commercial Buildings in China. *J. Therm. Sci.* **2019**, *28*, 1236–1251. [CrossRef]
32. Anastaselos, D.; Giama, E.; Papadopoulos, A.M. An Assessment Tool for the Energy, Economic and Environmental Evaluation of Thermal Insulation Solutions. *Energy Build.* **2009**, *41*, 1165–1171. [CrossRef]
33. Papadopoulos, A.M.; Giama, E. Environmental Performance Evaluation of Thermal Insulation Materials and Its Impact on the Building. *Build. Environ.* **2007**, *42*, 2178–2187. [CrossRef]
34. Pargana, N.; Pinheiro, M.D.; Silvestre, J.D.; De Brito, J. Comparative Environmental Life Cycle Assessment of Thermal Insulation Materials of Buildings. *Energy Build.* **2014**, *82*, 466–481. [CrossRef]
35. Deloitte Global Powers of Retailing. The Art and Science of Customers. 2017. Available online: https://www2.deloitte.com/content/dam/Deloitte/global/Documents/consumer-industrial-products/gx-cip-2017-global-powers-of-retailing.pdf (accessed on 11 September 2018).
36. Master Draw Arquitectura e Planeamento Lda Leroy Merlin Loulé. Available online: https://www.masterdraw.pt/project/leroy-merlin-loule/ (accessed on 31 October 2022).
37. European Commission Level(s) Common Framework | Product Groups Documents | Product Bureau. Available online: https://susproc.jrc.ec.europa.eu/product-bureau/product-groups/412/documents (accessed on 19 April 2022).
38. Andrews, C.J.; Krogmann, U. Explaining the Adoption of Energy-Efficient Technologies in U.S. Commercial Buildings. *Energy Build.* **2009**, *41*, 287–294. [CrossRef]
39. Irvine, S.J.C.; Rowlands-Jones, R.L. Potential for Further Reduction in the Embodied Carbon in PV Solar Energy Systems. *IET Renew. POWER Gener.* **2016**, *10*, 428–433. [CrossRef]
40. Zabalza Bribián, I.; Valero Capilla, A.; Aranda Usón, A. Life Cycle Assessment of Building Materials: Comparative Analysis of Energy and Environmental Impacts and Evaluation of the Eco-Efficiency Improvement Potential. *Build. Environ.* **2011**, *46*, 1133–1140. [CrossRef]
41. Santos Ferreira, A.S. Decarbonizing Strategies of the Retail Sector Following the Paris Agreement—27 Highest Revenue Retailers' CSR Reports. Available online: https://data.mendeley.com/drafts/rjtrcrps5p (accessed on 10 September 2022).
42. European Parliament Directive (EU) 2018/ 2002 on Energy Efficiency. Available online: https://eur-lex.europa.eu/legal-content/EN/TXT/PDF/?uri=CELEX:32018L2002&from=EN (accessed on 3 May 2019).
43. Ferreira, A.; Pinheiro, M.D.; de Brito, J.; Mateus, R. Decarbonizing Strategies of the Retail Sector Following the Paris Agreement. *Energy Policy* **2019**, *135*. [CrossRef]
44. Ferreira, A.; Pinheiro, M.D.; de Brito, J.; Mateus, R. Combined Carbon and Energy Intensity Benchmarks for Sustainable Retail Stores. *Energy* **2018**, *165*. [CrossRef]
45. Huang, Y.; Niu, J.L.; Chung, T.M. Study on Performance of Energy-Efficient Retrofitting Measures on Commercial Building External Walls in Cooling-Dominant Cities. *Appl. Energy* **2013**, *103*, 97–108. [CrossRef]
46. Kumanayake, R.; Luo, H.; Paulusz, N. Assessment of Material Related Embodied Carbon of an Office Building in Sri Lanka. *Energy Build.* **2018**, *166*, 250–257. [CrossRef]
47. Rissman, J.; Bataille, C.; Masanet, E.; Aden, N.; Morrow, W.R.; Zhou, N.; Elliott, N.; Dell, R.; Heeren, N.; Huckestein, B.; et al. Technologies and Policies to Decarbonize Global Industry: Review and Assessment of Mitigation Drivers through 2070. *Appl. Energy* **2020**, *266*, 114848. [CrossRef]
48. Cordero, A.S.; Melgar, S.G.; Márquez, J.M.A. Green Building Rating Systems and the New Framework Level(s): A Critical Review of Sustainability Certification within Europe. *Energies* **2019**, *13*, 66. [CrossRef]
49. One Click LCA. What Is Level(s)? Read about the New Framework for Sustainable Buildings. Available online: https://www.oneclicklca.com/levels-framework-for-sustainable-buildings/ (accessed on 6 September 2021).
50. European Commission Level(S). Available online: https://ec.europa.eu/environment/levels_en (accessed on 6 September 2021).
51. Kneifel, J. Life-Cycle Carbon and Cost Analysis of Energy Efficiency Measures in New Commercial Buildings. *Energy Build.* **2010**, *42*, 333–340. [CrossRef]

52. Hertwich, E.G.; Hammitt, J.K.; Pease, W.S. A Theoretical Foundation for Life-Cycle Assessment: Recognizing the Role of Values in Environmental Decision Making. *J. Ind. Ecol.* **2000**, *4*, 13–28. [CrossRef]
53. European Parliament Directive 2012/27/EU of the European Parliament and of the Council of 25 October 2012 on Energy Efficiency, Amending Directives 2009/125/EC and 2010/30/EU and Repealing Directives 2004/8/EC and 2006/32/EC Text with EEA Relevance. Available online: https://eur-lex.europa.eu/legal-content/EN/TXT/PDF/?uri=CELEX:02012L0027-20181224 (accessed on 28 April 2022).

MDPI
St. Alban-Anlage 66
4052 Basel
Switzerland
www.mdpi.com

Energies Editorial Office
E-mail: energies@mdpi.com
www.mdpi.com/journal/energies

www.ingramcontent.com/pod-product-compliance
Lightning Source LLC
Chambersburg PA
CBHW040742250926
51850CB00007BB/1383
9783725805167